AF781697

HANDBOOK OF GEOPHYSICAL EXPLORATION

SEISMIC EXPLORATION

VOLUME 26

MULTI-COMPONENT VSP ANALYSIS FOR APPLIED SEISMIC ANISOTROPY

HANDBOOK OF GEOPHYSICAL EXPLORATION

SEISMIC EXPLORATION

Editors: Klaus Helbig and Sven Treitel

PUBLISHED VOLUMES

1984 - Mathematical Aspects of Seismology. 2nd Enlarged Edition
(M. Båth and A.J. Berkhout)*
1984 - Seismic Instrumentation (M. Pieuchot) ISBN 0-08-036944-8
1984 - Seismic Inversion and Deconvolution (a) Classical Methods
(E.A. Robinson)*
1985 - Vertical Seismic Profiling (a) Principles. 2nd Enlarged Edition
(B.A. Hardage)*
1987 - Pattern Recognition & Image Processing (F. Aminzadeh)*
1987 - Seismic Stratigraphy (B.A. Hardage)*
1987 - Production Seismology (J.E. White and R.L. Sengbush)*
1989 - Supercomputers in Seismic Exploration (E. Eisner)*
1994 - Seismic Coal Exploration (b) In-Seam Seismics
(L. Dresen and H. Rűter) *
1994 - Foundations of Anisotropy for Exploration Seismics
(K. Helbig) ISBN 0-08-037224-4
1998 - Physical Properties of Rocks: Fundamentals and Principles of Petrophysics
(J.H. Schőn) ISBN 0-08-041008-1
1998 - Shallow High-Resolution Reflection Seismics (J. Brouwer and K. Helbig)
ISBN 0-08-043197-6
1999 - Seismic Inversion and Deconvolution (b) Dual-Sensor Technology
(E.A. Robinson) ISBN 0-08-043627-7
2000 - Vertical Seismic Profiling: Principles. 3rd Updated and Revised Edition
(B.A. Hardage) ISBN 0-08-043518-1
2001 - Seismic Signatures and Analysis of Reflection Data in Anisotropic Media
(I. Tsvankin) ISBN 0-08-043649-8
2001 - Wave Fields in Real Media: Wave Propagation in Anisotropic, Anelastic
and Porous Media (J. M. Carcione) ISBN 0-08-043929-2
2002 - Multi-Component VSP Analysis for Applied Seismic Anisotropy (C. MacBeth)
ISBN 0-08-042439-2
2002 - Nuclear Magnetic Resonance. Petrophysical and Logging Applications
(K.-J. Dunn, D.J. Bergman and G.A. LaTorraca) ISBN 0-08-043880-6

* *book out of print*

SEISMIC EXPLORATION

Volume 26

MULTI-COMPONENT VSP ANALYSIS FOR APPLIED SEISMIC ANISOTROPY

by

Colin MACBETH
Professor of Reservoir Geophysics
Reservoir Geophysics Group
Department of Petroleum Engineering
Heriot-Watt University
Riccarton, Edinburgh EH14 4AS
United Kingdom

2002
PERGAMON
An Imprint of Elsevier Science
Amsterdam – London – New York – Oxford – Paris – Shannon – Tokyo

ELSEVIER SCIENCE Ltd.
The Boulevard, Langford Lane
Kidlington, Oxford OX5 1GB, UK

8 2002 Elsevier Science Ltd. All rights reserved.

This work is protected under copyright by Elsevier Science, and the following terms and conditions apply to its use:

Photocopying
Single photocopies of single chapters may be made for personal use as allowed by national copyright laws. Permission of the Publisher and payment of a fee is required for all other photocopying, including multiple or systematic copying, copying for advertising or promotional purposes, resale, and all forms of document delivery. Special rates are available for educational institutions that wish to make photocopies for non-profit educational classroom use.

Permissions may be sought directly from Elsevier Science Global Rights Department, PO Box 800, Oxford OX5 1DX, UK; phone: (+44) 1865 843830, fax: (+44) 1865 853333, e-mail: permissions@elsevier.co.uk. You may also contact Global Rights directly through Elsevier=s home page (http://www.elsevier.com), by selecting 'Obtaining Permissions'.

In the USA, users may clear permissions and make payments through the Copyright Clearance Center, Inc., 222 Rosewood Drive, Danvers, MA 01923, USA; phone: (+1) (978) 7508400, fax: (+1) (978) 7504744, and in the UK through the Copyright Licensing Agency Rapid Clearance Service (CLARCS), 90 Tottenham Court Road, London W1P 0LP, UK; phone: (+44) 207 631 5555; fax: (+44) 207 631 5500. Other countries may have a local reprographic rights agency for payments.

Derivative Works
Tables of contents may be reproduced for internal circulation, but permission of Elsevier Science is required for external resale or distribution of such material.
Permission of the Publisher is required for all other derivative works, including compilations and translations.

Electronic Storage or Usage
Permission of the Publisher is required to store or use electronically any material contained in this work, including any chapter or part of a chapter.

Except as outlined above, no part of this work may be reproduced, stored in a retrieval system or transmitted in any form or by any means, electronic, mechanical, photocopying, recording or otherwise, without prior written permission of the Publisher.
Address permissions requests to: Elsevier Science Global Rights Department, at the mail, fax and e-mail addresses noted above.

Notice
No responsibility is assumed by the Publisher for any injury and/or damage to persons or property as a matter of products liability, negligence or otherwise, or from any use or operation of any methods, products, instructions or ideas contained in the material herein. Because of rapid advances in the medical sciences, in particular, independent verification of diagnoses and drug dosages should be made.

First edition 2002

Library of Congress Cataloging-in-Publication Data

MacBeth, Colin.
Multicomponent VSP analysis for applied seismic anisotropy / by Colin MacBeth.-- 1st ed.
p. cm. -- (Handbook of geophysical exploration. Seismic exploration, ISSN 0950-1401 ; v. 26)
Includes bibliographical references and index.
ISBN 0-08-042439-2 (hardcover)
1. Seismic waves. 2. Anisotropy. 3. Prospecting--Geophysical methods. I. Title. II. Handbook of geophysical exploration. Section I, Seismic exploration ; v. 26.

QE538.5 .M335 2002
622'.1592--dc21

2001058573

ISBN: 0-08-042439-2
ISSN: 0950-1401 (Series)

The paper used in this publication meets the requirements of ANSI/NISO Z39.48-1992 (Permanence of Paper).
Printed in The Netherlands.

Contents

Low-temperature deformation of the Upper Cretaceous Schrattenkalk, eastern Switzerland (sample 11, Groshong and others, 1984). Based on the twin strain, the rock has enjoyed a shortening of 6% at a temperature on the order of 150° C (from the vitrinite reflectance of about 3.0% in nearby rocks, using the well-standard curve of Bostick 1974). The fossil was discovered and photographed by Laurel Pringle-Goodell.

This picture is on the door of the EAP students room. It was originally hung there by David Campden in 1989. The caption read:

Rocks do not suffer deformation; they enjoy it.

Rob Knipe, 1982

PREFACE

Seismic anisotropy is no ordinary subject. Its complexity is often discouraging to those wishing to learn more, and yet its benefits can be clearly translated into a tangible form within the grasp of the general audience. For those embarking on the task of comprehending this subject, it is often useful to appreciate that the anisotropic world has been (and still is) traditionally divided into two types of people: those that treat anisotropy as a nuisance and develop correction procedures with the end objective of obtaining better reservoir images (processing focus); and those that use seismic anisotropy as a tool for providing lithology description (petrophysical focus). Recent research indicates that both viewpoints are now slowly merging as seismic anisotropy becomes more absorbed into mainstream geophysics. This book documents my experiences working with the latter philosophy. It demonstrates that seismic anisotropy is a predictive indicator unmatched by any other seismic technique. It is particularly powerful when employed in this context of deterministic analysis of vertical seismic profile data, providing an immense amount of useful information for calibration of surface seismic attributes from well and core scale information.

This book was written while the author was working at the Edinburgh Anisotropy Project, and summarizes the VSP work that was pursued in the project over the period 1988 to 1999. Thus, many of the diagrams have been generated by EAP's Shear-Wave Analysis Package (SWAP) software written by Phil Wild (now with Concept Systems). Synthetic seismogram examples were generated using another package in the staple diet of the Edinburgh anisotropist – ANISEIS, written by David Taylor of Edinburgh University. The data examples include most of the VSP collected and analysed by EAP, and donated by oil and service company sponsors.

The book assumes that the reader is acquainted with general acquisition and processing for VSP, and has some knowledge of seismic anisotropy and matrix algebra at university level. It is intended for use as a reference handbook for research students or geophysicists in the oil or service companies, principally for those who wish to design VSP surveys or to apply seismic anisotropy to their specific reservoir problems. The aim has been to treat the subject from cause to effect, developing a consistent and practical notation for universal application, and consequently simplifying many of the theories upon which this subject is based. Suggested books to complement this work include Hardage (1991) and the comprehensive treatment of seismic anisotropy by Helbig (1994). A background historical treatment of seismic anisotropy specifically for the Exploration Geophysicist is offered by MacBeth and Lynn (2000).

My thanks go to the many sponsor representatives who over the years have exchanged ideas, opinions and views within consortium meetings, and provided a tremendous stimulus to the subject. I am indebted to Stuart Crampin for first introducing me to this fascinating subject, and particularly for giving me my first job with EAP in 1988. I have enjoyed the many discussions we have had over the past years. I am also honored to have developed and discussed much of the material with students or staff who have passed through the doors of the British Geological Survey: Phil Wild (1988-1997), Enru Liu (Ph.D. 1989), David Campden (Ph.D. 1990), Iain Bush (Ph.D. 1991), Xiang-Yang Li (Ph.D. 1994), Xinwu Zeng (Ph.D. 1993), Gordon Holmes (Ph.D. 1995), Colin Slater (Ph.D. 1997), Brian Baptie (Ph.D. 1996), Hengchang Dai (Ph.D. 1995), Steve Horne (Ph.D. 1995), Frank Ohlsen (1996-1999), and Thomas Kühnel (Ph.D. 1998).

My special thanks are to my family for their support and tolerance of my long absences from home, allowing me the opportunity to complete this work.

Colin MacBeth
EAP Director (1994–1999),
Edinburgh Anisotropy Project,
British Geological Survey,
Edinburgh.

Now at
Reservoir Geophysics Group,
Department of Petroleum Engineering,
Heriot-Watt University,
Riccarton,
Edinburgh.

CHAPTER 1 – INTRODUCTION

One of the most durable concepts in seismic surveying is the replacement of a volume of small-scale heterogeneities that cannot be individually detected, by a single homogeneous medium. Indeed, by focusing on both the very small (pore scale) and the very large (blocks of rock defined by bed boundaries), the seismic method has worked remarkably well for many decades. Thus, the effect of porosity, pore fluids and their flow on seismic wave propagation can be substituted by the average seismic properties of velocity v, density ρ and quality factor Q. The resultant medium can produce a seismic signature sufficiently close to that actually measured to make a distinction difficult, albeit at the cost of introducing some non-uniqueness.

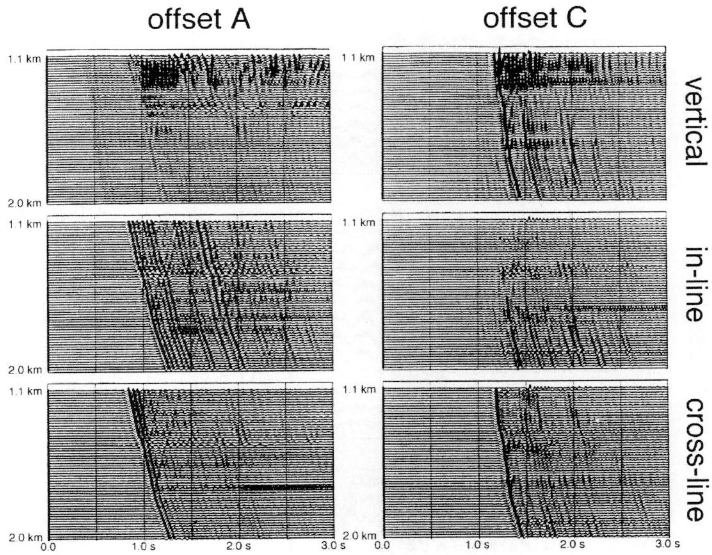

Figure 1.1. Two 3C recordings from a Paris Basin VSP experiment of 1983 run by a consortium of French companies. The horizontal vibrators are activated along a cross-line direction at both of these offsets. Offset A is at 272 m and offset C is 1443 m, with respective azimuths N80°E and N198°E. For offset A the vertical component is weak due to the nature of the source motion. The division of the shear-wave intensity between the in-line and cross-line components can indicate anisotropy, with the exact location in the overburden being uncertain (this interpretation assumes that the source or near-surface inhomogeneity do not introduce phase shifts). The in-line component at larger offset C is weak, with the shear-waves on the vertical trace again indicating anisotropy. A more detailed description of the particle motion analysis using polarization diagrams is given by Bush and Crampin (1991). It is shown that the full range of observations can be modeled by a combination of vertical cracks (mainly 2 %–3 % anisotropy, with a thin 12 % layer) and fine horizontal layering (4 %–5 % anisotropy). The cracks strike N30°W. The techniques outlined in the appendix help analyse these recordings. Data Courtesy of the British Geological Survey.

The benefits of such an approach are obvious: complicated sub-seismic geology can be analysed using simple processing tools, while retaining the impact of the complexity and a high level of description. The inevitable drawback is the sacrifice of some detail on the exact spatial distribution and its location – thus defining an ability to *detect* but not *resolve*. Seismic anisotropy exists as a natural extension of this fundamental concept, modified to involve heterogeneities with some degree of alignment that enhance the rock compliance along specific directions and thus create a directional dependence in the wave behavior (MacBeth 1994; Lynn 1996). Such heterogeneities could be fractures, small cracks, mineral alignments or sedimentary layering. The list of possible causes and resultant anisotropic symmetries is rather large, and might also include deformed pores, aligned sand grains, ripples on beds, cross-bedding, material gradients and many other physical features.

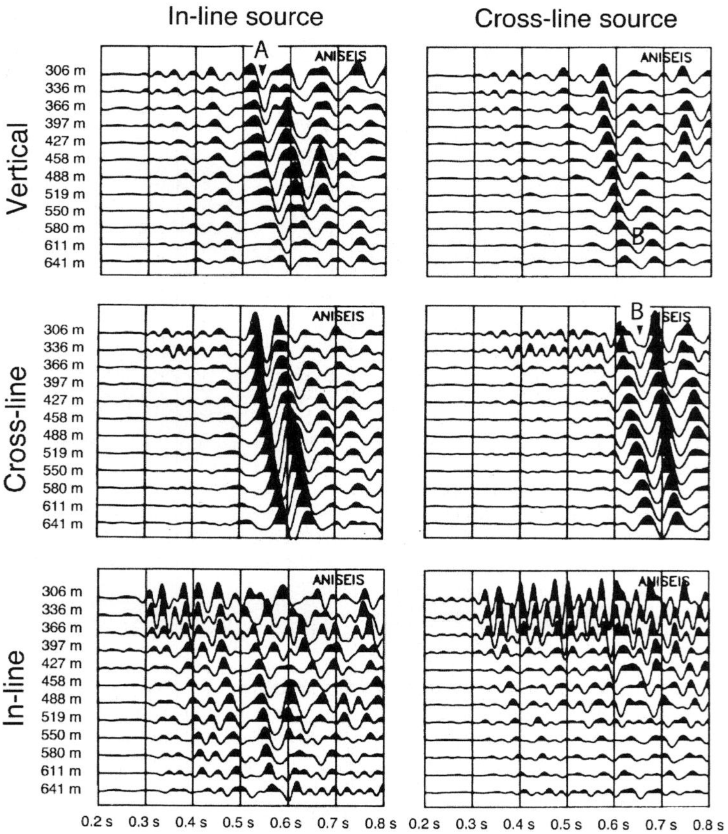

Figure 1.2. Six component (2×3C) data obtained with a horizontal vibrator at an offset of 518 m. The geophone components have been rotated into in-line and cross-line directions, and show an event A from the in-line source arriving 100 ms before event B from the cross-line source. Prima facie interpretation suggests an anisotropy of 30 %; however, Campden et al. (1990) show that the event does in fact arise due to P-SV conversion at a 10 m thick low-velocity surface layer. This is confirmed by a hammer seismics experiment. The cross-component energy still indicates anisotropy, but this could not be satisfactorily modeled and suggests further near-surface complications.

Surprisingly, most seismic observations to date reveal that the anisotropy may be roughly divided into two main types, both of which are classified as transverse isotropy (TI), but with different flavors (symmetries):

1 TIV – transverse isotropy with a vertical axis of symmetry (layering and shales); this symmetry has also been abbreviated as VTI (vertical transverse isotropy).

2 TIH – transverse isotropy with a horizontal axis of symmetry (cracks, fractures); this symmetry has also been abbreviated as HTI (horizontal transverse isotropy).

Table 1. Types of Transverse Isotropy

Category	Importance	Number of Publications	Anisotropy
Shales	high	large	TIV
Layering	uncertain	very large	TIV
Cracks	uncertain	large	TIH
Fractures	moderate	moderate	TIH

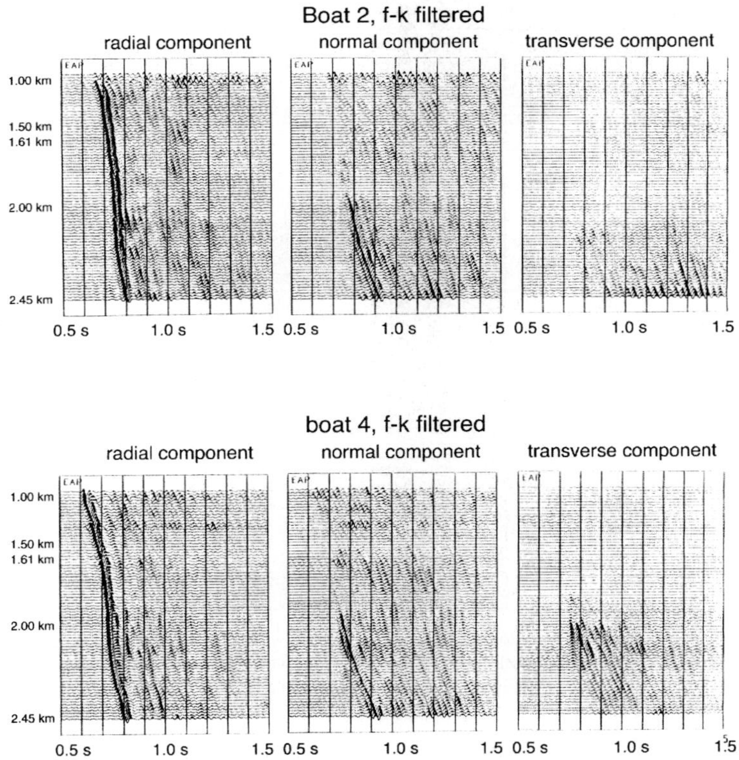

Figure 1.3. Processed three-component (3C) data from two of the four offset marine VSPs acquired in the Vulcan gas field in 1986, with offsets of 2180 m and 1830 m, respectively. The sources are airguns placed 9 m below the sea surface in water of depth 30 m. Note the strong converted shear-waves from the Zechstein caprock, which have been analysed by Campden (1990) revealing a strong cracks-related anisotropy of 10 %–2 % in the caprock.

Stress can also lead to a TI medium. However, the situation often more complicated, with the symmetry depending upon the applied stress field *and* the state of the initially unstressed rock.

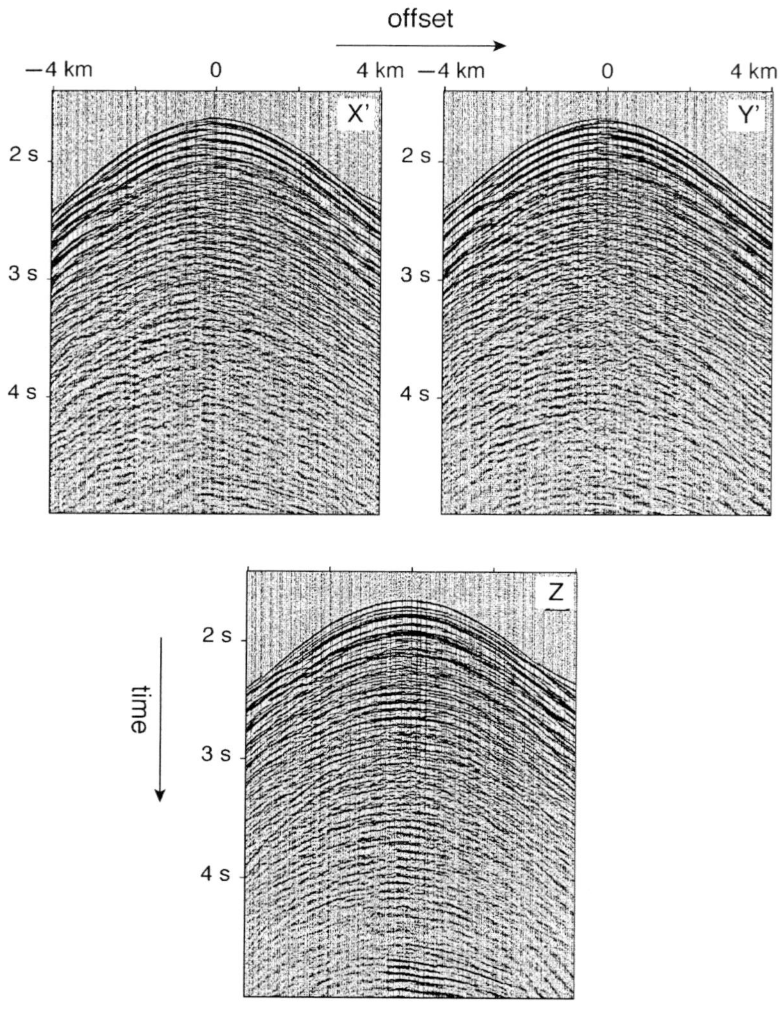

Figure 1.4. Three-component common receiver gather from a walkaway VSP acquired in the North Sea. *X'* and *Y'* are the horizontal components, and *Z* the vertical one. The sources are moved along a single line at fixed azimuth, creating a multi-offset VSP. This particular receiver is one of twelve in the downhole array placed in a deviated borehole. Data Courtesy of the British Geological Survey and Norsk Hydro.

It should be added that the TIV/TIH categorization above does not preclude the existence or prominence of other causative mechanisms, but reflects the simplest explanation for the effects observed and is thus chosen as the starting point. Literature and presentations at conferences also allow a general assessment of the way in which the seismic anisotropy is distributed with depth.

Table 2: Distribution of Anisotropy (in %) with Depth

	TIV	TIH
near surface	5–30 %	5–25 %
background sub-surface	unknown	less than 3 %
discrete (elevated) zones	5–40 %	up to 15 %

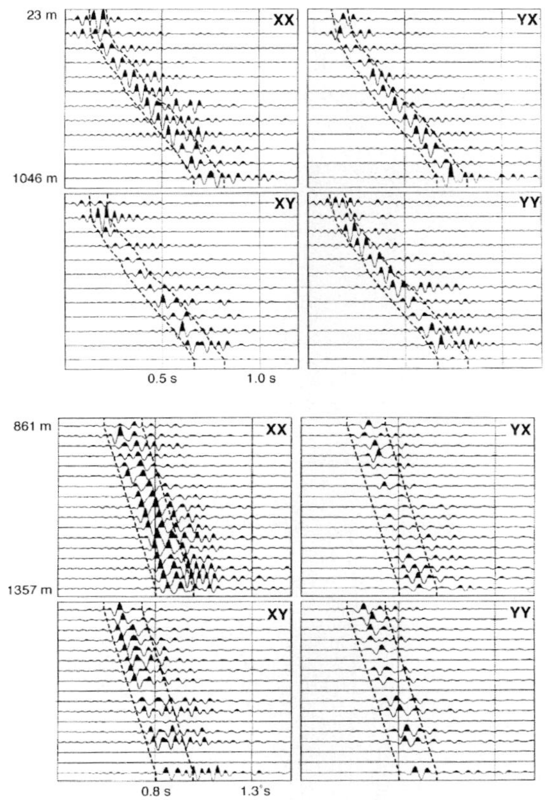

Figure 1.5. Four-component (4C) near-offset data recorded in a borehole close to Parkfield, California in 1987 as part of a sensor test. Each trace is labeled according to the source direction, followed by the receiver direction (see Figure 1.12). The location is within 1.4 km NE of the San Andreas fault in complicated geology. The vibrator offset is 100 m, with alignments being directed in-line and cross-line into a slightly (4°–12°) deviated well. Note that in both sections the direct arrival is distorted and shows a coda due to scattering. The symmetry between the off-diagonal data is necessary but not sufficient for anisotropy analysis. Despite these difficulties, some results have been obtained by Zeng and MacBeth (1992) indicating a 2 %–4 % anisotropy, possibly increasing with depth. Data Courtesy of the British Geological Survey.

Most subsurface TIV zones produce problems in isotropic imaging or mis-interpretations in AVO results. Nearly all can be explained by shales. The effect of fine layering remains largely uncertain and is still an area of active research. The origin of the elevated sub-surface TIH anisotropy is also uncertain, but in recent years some correlation has been observed between (i) open parallel fractures that permit fluid flow, (ii) anisotropy highs detected in seismic data, and (iii) the maximum horizontal compressive stress. Dipole sonic logging is also helping to understand the structure of these highs in more detail. VSP measurement techniques for assessing these anisotropies, their magnitude, location, and more specific departures from the rather broad generalizations above are discussed within this book. The description begins by outlining the different classes of heterogeneity style and their resultant anisotropies in Chapter 2, for a variety of cases for which mathematical solutions exist. These are then cast into a common framework for practical application throughout the remaining chapters. The book builds a quantitative description of the above, from which a more realistic assessment of the anisotropies can be drawn.

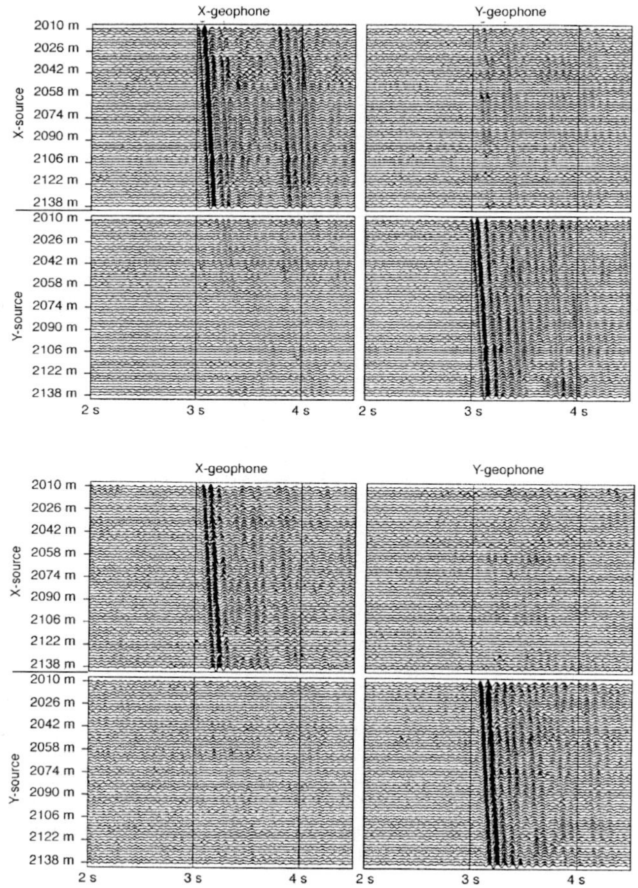

Figure 1.6. Bottom (near-offset) and top (offset of 645 m towards N308°E) four-component (4C) data recorded in the North Caucasus in 1993. The sources used are the electrodynamic VEIP-40 trucks. The formation is predominantly claystone. Negligible intensity on the off-diagonal components may indicate either isotropy or source polarizations aligned along the principal directions of the anisotropy. These and similar data are analysed by Slater (1997). Cross-correlation of the two principal traces results in an interpretation of inclined cracks with 20 % anisotropy.

Vertical seismic profiling (VSP) is an extra-ordinary technique for seismic surveying. It provides data close to the well, sensing the borehole environment by averaging rock volumes of 30 m or less. Due to the single ray path through the near-surface, frequencies are higher and bandwidths larger than in surface-to-surface seismics, thus the technique results in clearer data. This particular attribute has made VSP acquisition an invaluable tool for correlation with log data, because it reduces the risk in extrapolating these data to surface seismics. In fact, the data may almost be regarded as deterministic in nature: when rigorously examined they offer much insight into the physics of wave generation and propagation at depth. Such benefits have encouraged the use of VSP in the past for examining phenomena such as attenuation, permeability, fracture imaging, and the analysis of multiples. Such studies are outlined in the many excellent books on VSP acquisition and processing by Gal'Perin (1985), Simaan (1984), Balch and Lee (1984), and Hardage (2000). These benefits, together with modern instrumentation such as multi-level three-component tool arrays and multi-component downhole and surface sources that can be quickly and economically deployed, make VSP eminently suited for detailed anisotropy investigation. Indeed, it is should be emphasized that it is mainly through VSP that we can become convinced of the existence of seismic anisotropy.

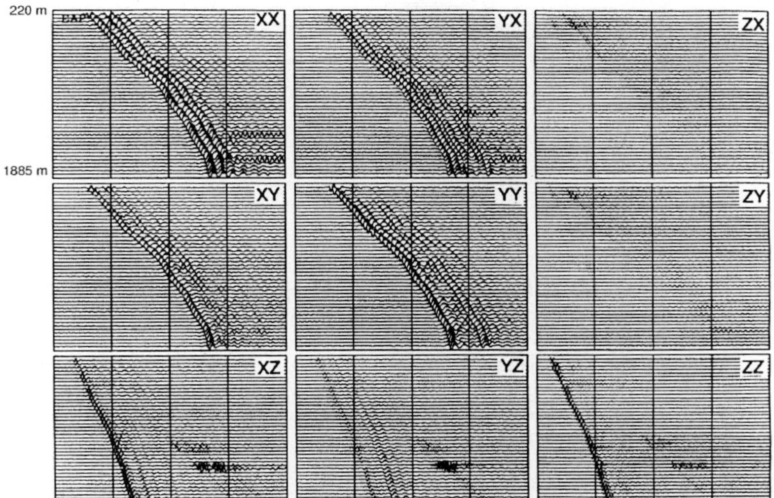

Figure 1.7 Nine-component (9C) matrix of traces recorded in the Paris Basin. A second VSP from the same area is discussed by Lefeuvre et al. (1992). Trace sets are labeled according to the convention used in Figure 1.5. Horizontal (*X* and *Y*) and vertical (*Z*) vibrators are offset at 365 m along an azimuth of S60°W. Most of the shear-wave anisotropy in this example is in the overburden above 500 m giving rise to a delay between the fast and slow shear-waves of 15 ms. As in all multi-component plots, true relative amplitudes are preserved between the components, and the multi-component traces are normalized at each depth. Data Courtesy of the British Geological Survey.

In the hydrocarbon industry VSP achieved popularity as a method for assessing and quantifying seismic anisotropy after evidence of azimuthal anisotropy in reflection data was published at the 1986 SEG Annual Meeting and the report of shear-wave anisotropy in VSPs at the SEG Research Workshop on Recording and Processing Vector Wavefield Data (Snowbird, Utah, August 1989). However, it should come as no surprise that in Former Soviet Union, where VSP had already been in use for over thirty years (Gal'Perin 1985), seismic anisotropy had been widely accepted and integrated into field practice in the geophysical community for some considerable time prior to its acceptance in the West (see for example, Puzyrev et al. 1985).

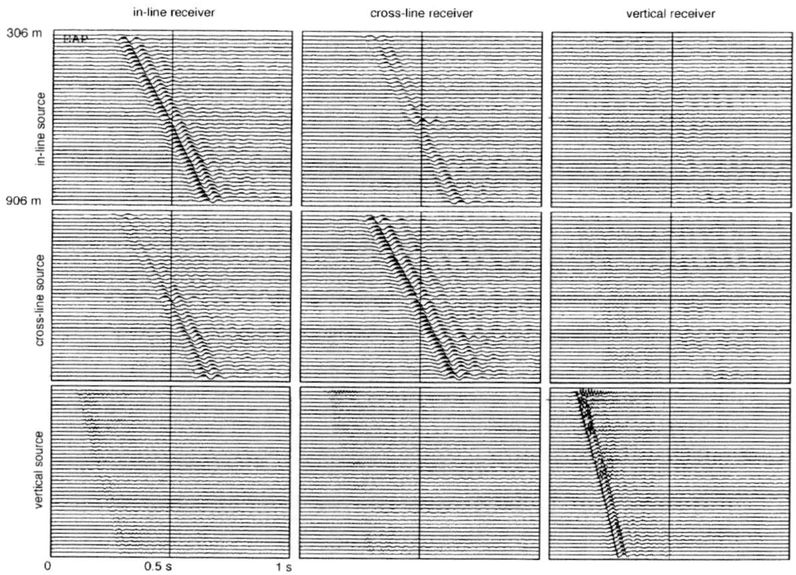

Figure 1.8. VSP data matrix from the Peel VSP shot at the CONOCO Borehole Test Facility, Kay County, Oklahoma. One of the finest examples of nine-component (9C) data. The off-diagonals are well balanced and nearly symmetrical. Sources are orthogonally aligned horizontal and vertical vibrators, with respective offsets of 65 m and 71 m at an azimuth of N122°E. Data have been rotated so that the geophone components lie along the in-line and cross-line directions. These data have been analysed together with another 9C VSP by Horne (1995) and Horne et al. (1997) to reveal inclined cracks with a vertical shear-wave birefringence of between 2 % and 7 %, larger in "reservoir" zones.

Once the technique was adopted, many further advances were made. For example, Alford (1986) described a technique to rotate four-component data (two horizontal source components, two horizontal receiver components) into other coordinate systems. Later, a classic study in integration by Queen and Rizer (1990) (CONOCO Borehole Test Facility, Oklahoma) demonstrated correlations between seismic anisotropy, fractures and stress based on analyses using cores and borehole logging measurements. Some of the early work aimed at improving the depth resolution by integrating the Alford rotation into a layer stripping technique (Winterstein and Meadows 1991a, 1991b, with applications over Cymric, Railroad Gap and the Lost Hills field in California). Another example of a higher-resolution technique was the propagator matrix method developed by Naville (1986), Nicoletis et al. (1988) and Lefeuvre et al. (1992) for the Paris Basin, France. Further improvements in the performance were afforded by the generalized downgoing deconvolution for near-surface correction (MacBeth et al. 1995), and corrections for tool spin and coupling (MacBeth et al. 1994). Miller and Spencer (1994) applied the phase slowness technique of White et al. (1983) to a high-quality multi-level, wide aperture marine walkaway VSP in the North Sea, opening up anisotropy analysis to the offshore environment. Other developments with walkaway VSP used AVO (Leaney 1994, Armstrong et al. 1995). Further advances transferred from surface seismics include the use of azimuth-dependent AVO (Lynn et al. 1995, MacBeth et al. 1999), moveout, interval velocities (Chang and Gardner 1997), and converted waves (MacBeth and Li 1999).

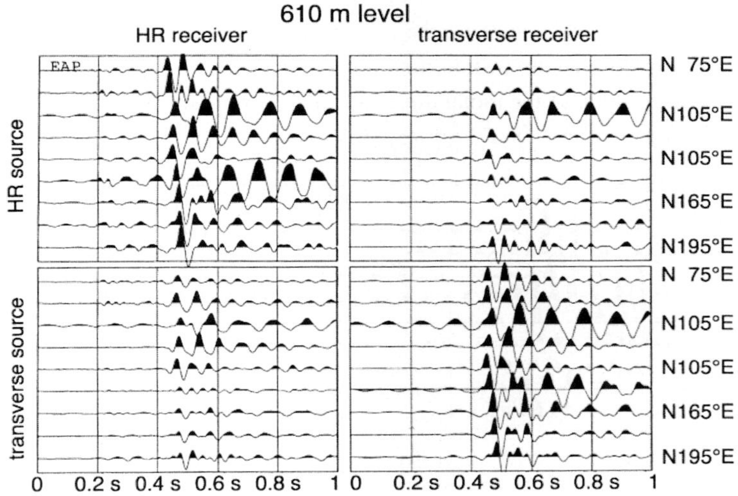

Figure 1.9. Six component (6C) azimuthal VSP recorded at the CONOCO Borehole Test Facility Kay County, Oklahoma. Vibrator sources are arranged at a constant offset of 290 m, and transmit into a receiver at 610 m depth. The azimuthal increment of each offset is 15°. Note the symmetry between the first shear-wave arrivals on the off-diagonals indicating that anisotropy analysis is possible. These, and the remaining levels, have been analysed by Horne and MacBeth (1995) to reveal a line singularity and a crack related anisotropy of 3 % with inclined cracks striking N72°E. A model incorporating a point singularity also fits the data.

It is instructive to examine Figures 1.1 to 1.11, which illustrate a variety of multi-component VSP examples drawn from the period 1983 to 1996. The examples range from three-component downhole recordings of a single impulsive source to full nine-component survey using three orthogonal source motions generated using the vibroseis technology. Orientation of the three-component receiver package is specified by the alignment of each sensor along the axes of an orthogonal Cartesian system. There are several coordinate systems in common use

a) *X-Y-Z* or *IL-CL-V*. This system is defined by the in-line direction (*X* or *IL*) that connects the projections of source and receiver on a horizontal plane. The horizontal direction perpendicular to this axis is the cross-line direction (*Y* or *CL*). It is taken to be positive if the rotation from *X* to *Y* is in the same direction as a rotation from North to East. The vertical (Z or V) points downward, thus completing a right-handed set. This acquisition-related coordinate system can become problematic when there is more than one source offset (see below).

b) X'-Y'-Z or H_1-H_2-V. Most sensor packages are gimbal-mounted, so that the axis of one sensor points vertically downward. The orientation of the two horizontal sensors remains arbitrary.

c) *R-T-N*. This system is defined according to the incident wave. Mathematical rotation of the sensor package into this coordinate system is often employed to separate the wave types. The radial (R) direction is aligned with the ray path, while T is horizontal and transverse. Normal (N) direction lies in the vertical plane through source and receiver (the *sagittal plane*) and completes the orthogonal set. Occasionally the adjective *radial* is incorrectly applied to the in-line direction (as in item a). The in-line direction is the horizontal projection of the radial direction (*HR*). A rotation about the Y-axis transforms the *R-T-N* system to the *HR-T-V* (or *IL-CL-V*) system.

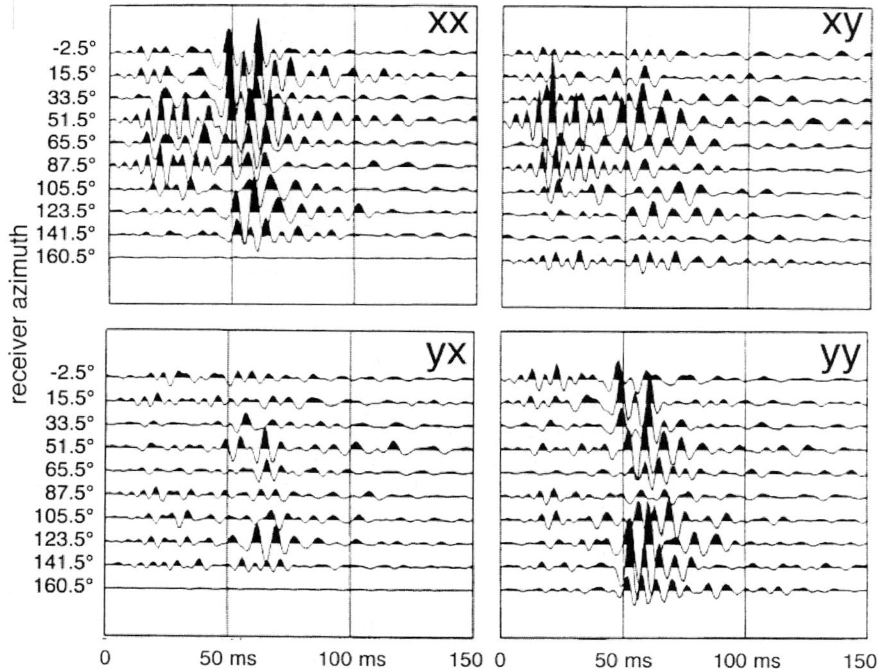

Figure 1.10. Four component (4C) common source gathers recorded at a source depth of 23.5 m and geophone offset of 7.62 m for an azimuthal "reverse VSP" shot at the CONOCO Borehole Test Facility. The CONOCO downhole source creates polarized shear-waves by a rotating eccentric mass. The source and receiver have been rotated parallel (x) and perpendicular (y) to the presumed fracture orientation of N70°E. Note that the lower-case labels are used to indicate components directed parallel and perpendicular to the strike of the fractures It is difficult to distinguish noise from signal on the off-diagonals, suggesting that the fracture orientation is correct. The systematic amplitude and travel-time variations with azimuth are discussed in detail by Liu et al. (1993), and relate to large-scale fractures.

Single three-component recordings are the earliest form of seismic anisotropy measurement, for which particular emphasis was placed on the shear-waves. Thus, the phenomenon of shear-wave splitting (Crampin 1984) was analysed. Here, shear-wave energy is split upon transmission through a (fracture or crack-related) TIH rock into two distinct arrivals, with slightly different velocities and very different polarizations. Initially, the only way of analysis was by interpreting the complicated morphology of the shear-wave signature (techniques used in the past are described in the appendix).

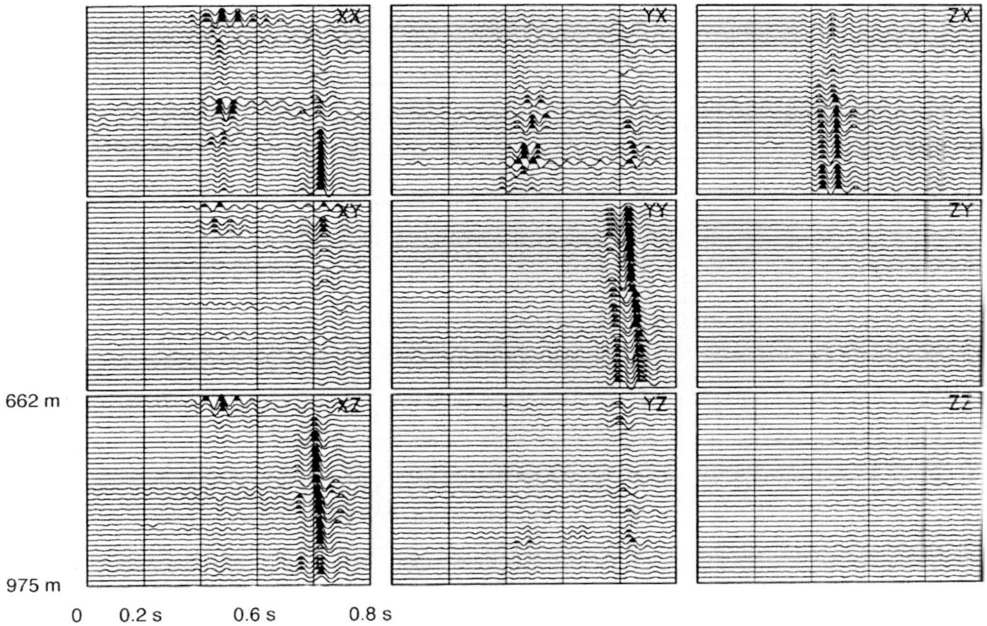

Figure 1.11. An unusual 9C VSP from the Iatan East Howard field in Texas. Acquisition is with an OMNIPULSE source, offset 839 m with an azimuth of N85°E from the wellhead. These data have been analysed in detail by Baptie (1993) who explained the switch of the intensity to the cross-diagonal as due to bending of the ray path. Analysis reveals an average anisotropy of several percent.

Due to the strong connection between fractures and shear-waves (see Chapter 3 for details), shear-wave splitting remains at the focus of our attention. However, it should be added that recent research has also shown that a "compressional wave parameter" is equally important. This parameter is accessed by utilizing multi-offset sources and multiple level tools (Figure 1.4, Chapter 7). Additional information on seismic anisotropy is provided by acquisition geometries incorporating different line azimuths or a range of azimuths at a fixed offset (azimuthal VSP or walkaround, Figure 1.9) opening up possibilities to access amplitude versus offset/direction or moveout variations (see Chapter 7 and 8).

Although downhole recording can now guarantee high fidelity vector data, it should be emphasized that the overlap provided by combining different sources in the nine-component VSP (Figures 1.7, 1.8, and 1.11) contributes substantially to the robustness of the method.

Indeed, the extra directional sources ensure firstly that all wave types are adequately represented, and secondly that a known proportion of their energy is transmitted into each survey direction. Individual source directions in these examples are usually in-line, cross-line and vertical. This convention can be adhered to when many source points are arranged along a line, but the nomenclature gets into difficulties when the well is deviated (so that some receiver positions do not lie in the vertical plane through the source line). In such cases it is best to rotate into the *HR-T-V* system.

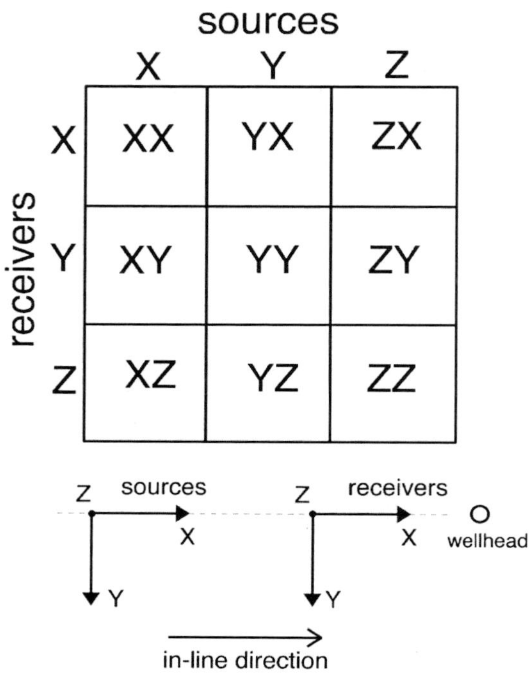

Figure 1.12. Alignment convention for a multi-component VSP. The *X* direction is in all cases defined as the in-line direction (pointing towards the wellhead), with the cross-line direction (transverse to the wellhead line) preserving the right handed coordinate system. The *Z* axis is directed downwards. With source motion and receiver components in this acquisition system, a consistent and straightforward anisotropy analysis can be made.

Tatham and McCormack (1991) give a good description of the special equipment required to generate the source motions and guarantee directionality. The operational principle behind this acquisition may be understood by considering an isotropic earth. It is shown that the particle motion for the compressional waves is along the direction of propagation, those of the shear waves (SV or SH) perpendicular to the direction of propagation. Consequently, the orientation of the source influences the intensity of each type of wave.

To get the best out of each wave type, each should be separately excited by an appropriate source. For example, for a near-offset VSP, we should use a vertical (*Z*) impulse for compressional waves and a horizontal impulse for the shear-waves – the horizontal impulse could be along the *X* or *Y*-axis. In a similar way, reception of the waves can be achieved along the three specific directions of motion (*X*, *Y* or *Z*). It is thus possible to generate in one of three directions and receive in one of three directions, thus a nine-component data matrix can be built up (Figure 1.12).

For an isotropic or TIV medium, the 9C near-offset data matrix should have only significant diagonal components, consisting of the P- and S-waves generated directly by the source. For an anisotropic medium with another symmetry system, the *XY* and *YX* components are also non-zero due to shear-wave splitting. They are linked to each other and to the principal directions by the vectorial geometry. However, it should be noted that misunderstood source motion (Figure 1.2), inhomogeneities (Figure 1.5), near-surface anisotropy (Figure 1.7), or inaccurate source-receiver alignment will result in a departure from this picture and create non-zero off-diagonal components. Symmetry helps as a first step towards assessing these acquisition problems, but this is not sufficient for anisotropy analysis (Chapter 6). Thus the data must first be conditioned by a number of pre-processing steps (Chapter 4) to ensure the quality of the final result and to facilitate the application of the matrix techniques that form the basis of multi-component VSP processing (Chapters 5 and 5). The major aim of the book is to take the reader through the steps involved in this process, towards the final product of robust seismic anisotropy measurements.

CHAPTER 2 – ANISOTROPIC REPLACEMENT MEDIA

The VSP techniques described in the following chapters detect the directional behavior of seismic signatures caused by preferred orientation of fine-scale features in the rock fabric. As this directionality can also be explained by elastic waves propagating through a homogeneous anisotropic medium, seismic anisotropy provides an effective downscaling (or inversion) tool, provided the type of heterogeneity is known. The starting point for anisotropy analysis is the generalized Hooke's law, and in particular the symmetric, fourth-rank stiffness tensor relating the second rank stress and strain tensors. Indeed, anisotropic wave behavior depends upon the material symmetry exhibited through the 21 independent stiffnesses (elastic moduli) in this tensor together with their relative magnitudes. A detailed treatment is given by Helbig (1994), and the relation to reflectivity and moveout in Tsvankin (2001).

The description of elastic behavior is simpler if the fourth-rank stiffness tensor **C** is mapped onto a symmetric 6×6 matrix **c**. Each element c_{ij} $(i,j \in \{1,\ldots,6\})$ is related to the elements C_{pqrs} of the tensor by the "Voigt convention" (see, e.g., Voigt 1928 or Helbig 1994). There are only a few isolated cases where the use of the fourth-rank tensor is necessary. To fully appreciate the downscaling process, its stability and inherent non-uniqueness, details of the stiffness tensor must be constructed from mathematical idealizations of specific rock features. With this aim in mind, this chapter presents formulae for the stiffness matrix in four main categories of fine-scale heterogeneity in rocks: cracks, fractures, sedimentary layering, and the intrinsic structure of shales.

Cracks and fractures give rise to a replacement medium that is transversely isotropic with a horizontal axis of symmetry TIH, and a stiffness matrix of the form

$$\mathrm{c} = \begin{pmatrix} c_{11} & c_{12} & c_{12} & 0 & 0 & 0 \\ c_{12} & c_{33} & c_{33}-2c_{44} & 0 & 0 & 0 \\ c_{12} & c_{33}-2c_{44} & c_{33} & 0 & 0 & 0 \\ 0 & 0 & 0 & c_{44} & 0 & 0 \\ 0 & 0 & 0 & 0 & c_{66} & 0 \\ 0 & 0 & 0 & 0 & 0 & c_{66} \end{pmatrix}. \tag{2.1}$$

Such media give wave signatures that vary as a function of azimuth; correspondingly, they are referred to as azimuthally anisotropic.

On the other hand, shales or horizontally stratified materials result in a replacement medium that is transversely isotropic with a vertical symmetry axis (TIV), and a corresponding stiffness matrix

$$\mathbf{c} = \begin{pmatrix} c_{11} & c_{11}-2c_{66} & c_{13} & 0 & 0 & 0 \\ c_{11}-2c_{66} & c_{33} & c_{13} & 0 & 0 & 0 \\ c_{13} & c_{13} & c_{33} & 0 & 0 & 0 \\ 0 & 0 & 0 & c_{55} & 0 & 0 \\ 0 & 0 & 0 & 0 & c_{55} & 0 \\ 0 & 0 & 0 & 0 & 0 & c_{66} \end{pmatrix}. \tag{2.2}$$

For this group of materials, wave signatures vary only with incidence angle and are thus azimuthally isotropic. It should be noted that inclined cracks/fractures, dipping shales or layers, cracks/fractures intersecting shale/layering or other sets of cracks/fractures, in addition to stress, all result in lowering of the material symmetry (a good description of anisotropy symmetries is given by Winterstein 1990). Such systems could be orthorhombic

$$\mathbf{c} = \begin{pmatrix} c_{11} & c_{12} & c_{13} & 0 & 0 & 0 \\ c_{12} & c_{22} & c_{23} & 0 & 0 & 0 \\ c_{13} & c_{23} & c_{33} & 0 & 0 & 0 \\ 0 & 0 & 0 & c_{44} & 0 & 0 \\ 0 & 0 & 0 & 0 & c_{55} & 0 \\ 0 & 0 & 0 & 0 & 0 & c_{66} \end{pmatrix}, \tag{2.3}$$

monoclinic, with one possible form

$$\mathbf{c} = \begin{pmatrix} c_{11} & c_{12} & c_{13} & 0 & 0 & c_{16} \\ c_{12} & c_{22} & c_{23} & 0 & 0 & c_{26} \\ c_{13} & c_{23} & c_{33} & 0 & 0 & c_{36} \\ 0 & 0 & 0 & c_{44} & c_{45} & 0 \\ 0 & 0 & 0 & c_{45} & c_{55} & 0 \\ c_{16} & c_{26} & c_{36} & 0 & 0 & c_{66} \end{pmatrix}, \tag{2.4}$$

or even triclinic (with 21 significant stiffnesses). In these cases the stiffness matrix is clearly fuller, with calculation and visualization of the wave properties more challenging. This chapter reparameterizes the stiffness matrices above by introducing a two-parameter system for the cracks and fractures (e_n and e_t), and a three-parameter system for shales and layering ($\varepsilon, \delta,$ and γ). These two groups are sufficient to simulate the majority of anisotropies in geologic media. Application of the multi-component VSP technique to these different anisotropy categories is the primary objective of this book, with cracks/fractures forming the focus in Chapters 4, 5, and 6. Chapters 7 and 8 are more relevant to shales/layering.

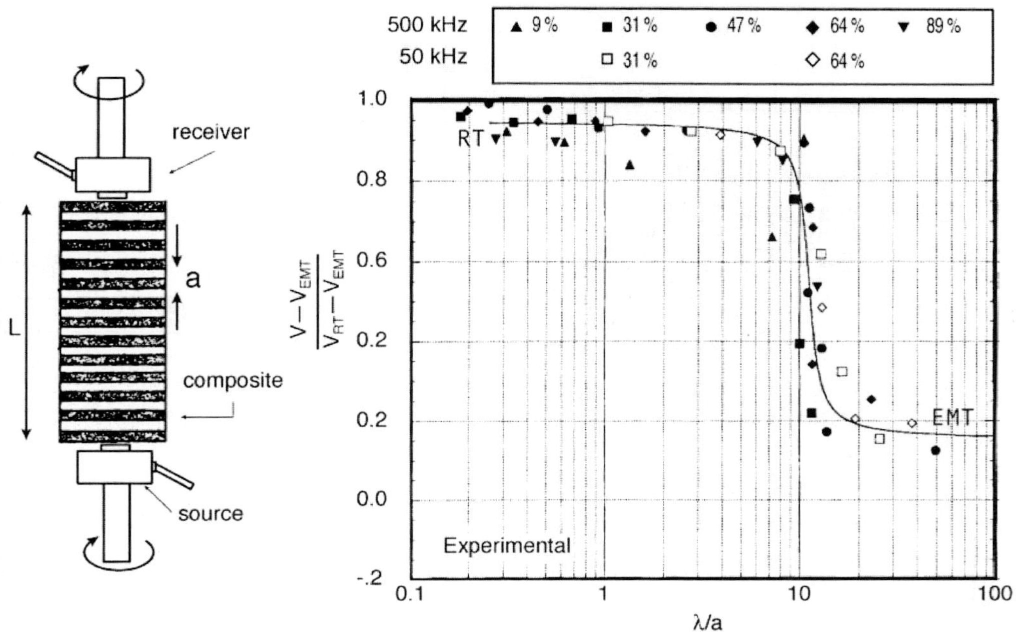

Figure 2.1. Results from a laboratory study by Marion et al. (1994) to test the transition between short wavelength (ray theory) and long wavelength (equivalent medium theory) predictions for velocity. P-wave propagation is normal to a cyclic composite of steel and plastic discs. The ratio of wavelength λ to thickness a is varied, while the total thickness L of the composite remains fixed. Measurements are shown for two peak frequencies of the transducers and for several different volume fractions of steel. The coice of materials give $V_{EMT} < V_{RT}$.

Scale is the major controlling factor in understanding the applicability of the replacement media concept (MacBeth 1994, Lynn 1996). This can be appreciated by considering vertical wave propagation through a horizontally layered sequence. If the layers are several times thicker than the incoming wavelength, the wave spends several complete cycles of its motion in each layer. The velocity through the sequence is obtained by dividing the total path length by the sum of the travel-times in the individual layer segments. Thus the "effective velocity" is the harmonic average of the constituent velocities. On the other hand, if the layer thickness is only a fraction of a wavelength, the wave spends time in several layers during its cycle of motion. These layers are now treated as one homogeneous medium. The velocity is now related to the square root of the harmonic average of the elastic stiffnesses.

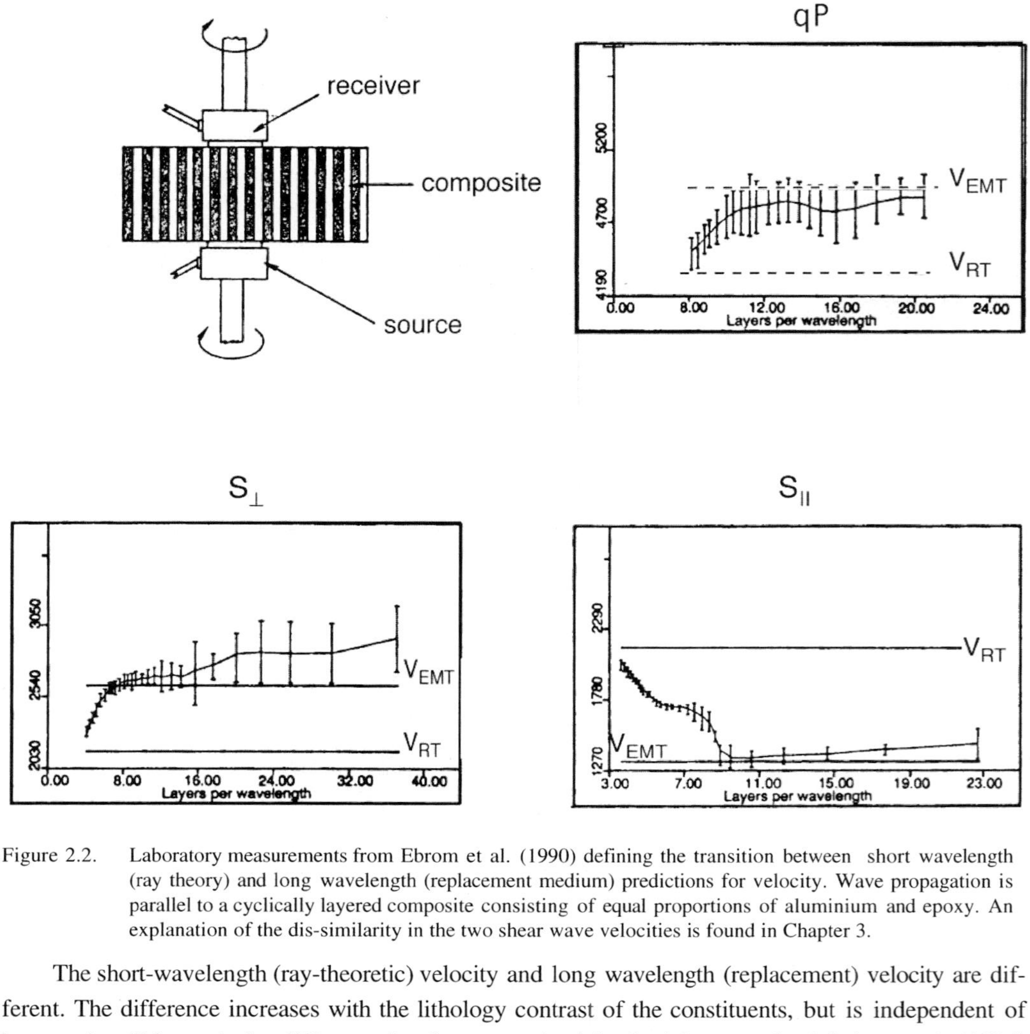

Figure 2.2. Laboratory measurements from Ebrom et al. (1990) defining the transition between short wavelength (ray theory) and long wavelength (replacement medium) predictions for velocity. Wave propagation is parallel to a cyclically layered composite consisting of equal proportions of aluminium and epoxy. An explanation of the dis-similarity in the two shear wave velocities is found in Chapter 3.

The short-wavelength (ray-theoretic) velocity and long wavelength (replacement) velocity are different. The difference increases with the lithology contrast of the constituents, but is independent of layer order. This particular difference has been examined in the laboratory by Marion et al. (1994) (Figure 2.1), and also tested for propagation parallel to a layered medium by Ebrom et al. (1992) (Figure 2.2). Both find that a narrow transition zone exists between these two velocities, in which the waves are attenuated and dispersed and where the scattering effects depend strongly upon layer order. The approximate position of this zone is generally independent of composition and impedance contrasts. Literature on this topic predicts a range of layer thicknesses between one fifth and one fifteenth of a wavelength.

The principles above also translate to cracks and fractures. For cracks this implies that their size and separation must be smaller than the probing seismic wavelength. Crack scattering has been tested in the long wavelength limit using 2D numerical experiments (Kachanov (1992), Davis and Knopoff (1995), Muijres (1998), and Kelner, Bouchon and Coutant (1999)). For a VSP peak frequency of

35 Hz and a typical velocity of 3.5 km/s, the long-wavelength limit implies that features of a size of less than about 6 m are averaged into a replacement medium. This therefore includes small-scale sandstone architecture such as dunes, sheets, cross-beds (that possess substantial permeability contrast and thus presumably elastic impedance contrast), pore-scale phenomena, shale minerals and platelets, in addition to grain-sized cracks and fracture apertures.

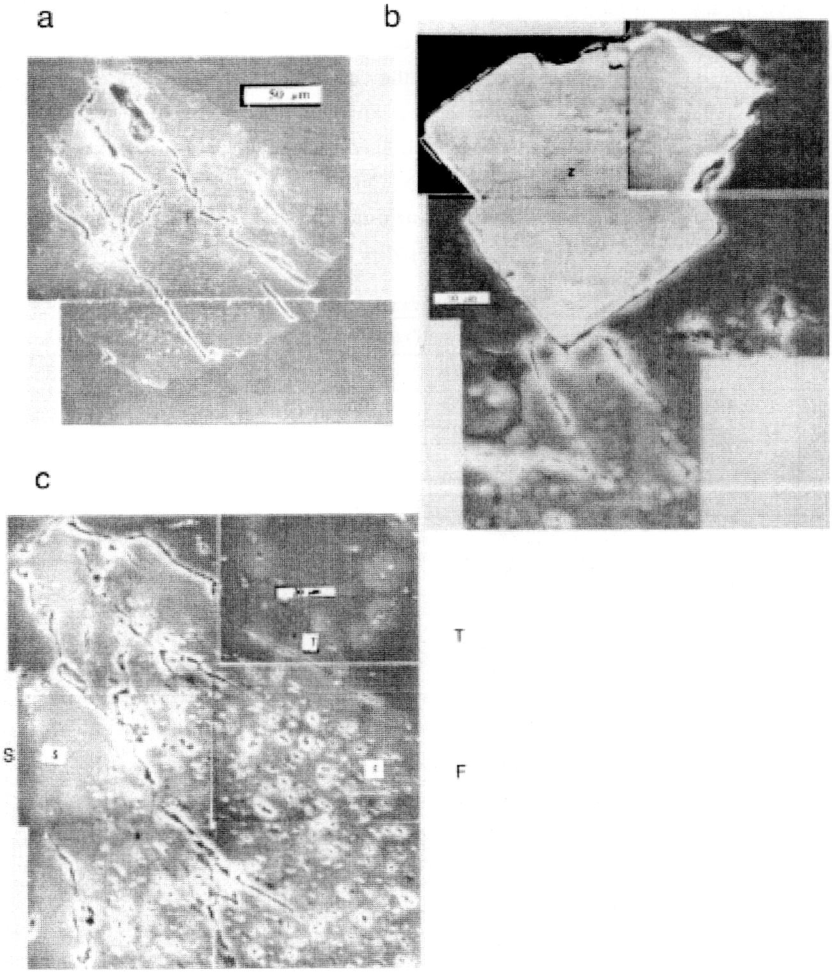

Figure 2.3. Figures extracted from Wong et al. (1989) displaying cracks at the microscopic scale, imaged using the scanning electron microscope. Specimens are from Rutland quartzite with grain sizes of 300 microns, interconnected porosity of 0.5%, total porosity of 0.13% and crack apertures of 0.5 microns. (a) intra-granular cracks with wide apertures in a feldspar grain, occuring principally along the cleavage planes. The surrounding quartz grains are relatively free of cracks. (b) cracking along a grain boundary between zircon and quartz; (c) extensive cracking in feldspar which can be related to different stages of crack healing. Open grain boundary cracks (T), partially healed arrays of cracks (F), and completely healed cracks (S). Courtesy of the American Geophysical Union (AGU).

2.1 CRACKS

Cracks in rocks appear as small flattened in intra- or inter-granular defects (Figure 2.3). They are generally caused by geochemical interactions, or by or thermal loading. They range in dimension from microns up to several millimeters, depending upon the particular lithology and origin. Typical thickness-to-length ratios ("aspect ratio") are 0.001 to 0.1 (resembling flattened pores). Cracks are important as they relate to the mechanical integrity and flow characteristics of the rock. They should be considered when predicting the effect of stress changes on the rock properties, as they are more compliant to loading than the pores. Significantly, they contribute no more than a few percent to the overall porosity (typically much smaller), and hence cannot generally be detected by their effect on seismic impedance. On the other hand, aligned crack give a very large anisotropy response, as can be seen in the table below.

Table 2.1. Aspect ratio and porosity combinations for gas-filled crack, all of which give a velocity anisotropy of 15 percent.

P-wave response

Aspect Ratio	0.001	0.01	0.1
Porosity	0.2 %	2 %	4 %

S-wave response

Aspect Ratio	0.001	0.01	0.1
Porosity	0.005 %	0.05 %	0.5 %

This strong connection between mechanism and seismic effect is why many researchers contend that seismic anisotropy can help to monitor hydrocarbon production or even predict earthquakes. Evidence of crack alignments and the resultant seismic anisotropy have been shown in laboratory studies for a variety of lithologies ranging from sandstones (Sayers et al. 1990), limestones (Gupta 1973) and granite (Nur and Simmons 1969). There are numerous other laboratory examples of crack-induced seismic anisotropy throughout the literature.

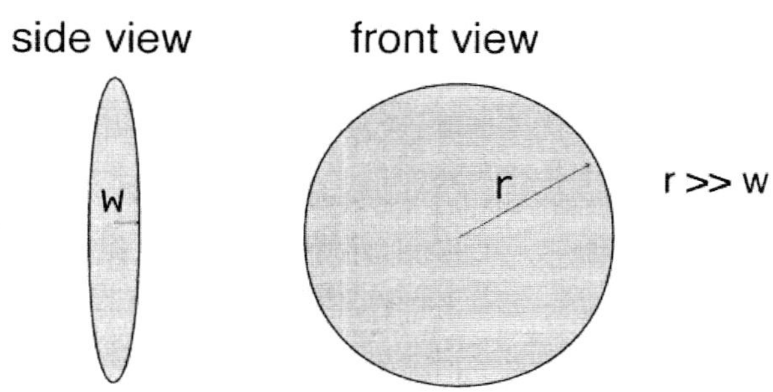

Figure 2.4. The oblate spheroid is used to model a representative crack or pore, thus making the mathematics for the replacement medium tractable. This 'Hudson crack' is an ellipsoid of revolution, with a circular cross-section and a small width or thickness. The aspect ratio is defined as the ratio of half-width, w, to radius of the crack face, r.

Figure 2.5. This modern lava specimen shows vesicles that closely approximate the idealized Hudson crack. These oblate spheroidal inclusions display preferential alignments along the flow direction. Note the change in the density of vesicles near the top of the flow, and the rubble top. The high background velocities would render this particular medium highly anisotropic. Courtesy of Dave Long, British Geological Survey

This section is based mainly on Hudson (1980) and Hudson (1981). These papers provide essential references to the effects of various geometries and fills. The specific contribution of Hudson (1980) was to extend the work of Garbin and Knopoff (1975) to second-order accuracy, thus including crack-crack interactions, in addition to formulae for an arbitrary shape and spatial probability density. Analytic formulae are provided for cracks modeled as oblate spheroids (flattened spheres – often inaccurately called 'penny-shape' crack, see Figures 2.4 and 2.5). The formulae are restricted to dilute concentration (small porosity), random spatial distribution, aspect ratios smaller than 0.3, and fluids that remain isolated in each crack. They have been verified under laboratory conditions by Assad et al. (1990) and Peacock (1991).

2.1.1 Volume distribution of aligned vertical cracks (TIH)

A weak concentration of aligned cracks in an otherwise isotropic matrix generally weakens the rock differently along and perpendicular to the plane of the crack. This reduction in the background stiffnesses $\mathbf{c}_b$ is proportional to the overall crack porosity Φ, their individual compliance and the properties of the background material. The resultant elastic stiffness matrix incorporating a first-order perturbation (corresponding to the first-order term in number density given by Hudson (1980)) can be written in the form

$$\mathbf{c} = \begin{pmatrix} (\lambda+2\mu)(1-e_n) & \lambda(1-e_n) & \lambda(1-e_n) & 0 & 0 & 0 \\ \lambda(1-e_n) & (\lambda+2\mu)(1-\zeta^2 e_n) & \lambda(1-\zeta e_n) & 0 & 0 & 0 \\ \lambda(1-e_n) & \lambda(1-\zeta e_n) & (\lambda+2\mu)(1-\zeta^2 e_n) & 0 & 0 & 0 \\ 0 & 0 & 0 & \mu & 0 & 0 \\ 0 & 0 & 0 & 0 & \mu(1-e_t) & 0 \\ 0 & 0 & 0 & 0 & 0 & \mu(1-e_t) \end{pmatrix}, \tag{2.5}$$

where λ and μ are the Lamé constants and $\zeta = 1 - 2\beta^2/\alpha^2 = \lambda/(\lambda+2\mu)$. α and β are the P and S wave velocities, respectively. It will be seen later in this section that this expression can be used to accommodate a number of crack- and fracture styles, and is the *generic* form for the crack and fracture-related anisotropy. The dimensionless scalar parameters e_n and e_t can be interpreted as a measure of the overall crack compliance relative to the background material. They are directly proportional to the crack porosity Φ, and they are the response of the crack to an incident stress field. Together with the normal and tangential boundary conditions they result in, respectively, the "response factors" b_n and b_t, that are proportional to the compliance of a representative crack. These factors account for the crack stiffening due to either material fill of the crack or to an increase in aspect ratio. The contributions to e_n and e_t from the porosity Φ and the response factors b_n and b_t are expressed as

$$e_n = \Phi b_n \tag{2.6}$$

and

$$e_t = \Phi b_t \;. \tag{2.7}$$

These equations immediately reveal the inherent ambiguity in interpretation of the crack properties: different combinations of porosity Φ and response factors b_n and b_t give the same perturbation. The cracked medium can thus be rendered anisotropic by virtue of a large number of slightly compliant crack, or a smaller number of more compliant ones. An additional constraint from core or log analyses is always required to calculate these quantities individually. The porosity for the specific crack geometry used by Hudson (1981) is given by $\Phi = \frac{4}{3}N\pi\bar{r}^3 a$, where a is the aspect ratio (minor axis divided by major axis), $\bar{r}$ the mean radius, and N the number of crack per unit volume. The exact relationship between fill and geometry depends upon the type of crack. For the Hudson crack, the response factor b_n differentiates between gas fill and liquid fill:

$$b_n^{gas} = \frac{1}{\pi a \eta(1-\eta)}, \qquad \eta = \frac{\beta^2}{\alpha^2}, \tag{2.8}$$

and

$$b_n^{liq} = \frac{b_n^{gas}}{1+\eta_{liq} b_n^{gas}}, \qquad \eta_{liq} = \frac{\rho_{liq}\alpha_{liq}^2}{\rho\alpha^2} \;. \tag{2.9}$$

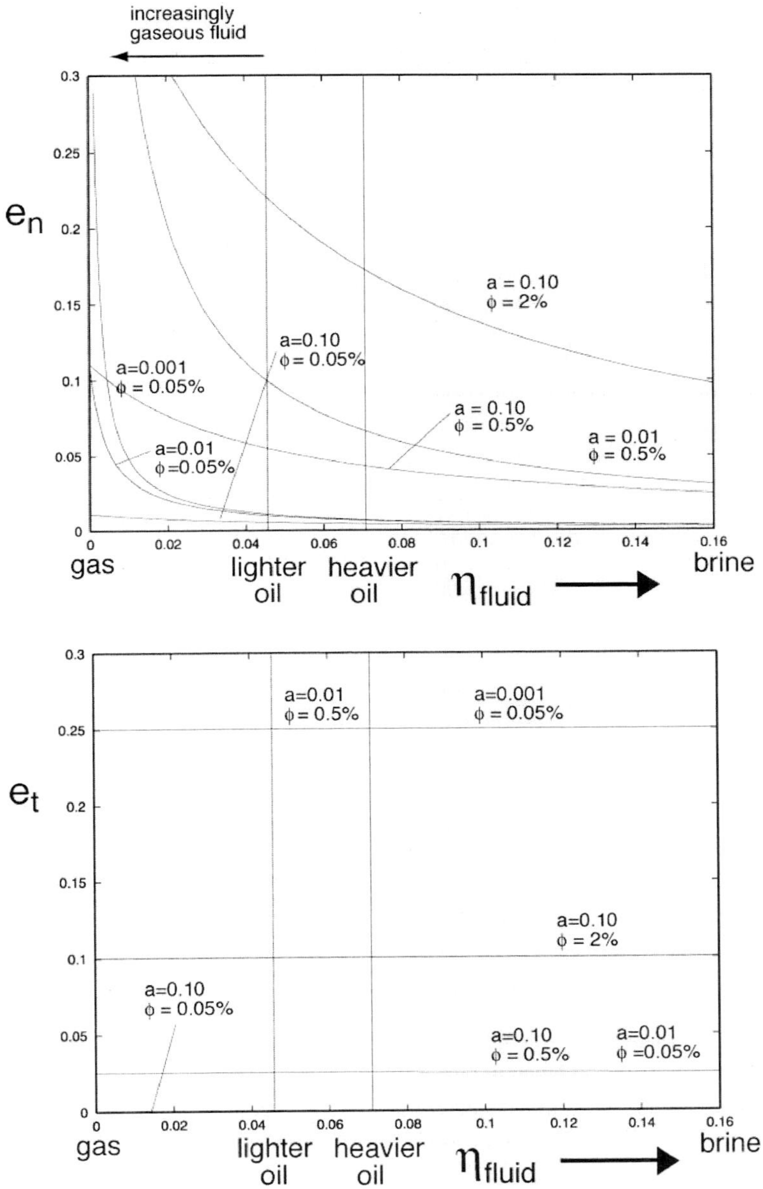

Figure 2.6. Variation in normal and tangential crack indices e_n and e_t, respectively, for continuously changing fill parameter η_{fluid} (= $\rho_{fluid}\alpha^2_{fluid}/\rho\alpha^2$) and different aspect ratios a. The fluid properties are derived for a reservoir pressure of 39MPa and a temperature of 112°C using Batzle and Wang (1992). The vertical lines demarcate conditions appropriate to most produced hydrocarbons. The sandstone matrix has α = 3080 m/s, and ρ = 2.25 gcm^{-3}. The crack index e_t is invariant with changing fluid fill according to the Hudson theory, but varies with aspect ratio and porosity. Different crack porosities Φ are considered, with the largest being the upper limit for the Hudson (1980, 1981) theory.

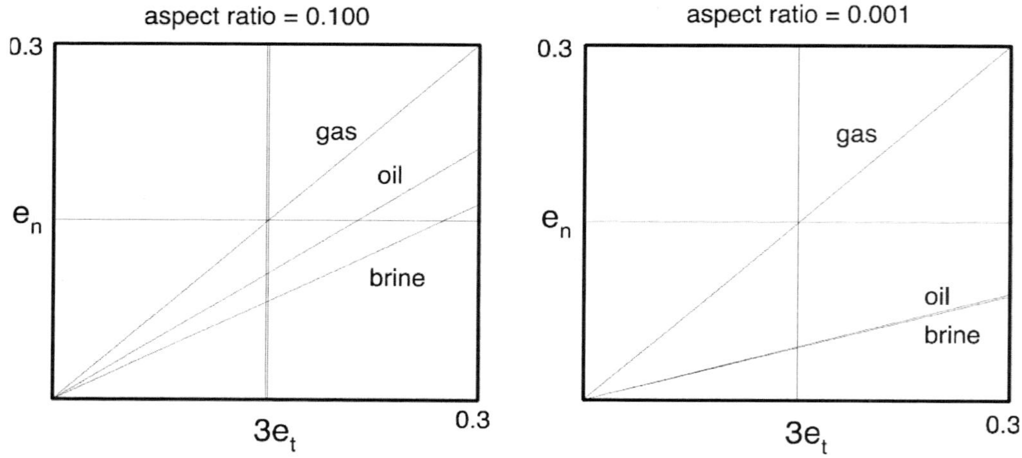

Figure 2.7. The link between normal and tangential fracture/crack indices for three different fills and two different aspect ratios using Hudson's expressions for a volumetric distribution of cracks. The lines are generated by varying porosity, but fixing $\alpha/\beta = 2$. Note the predicted insensitivity to the saturant for thin crack.

Equation (2.9) is equivalent to adding the crack and fill stiffnesses. The formula is obtained when both a and η_{liq} are small, and reverts to (2.8) when $\eta_{liq} \to 0$. The density of the liquid is ρ_{liq}, the density of the background material is ρ, and the P-wave velocity in the liquid is α_{liq}. Variations of e_n with crack fill and aspect ratio are shown in Figure 2.6. The curves are based on equations (2.6) and (2.9) for a continuous range of η_{fluid} (defined using η_{liq} for increasingly gaseous constituents), specifying fill conditions from gas through to oil and then brine. One common feature is a rapid increase in e_n with increasing gas saturation; this becomes more pronounced for thin crack due to a sudden reduction in crack stiffness. There is little difference between oil-filled and brine-filled crack with small aspect ratio, but the saturant can be discriminated for larger aspect ratios. A correction to (2.9) was given by Thomsen (1995) to provide more accuracy when dealing with larger values of η_{liq}:

$$\eta_{liq} \to \frac{\eta_{liq}}{1-\eta_{liq}} . \tag{2.10}$$

This ensures that with increasing incompressibility $b_n^{liq} = 0$ for $\eta_{liq} = 1$. The factor corrects an error of 10 % for brine-filled crack in chalk.

The response factor b_t for the aligned Hudson crack does not distinguish between gas or liquid fill:

$$b_t^{fluid} = \frac{4}{\pi a(3-2\eta)} , \tag{2.11}$$

and thus e_t depends only on the crack porosity (Figure 2.6). This particular result from the Hudson theory has not been fully tested (see Chapter 9 for further considerations). For crack of Hudson-type, the ratio of e_n to e_t is approximately 3 for $\alpha/\beta = 2$ – cracked media are more compliant under compression than under shear.

Figure 2.7 shows the relationship between e_n and e_t, and its effect on the discrimination ability for different crack geometries. The expressions of Hudson (1981) also accommodate a solid fill, provided it is softer than the background matrix (for example, clay). This adjusts the tangential response factor:

$$b_t^{sol} = \frac{b_t^{fluid}}{1+\eta_{sol} b_t^{fluid}} , \tag{2.12}$$

where $\eta_{sol} = \rho_{sol}\beta_{sol}^2 / \left(\rho\beta^2\right)$, and β_{sol} is small but finite (for example, with a clay fill, β_{sol} = 400 m/s, ρ_{sol} = 1.5, and hence $\eta_{sol} \approx 0.1$). The expression is valid provided $\eta_{sol} \ll 1$ but larger than the aspect ratio a. For very thin crack equation (2.12) reduces to

$$b_t^{sol} = \frac{1}{\eta_{sol}} . \tag{2.13}$$

2.1.2 Accuracy of the crack expressions

It is expected that there should be an upper limit for the magnitude of e_n and e_t used in the expressions, since they are based on single-scattering theory. This in turn limits the range of aspect ratios, fills and porosities that can be accurately predicted. Hudson (1980) provided an extension to these formulae by including a second-order contribution that accounts for crack-crack interactions. This additional term predicts

$$e_n \rightarrow e_n - \tfrac{1}{15}\left(15\zeta^2 + 28\zeta\eta + 28\eta^2\right)e_n^2 \approx e_n - 0.6{e_n}^2 , \tag{2.14}$$

and

$$e_t \rightarrow e_t - \tfrac{2}{15}(3+2\zeta)e_t^2 \approx e_t - 0.5{e_t}^2 , \tag{2.15}$$

with the approximations holding for $\alpha/\beta = 2$. These expressions are plotted in Figures 2.8, where the effect can be seen as a reduction in the rate of change of the overall crack compliance with increasing individual crack compliance (more crack give lower overall compliance than their straight summation). Calculations show that the first-order term gives a 10 % error in the velocity prediction when e_n and e_t are about 0.3; thus this is the limit of applicability. The equations also indicate that the form of (2.5) remains appropriate, with predictions now accurate for large crack compliances. Thus, the second-order correction is only relevant when the measured fracture parameter requires interpretation based on the microstructural models.

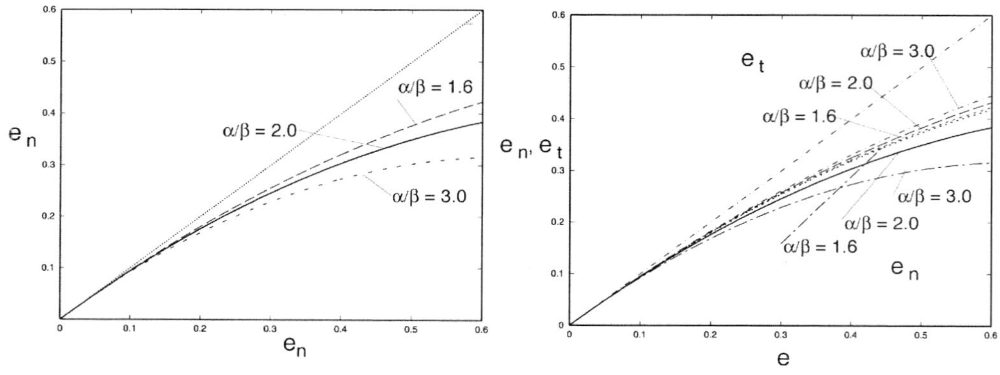

Figure 2.8. Left: divergence between the second order corrected fracture index (e'_n) and the first order term (straight line). The effect of the correction is to reduce the overall impact of larger e_n values. Right: both normal and tangential indices, second order corrected and evolving with a common value e. This highlights the differences in the two approximations for different background media.

There is an additional limit on the aspect ratio, that causes a breakdown in the representation of (2.5) for $a \approx 0.3$ (Schoenberg and Douma 1988). For aspects ratios this large, recourse must be made to numerical computation based on Nishizawa (1982) or Hudson (1994). An interesting prediction is that for prolate (needle-shaped) pores $c_{44} < c_{66}$ (Horne, MacBeth and Liu 1997).

2.1.3 Volume distribution of randomly oriented crack (isotropic)

Cracks oriented at random reduce the strength of the rock but the stiffness matrix remains isotropic:

$$c = \begin{pmatrix} (\lambda+2\mu)(1-e_1) & \lambda(1-e_2) & \lambda(1-e_2) & 0 & 0 & 0 \\ \lambda(1-e_2) & (\lambda+2\mu)(1-e_1) & \lambda(1-e_2) & 0 & 0 & 0 \\ \lambda(1-e_2) & \lambda(1-e_2) & (\lambda+2\mu)(1-e_1) & 0 & 0 & 0 \\ 0 & 0 & 0 & \mu(1-e_3) & 0 & 0 \\ 0 & 0 & 0 & 0 & \mu(1-e_3) & 0 \\ 0 & 0 & 0 & 0 & 0 & \mu(1-e_3) \end{pmatrix}, \tag{2.16}$$

where

$$e_3 = \frac{2\Phi}{15}(2\eta b_n + 3b_t), \tag{2.17}$$

$$e_1 = \frac{\Phi}{3}\left(\frac{(3-4\eta)^2}{3}b_n + 4\eta e_3\right), \tag{2.18}$$

and

$$e_2 = \frac{\Phi}{3\zeta}\left(\frac{(3-4\eta)^2}{3}b_n - 2\eta e_3\right), \tag{2.19}$$

and response factors b_n and b_t are defined as above. Interestingly, these equations predict that both P and shear waves should be affected by the crack fill and geometry. This crack-specific result should be contrasted with predictions from Gassmann's equations (for isotropic pores) for which the shear modulus is independent of the fluid content.

For $\alpha/\beta = 2$ the relations in (2.17), (2.18) and (2.19) can be approximated by

$$e_1 \approx \Phi\left(\frac{b_n}{2} + \frac{b_t}{7}\right), \tag{2.20}$$

$$e_2 \approx \Phi\left(\frac{b_n}{5} - \frac{b_t}{45}\right), \tag{2.21}$$

and

$$e_3 \approx \Phi\left(\frac{b_n}{7} + \frac{b_t}{2}\right). \tag{2.22}$$

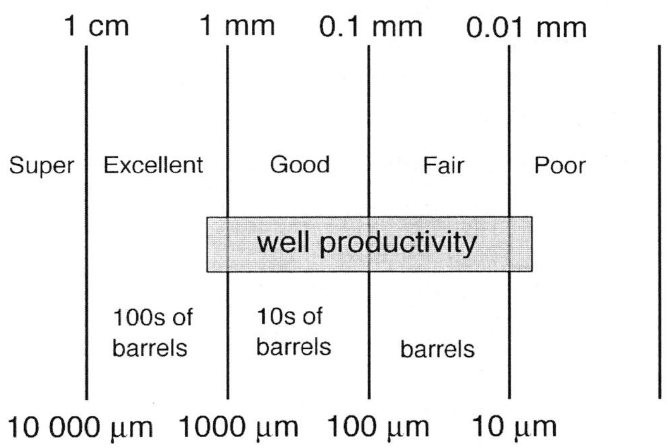

Figure 2.9. Correlation between fracture aperture and productivity for Austin Chalk, Texas Diagram after Lynn (1996).

2.2 FRACTURES

Natural fractures are complicated macroscopic planar discontinuities in the rock, caused by deformation or physical diagenesis. They range in variety from shear fractures that have had displacements parallel to their surfaces (and often form a conjugate pair with two intersecting planes) to joints that have experienced tensional/extensional displacement perpendicular to the fracture surface. The apertures of joints are generally on the millimeter scale, whereas a shear displacement can have an effective hydraulic aperture of several orders of magnitude lower. Typical aperture sizes and their link to production in the Austin Chalk in Texas are shown in Figure 2.9. Fractures generally possess a surface morphology of complicated shape (Figure 2.10), characterized by the stress conditions that occurred during formation (Ameen 1995). Subsequent lateral movement has shifted the relative positions of these surfaces so that they make contact at a number of asperity points.

Figure 2.10. An example of complicated fracture surface morphology. The science of tracing back the roughness to the generative stress mechanism is called "Fractography" (Ameen 1995, Special Publication no. 92. Printed courtesy of the Geological Society, London).

Fractures associated with faults can have complicated gouges that include such as weakened solid material, fluids, or smaller-scale parallel fracturing. In addition, vugs may develop as a consequence of dissolution, giving rise to spheroidal or oblate shaped inclusions with often significant porosity (Figure 2.11). In sedimentary rocks the fracture orientations are often determined by the local stress; fractures open at depth are oriented normal to direction of minimum in-situ stress. Fracture spacing can vary from a few centimeters to some meters or more, and can change with bedding thickness and lithology (Figure 2.11). In general, brittle rocks (e.g., dolomites) have more closely spaced fractures than those with a low percentage of brittle constituents (e.g., limestone) (Nelson 1985). Permeabilities as low as 0.1 mDarcy and as high as 1000 mDarcy are possible in oriented sets, while the fracture porosity Φ spans a range from a fraction of 1 % to several 10 %.

Due to their wide diversity, fractures can form primary, secondary, or even tertiary controls on reservoir flow. A good description of natural fractures is to be found in Aguilera (1980) with emphasis on reservoir engineering aspects, and Nelson (1985) with emphasis on geological aspects.

Single fractures cutting through the rock mass can be visualized theoretically as two interlocking, approximately planar surfaces with lateral dimensions larger than a seismic wavelength. As the aperture or gouge dimensions are much smaller than the wavelength, this fracture model can be further reduced to a single interface, with a distribution of boundary conditions over this surface now representing the different possible conditions listed above. For this model, the stress is assumed to remain continuous across the interface, and throughout the fractured volume, but strain and displacement are discontinuous. The magnitude of the discontinuity depends upon the strength of the bond between the two surfaces. A loosely bonded slip interface is called a "Somigliano dislocation" (Murty 1976).

Upon the further assumption that a linear relation exists between the slip and the stress components, the mean displacement discontinuity $[\mathbf{u}]$ at the interface can be written in terms of mean stress σ in the form given by Schoenberg (1980) as

$$[\mathbf{u}] = \mathbf{Z}\sigma \tag{2.23}$$

where the degree of fracture bonding is now expressed by a 3×3 compliance tensor **Z**. This linear slip model has been successfully tested in physical modeling studies by Pyrak-Nolte, Myer and Cook (1990), Hsu and Schoenberg (1993), and more recently by Haugen and Schoenberg (1995).

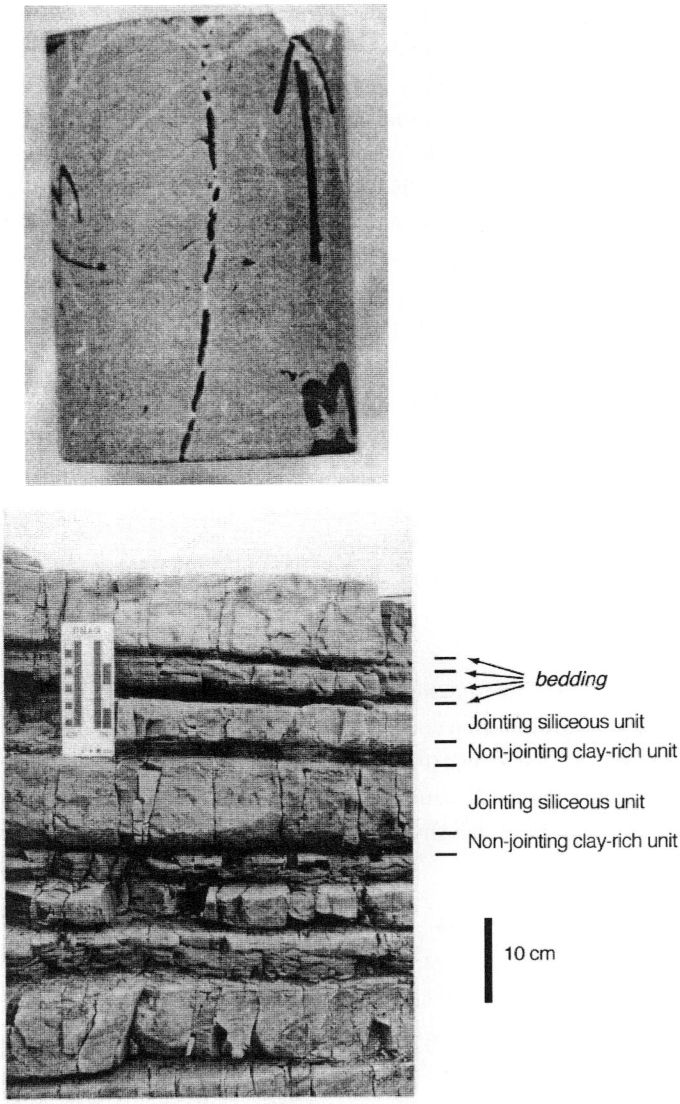

Figure 2.11. Top: vuggy fractures in a carbonate core, extracted from Nelson (1985). Bottom: joints confined to layering, showing the correlation between thickness and spacing (Ameen 1995, Special Publication no. 92. Printed courtesy of the Geological Society, London).

For most practical purposes it is sufficient to consider the fracture interface to possess features that are rotationally symmetry about its normal. Thus the compliance tensor can be reduced to

$$\mathbf{Z} = \begin{pmatrix} Z_n & 0 & 0 \\ 0 & Z_t & 0 \\ 0 & 0 & Z_t \end{pmatrix}, \tag{2.24}$$

where Z_n and Z_t are, respectively, the normal and tangential compliance of the fracture, i.e., the degree to which the fracture yields when subject to the corresponding stress. Positive/negative normal stress corresponds to pulling apart/pushing together the sides of the fracture, tangential stress to sliding the fracture surfaces. Both compliances must satisfy $0 \leq Z \leq 1$.

The model permits computation of the overall compliance for the fractured medium by adding the excess from each fracture in a representative volume of rock to the compliance of the rock matrix. Schoenberg and Sayers (1995) show that for an isotropic background medium and fracture parallel to the 23-plane (normal to the 1-axis) this process yields

$$\mathbf{s} = \begin{pmatrix} \frac{\lambda+\mu}{\mu(3\lambda+2\mu)} + Z_n & -\frac{\lambda}{2\mu(3\lambda+2\mu)} & -\frac{\lambda}{2\mu(3\lambda+2\mu)} & 0 & 0 & 0 \\ -\frac{\lambda}{2\mu(3\lambda+2\mu)} & \frac{\lambda+\mu}{\mu(3\lambda+2\mu)} & -\frac{\lambda}{2\mu(3\lambda+2\mu)} & 0 & 0 & 0 \\ -\frac{\lambda}{2\mu(3\lambda+2\mu)} & -\frac{\lambda}{2\mu(3\lambda+2\mu)} & \frac{\lambda+\mu}{\mu(3\lambda+2\mu)} & 0 & 0 & 0 \\ 0 & 0 & 0 & \frac{1}{\mu} & 0 & 0 \\ 0 & 0 & 0 & 0 & \frac{1}{\mu} + Z_t & 0 \\ 0 & 0 & 0 & 0 & 0 & \frac{1}{\mu} + Z_t \end{pmatrix}, \tag{2.25}$$

where Z_n and Z_t are defined to be the sum for all of the fractures. This can be seen as an enhancement of the normal compliance in the direction normal to the fracture plane, and of the tangential compliance along the two other orthogonal directions (the 2- and 3-axes).

After computing (2.25), the elastic stiffnesses of the composite can be recovered by inverting the compliance matrix **s**. In the stiffness domain, the general form of these equations is identical to that of (2.5) upon the substitution

$$e_n \rightarrow \frac{(\lambda+2\mu)Z_n}{1+(\lambda+2\mu)Z_n} \approx (\lambda+2\mu)Z_n \tag{2.26}$$

and

$$e_t \rightarrow \frac{\mu Z_t}{1+\mu Z_t} \approx \mu Z_t \ . \tag{2.27}$$

The approximation holds for small e_n and e_t (e_n, $e_t < 0.30$ is reasonable). Thus, e_n and e_t can also be interpreted as the ratio of excess fracture compliance to the background rock compliance.

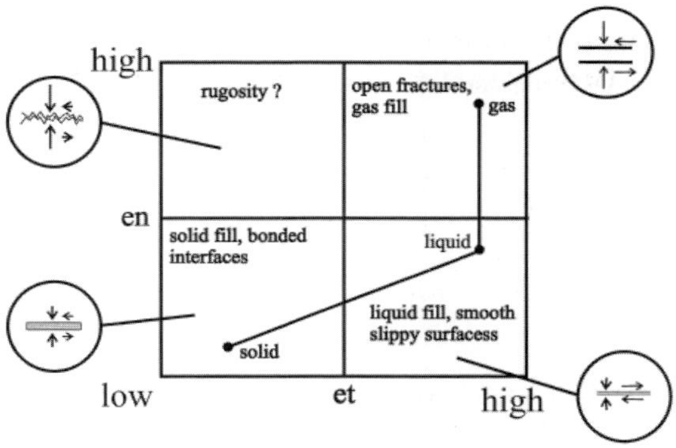

Figure 2.12. Qualitative assessment of the relative compliances for different physical conditions at the fracture interface. The line of transition from gas to liquid and then to solid is the prediction for crack using Hudson's theory.

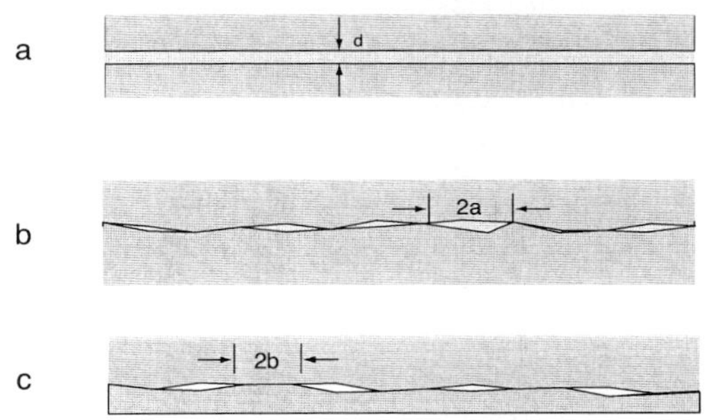

Figure 2.13. A schematic cross-section of three very different styles of fracture model. (a) a weak infill with constant aperture; (b) a distribution of weakness patches; (c) a distribution of welded contact points in an otherwise open fracture. After Hudson et al. (1996).

The linear slip model is sufficiently general to include a wide range of possible fracture interface conditions, and hence gouge or asperity conditions. For example, a welded interface is represented by zero compliance, whereas an increase in compliance signifies a reduction in the number of asperity points, a larger aperture, smoother surface, or a change to a more gaseous fill (Figure 2.12). An example is shown by Hsu and Schoenberg (1993) for rough lucite plates pressed together, for which Z_t/Z_n decreases with increasing pressure. Thus, the generic and physical nature of this model, and the link to the crack models in (2.5), encourage estimation of e_n and e_t as the primary crack- or fracture parameters in multi-component seismics. The non-uniqueness of the interpretation presents the most challenging aspect of this approach. This inversion process requires expressions for e_n and e_t for specific microstructural models, and must ultimately be guided by hard information from core- and log analyses. Three specific fracture models (Figure 2.13) are described in the following sections, and correspond to the fracture models proposed by Hudson and Liu (1999).

2.2.1 Thin layer of weak material

This model was considered by Morland (1974), Hudson (1981), and Fehler (1982). Morland (1974) considers joints with an aperture Δd and a regular spacing of d in an isotropic material (Figure

2.13a). The properties of the gouge material are ρ_g, α_g, and β_g. This gives $1/d$ fractures per unit length, a fractional gouge volume $\Delta V = \Delta d/d$ and fracture parameter

$$e_n = \Delta V / \eta_n \ , \qquad \eta_n = \rho_g \alpha_g^2 / \rho \alpha^2 \ , \tag{2.28}$$

and

$$e_t = \Delta V / \eta_t \ , \qquad \eta_t = \rho_g \beta_g^2 / \rho \alpha^2 \ . \tag{2.29}$$

The equations hold for $\Delta d/d \ll 1$ and $\Delta d/d\eta_n \ll 1$.

2.2.2 Rough surface in contact at a small number of discrete points or asperities

The extent to which the fracture surfaces touch depends largely upon the relative movement of the initially exactly fitting surfaces during and after initiation. Subsurface forces or geochemical bonds press surface asperities together, so that incident stresses can be transmitted across the fracture interface. For the purposes of seismic wave propagation, the spots (Figure 2.13b) behave as if they were welded. This model is in essence the "bed-of-nails" model proposed by Gangi (1978) and considered by White (1983). The latter idealized it for circular Hertzian contacts. For fracture surfaces stiffly welded at only a discrete number of circular spots of mean radius $\bar{r}$ giving a total contact area that is a small fraction A_f of the total area, the fracture parameters can be obtained from Hudson et al. (1997) for gas fill (dry)

$$e_n^{gas} = \frac{\pi \bar{r} / d}{4A_f \eta(1-\eta)} \frac{1}{1+\sqrt{4A_f / \pi}} \ , \tag{2.30}$$

and

$$e_t^{gas} = \frac{\pi \bar{r}/d \ (3-2\eta)}{8A_f \eta(1-\eta)} \frac{1}{1+\sqrt{4A_f / \pi}} \ , \tag{2.31}$$

with $A_f < 0.2$; d is the spacing.

As the fractional area tends to zero, the fractures become infinitely compliant and thus they no longer transmit the seismic waves. Although the theory does not consider fluids, Hudson and Liu (1999) suggest adding the fill and fracture plane stiffnesses for each fracture in the set of $1/d$ per unit length to obtain

$$e_n^{fill} = \frac{e_n^{gas}/d}{1+\eta_n e_n^{gas}} \ , \tag{2.32}$$

and

$$e_t^{fill} = \frac{e_t^{gas}/d}{1+\eta_t e_t^{gas}} \ , \tag{2.33}$$

where η_n and η_t for the fluid or solid constituents are defined as in Section 2.2.1. These formulae do not explicitly depend upon aperture and hence fracture porosity. Indeed, for this situation an additional relationship is required to connect the geometry of the contact points and their deformational properties with the contact radius r.

2.2.3 Plane surfaces in welded contact except at small patches of weakness

Most of the fracture interface can be in welded contact, but interspersed with small patches of non-welded contact or weakness (Figure 2.13c). The solution for a medium consisting of planar distributions of spots of voids with small aspect ratio is given by Hudson et al. (1996b). The relationship can be expressed as the boundary condition terms b_n^{gas} and b_t^{gas} used in (2.8) and (2.11) for calculating the volumetric result

$$e_n = \frac{\Phi b_n^{gas}}{\eta_n b_n^{gas} + 1 - \frac{3}{4} A_O^{3/2}}, \tag{2.34}$$

where A_O is a (small) fraction of the total area that has an open aperture on the fracture. With the mean thickness Δd of the region, and the fracture spacing d, the available porosity Φ can be approximated by $A_O \Delta d/d$. A similar expression exists for the tangential fracture parameter for solid fill:

$$e_t = \frac{\Phi b_t^{gas}}{\eta_t b_t^{gas} + 1 - \frac{3}{4} A_O^{3/2}}. \tag{2.35}$$

These expressions are characterized by the second-order dependence on the fractional area A_O. For very small values of A_O, neglecting the last term in the denominators gives the same result as the volume distribution in (2.9) and (2.12). However, for larger values of A_O the higher-order expressions differ significantly from those in Section 2.1.2. Intuitively, this makes sense: the compliance of fracture planes increases with the density of the voids, but they remain confined to the planes and hence still have an established spacing. Indeed, the theory accounts for interactions between the voids on the same surface, but neglects interactions between individual voids on different fracture surfaces (Hudson and Liu 1999).

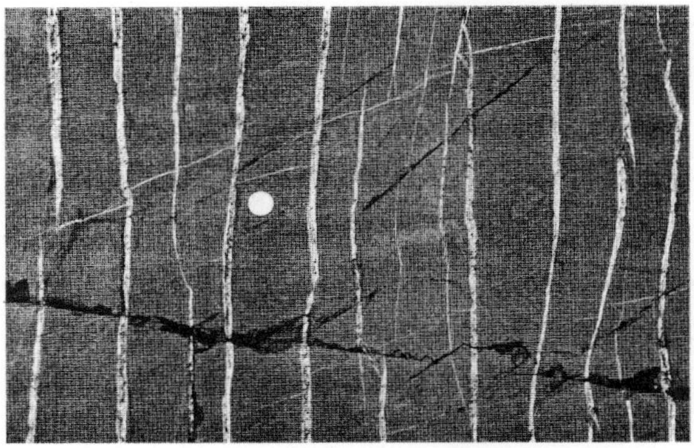

Figure 2.14. A regular array of tensile fractures exposed on a bedding plane in Carboniferous sandstone at Millook, North Cornwall. The alignment of the calcite filled veins indicates that a significant differential stress existed during fracturing. After Ameen 1995, Special Publication no. 92. Printed courtesy of the Geological Society, London.

2.2.4 Fractures filled with a hard solid material

Mineral-filled fractures are very common in all rock types, and their sealing capacity creates major problems in hydrocarbon production. The thin cementing material can be a quartz or carbonate, both usually much harder than the host rock (Figure 2.14). Mineral-filled rock fractures can be represented as thin stiff elements. Using Schoenberg et al. (1996) it is possible to write down the resultant stiffness matrix directly, with additional contributions from the stiffnesses, S_n and S_t due to each thin mineral layer

$$c = \begin{pmatrix} (\lambda+2\mu) & \lambda & \lambda & 0 & 0 & 0 \\ \lambda & (\lambda+2\mu)(1+m_n) & \lambda & 0 & 0 & 0 \\ \lambda & \lambda & (\lambda+2\mu)(1+m_n) & 0 & 0 & 0 \\ 0 & 0 & 0 & \mu(1+m_t) & 0 & 0 \\ 0 & 0 & 0 & 0 & \mu & 0 \\ 0 & 0 & 0 & 0 & 0 & \mu \end{pmatrix}, \tag{2.36}$$

where

$$m_n = \frac{S_n}{\lambda+2\mu} \tag{2.37}$$

and

$$m_t = \frac{S_t}{\mu}, \tag{2.38}$$

with $S_n = 2S_t > 0$. An additional constraint is

$$0 < \eta = \frac{\mu}{\lambda+2\mu} \le \frac{1}{2}. \tag{2.39}$$

This gives the expected TIH medium with a response similar to the compliance case, but now the matrix does not fit the generic form of (2.5). For a calcite fill ($\rho = 2.5$ g/cm^2 and $\alpha = 4.5$ km/s, $\beta = 2.25$ km/s) in a sandstone background ($\rho = 2.1$ g/cm^2 and $\alpha = 4.0$ km/s, $\beta = 1.5$ km/s) occupying 1 % of the volume (10 fractures of 1 mm aperture in 1 m of rock), S_n and S_t are about 2 %. To make this contribution significant requires a background stiffness larger than the ratio of the spacing to aperture. Although open fractures can have a compliance many orders of magnitude larger than the compliance of intact rock, a thin mineral layer cannot be orders of magnitude stiffer than the background rock. It is unlikely that these fractures will induce a significant anisotropic response.

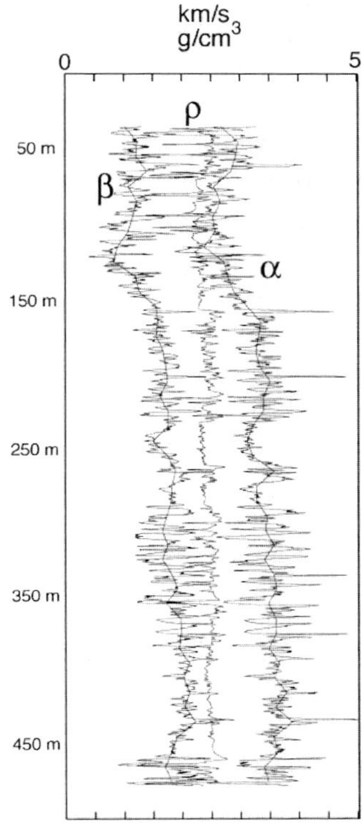

Figure 2.15. Sedimentary sequence for a massive sandstone formation, showing strong fluctuations. The compressional, shear and density logs are measured in a shallow borehole in Kincardine, Scotland. Data courtesy of the British Geological Survey.

2.3 SEDIMENTARY LAYERING (TIV)

This is the best known and most frequently discussed form of seismic anisotropy. Its importance is underlined by the fine layered sequences evident in well logs collected in almost every geographical location throughout the world. Fluctuations are observed due to the interleaving of different types of sandstone (Figure 2.15), sand-shale sequences, but also in carbonates. Large scale fluctuations are also observed in basalt sequences, where the interbedding can cause severe attenuation problems. This style of seismic anisotropy owes its popularity to the fact that it provides a way of upscaling log measurements in a more sophisticated way than by a crude block averaging procedure. In particular, the work of Backus (1962) is much referenced as it provides the basic building blocks for constructing the TIV medium from a number of known constituents. A detailed summary of all aspects of this subject and important developments since Backus (1962) can be found in Helbig (1994).

2.3.1 General parameterization

A simple set of parameters for practical analysis in TI media was suggested by Thomsen (1986). Thomsen introduced three parameters, ε, δ and γ that could be manipulated, remembered and compared in seismic surveys and the laboratory, and developed into processing solutions. In seismic exploration, these are the most commonly used parameters of any anisotropic nomenclature, and Thomsen (1986) is one of the papers most frequently cited. The parameters are based on a re-organization of the elastic stiffnesses that simplifies the algebra and offers an insight into the physical relationship between the wave properties and seismic anisotropy (see Chapter 3). They are introduced here to establish a framework that connects all forms of TIV anisotropy. The Thomsen parameters are defined as

$$\varepsilon = \frac{c_{11} - c_{33}}{2c_{33}} , \tag{2.40}$$

$$\gamma = \frac{c_{66} - c_{55}}{2c_{55}} , \tag{2.41}$$

and

$$\delta = \frac{(c_{13} + c_{55})^2 - (c_{33} - c_{55})^2}{2c_{33}(c_{33} - c_{55})} . \tag{2.42}$$

The elastic stiffnesses are completely specified when the vertical velocities are also given (i.e., when c_{33} and c_{44} are known). Although proposed originally for weak anisotropy, these relations can be used to specify anisotropy of any strength. Starting with the vertical velocities, c_{11} and c_{66} are found from ε and γ with the help of (2.40) and (2.41). In turn, c_{13} can then be determined from δ. It will be seen in Chapter 3 that ε and γ are directly related to the notion of the magnitude (or strength) of the P-wave and S-wave anisotropy, respectively. However, δ is more difficult to understand but lithologically more important. Chapter 3 shows that the difference $\varepsilon - \delta$ relates to the anellipticity of the phase slowness. The table of TI measurements by Thomsen (1986) shows that in most rocks ε and γ lie in the range 0 to 0.3 and δ in the range –0.1 to 0.2. For very small δ such that

$$\delta = \frac{c_{13} + 2c_{55} - c_{33}}{2c_{33}} , \tag{2.43}$$

it is possible to write the stiffness matrix in terms of these parameters

$$\mathbf{c}^{TIV} = \mathbf{c}_0 + 2\begin{pmatrix} (\lambda_0 + 2\mu_0)\varepsilon & (\lambda_0 + 2\mu_0)\varepsilon - 2\mu_0\gamma & (\lambda_0 + 2\mu_0)\delta & 0 & 0 & 0 \\ (\lambda_0 + 2\mu_0)\varepsilon - 2\mu_0\gamma & (\lambda_0 + 2\mu_0)\varepsilon & (\lambda_0 + 2\mu_0)\delta & 0 & 0 & 0 \\ (\lambda_0 + 2\mu_0)\delta & (\lambda_0 + 2\mu_0)\delta & 0 & 0 & 0 & 0 \\ 0 & 0 & 0 & 0 & 0 & 0 \\ 0 & 0 & 0 & 0 & 0 & 0 \\ 0 & 0 & 0 & 0 & 0 & \gamma\mu_0 \end{pmatrix} , \tag{2.44}$$

where the isotropic background matrix $\mathbf{c}_0$ is given by the vertical velocities, i.e., $\lambda_0 + 2\mu_0 = \rho c_{33}$ and $\mu_0 = \rho c_{55}$.

2.3.2 Periodic layering with two constituents

Stratigraphic sections frequently possess cyclic or quasi-cyclic layering due to alternating depositional mechanisms (Ebrom et al. 1992). Examples of this include interbedded sand/shale, limestone/shale, or sand/coal. Riznichenko (1949), Thompson (1950), Postma (1955), Helbig (1958) and Anderson (1961) have developed expressions for the specific case of two material constituents. Postma (1950) gives the result for layers with elastic parameters λ_1, μ_1, ρ_1, and λ_2, μ_2, ρ_2, and relative proportions P_1 and P_2 (ratio of individual thickness to total thickness):

$$c_{11} = \frac{K_1 K_2 + 4P_1 P_2(\mu_1 - \mu_2)(L_1 - L_2)}{K_1 P_1 + K_2 P_2} , \tag{2.45a}$$

$$c_{13} = \frac{\lambda_1 K_2 P_1 + \lambda_2 K_1 P_2}{K_1 P_1 + K_2 P_2} , \tag{2.45b}$$

$$c_{33} = \frac{K_1 K_2}{K_1 P_2 + K_2 P_1} , \tag{2.45c}$$

$$c_{55} = \frac{\mu_1 \mu_2}{\mu_1 P_2 + \mu_2 P_1} , \tag{2.45d}$$

$$c_{66} = \mu_1 P_1 + \mu_2 P_2 , \tag{2.45e}$$

with

$$K_i = \lambda_i + 2\mu_i \tag{2.45f}$$

and

$$L_i = \lambda_i + \mu_i \ . \tag{2.45g}$$

The average density is given by

$$\rho = \rho_1 P_1 + \rho_2 P_2 \ . \tag{2.45h}$$

For this periodic medium, the Thomsen parameters are

$$\varepsilon = \frac{4P_1 P_2(\mu_1 - \mu_2)(L_1 - L_2)}{2K_1 K_2} , \tag{2.46}$$

$$\delta = \varepsilon - \frac{K_1 K_2}{K_1 P_1 + K_2 P_2}\left(\frac{K_1 K_2}{K_1 P_1 + K_2 P_2} - \frac{\mu_1 \mu_2}{\mu_2 P_1 + \mu_1 P_2}\right)\left(\frac{4P_1 P_2(\mu_1 - \mu_2) L_1 L_2}{(\mu_2 P_1 + \mu_1 P_2)(K_2 P_1 + K_1 P_2)}\right) , \tag{2.47}$$

and

$$\gamma = \tfrac{1}{2}\left(1 - P_1 P_2 \frac{(\mu_1 - \mu_2)^2}{\mu_1 \mu_2}\right) . \tag{2.48}$$

γ is always positive and depends quadratically on the contrast of the shear stiffnesses. Moreover, $\varepsilon \geq \delta$.

Most TIV media of practical importance can be modeled by (i.e., are long-wave equivalent to) periodic layering with two constituents. However, there are obviously TIV media that cannot be modeled in this way: e.g., any TIV medium with $c_{66} < c_{55}$ would have $\gamma < 0$ and thus violate (2.48). There are a number of inequalities that have to be satisfied for a medium to be "representable" by periodic layer sequences (see, e.g., Backus 1962). Backus (1962) has also shown that even among the TIV media that satisfy these inequalities, there is a class that requires more than two – but at most three – constituent materials.

2.3.3 Periodic layering with an arbitrary number of isotropic constituents

Backus (1962) developed more general averaging formulae for either isotropic or TIV constituents. These formulae determine replacement stiffnesses from algebraic combinations of averages of the constituents' elastic properties. The following expressions are based on the summary of Folstad and Schoenberg (1992) for isotropic layering:

$$c_{11} = 4c_{66} - 4\langle \eta\mu \rangle + \frac{c_{13}^2}{c_{33}} , \tag{2.49a}$$

$$c_{13} = \left(1 - 2\langle \eta \rangle\right)\, c_{33} , \tag{2.49b}$$

$$c_{33} = \left\langle \frac{\eta}{\mu} \right\rangle^{-1} , \tag{2.49c}$$

$$c_{55} = \left\langle \frac{1}{\mu} \right\rangle^{-1} , \tag{2.49d}$$

and

$$c_{66} = \langle \mu \rangle , \tag{2.49e}$$

where the angular brackets indicate the "thickness-weighted arithmetic average". These equations should be evaluated in reverse order. The equations give the general condition $c_{66} \geq c_{55}$. The "rule of thumb" $c_{11} \geq c_{33}$ holds for most layer sequences, but is no *mathematical* limit: there are sequences that lead to replacement media with $c_{11} < c_{33}$. However, it can be shown that always $c_{11} > c_{33}/4$ (see, e.g., Helbig 1994, Section 9.9).

A constant $\eta = (\rho\beta^2)/(\rho\alpha^2)$ gives $\delta = 0$. If the shear stiffness $\mu = \rho\beta^2$ is the same throughout the layer sequence, the medium is isotropic in spite of fluctuations in α.

The equations can be simplified by considering small zero-mean fluctuations $\Delta\mu/\mu$ and $\Delta\eta/\eta$ about average values μ_0 and η_0 with standard deviations (auto- and cross correlations)

$$\sigma_\mu = \left\langle \left(\frac{\Delta\mu}{\mu} \right)^2 \right\rangle , \tag{2.50}$$

$$\sigma_\eta = \left\langle \left(\frac{\Delta\eta}{\eta} \right)^2 \right\rangle , \tag{2.51}$$

and

$$\sigma_{\mu\eta} = \left\langle \left(\frac{\Delta\mu\Delta\eta}{\mu\eta} \right)^2 \right\rangle . \tag{2.52}$$

The stiffness matrix can be written to first order in these quantities

$$\mathbf{c} = \begin{pmatrix} (\lambda_0 + 2\mu_0)(1-e_1) & \lambda_0(1-e_4) & \lambda_0(1-e_2) & 0 & 0 & 0 \\ \lambda_0(1-e_4) & (\lambda_0 + 2\mu_0)(1-e_2) & \lambda_0(1-e_2) & 0 & 0 & 0 \\ \lambda_0(1-e_2) & \lambda_0(1-e_2) & (\lambda_0 + 2\mu_0)(1-e_2) & 0 & 0 & 0 \\ 0 & 0 & 0 & \mu_0(1-e_3) & 0 & 0 \\ 0 & 0 & 0 & 0 & \mu_0(1-e_3) & 0 \\ 0 & 0 & 0 & 0 & 0 & \mu_0 \end{pmatrix}, \tag{2.53}$$

where

$$e_1 = (1-4\eta)\sigma_{\mu\eta}^2 - (1-2\eta)^2\sigma_\mu^2 , \tag{2.54}$$

$$e_2 = \sigma_\mu^2 - \sigma_{\mu\eta}^2 , \tag{2.55}$$

$$e_3 = \sigma_\mu^2 , \tag{2.56}$$

and

$$e_4 = \frac{e_1}{1-2\eta} . \tag{2.57}$$

This leads to the Thomsen parameters

$$\gamma = \sigma_\mu^2 , \tag{2.58}$$

$$\varepsilon = 2\eta\sigma_{\mu\eta}^2 - \left(1 + 2\eta - 2\eta^2\right)\sigma_{\mu\eta}^2 \ , \tag{2.59}$$

and

$$\delta = 0 \ . \tag{2.60}$$

The last equation shows that Thomsen's δ senses the higher statistical moments of the layer fluctuations. Berryman et al. (1997) show that with large fluctuations in the Lamé-constant λ and constant density ρ, $-\frac{3}{8} \le \varepsilon \le \frac{1}{2}\left(\langle\alpha^2\rangle\langle\alpha^{-2}\rangle - 1\right)$. The smaller positive and all negative values of ε occur in media with large fluctuations in λ. δ can be positive or negative, with both signs equally likely for periodic media with two constituents. Large fluctuations in λ give a positive δ. These results suggest that positive ε and δ are indicative of fluids.

2.3.4 Periodic layering with an arbitrary number of TIV constituents

For a stack of TIV layers, it has been shown that

$$c_{11} = \left\langle c_{11} - \frac{c_{13}^2}{c_{33}} \right\rangle + \left\langle \frac{1}{c_{33}} \right\rangle^{-1} \left\langle \frac{c_{13}}{c_{33}} \right\rangle^2 , \tag{2.61a}$$

$$c_{33} = \left\langle \frac{1}{c_{33}} \right\rangle^{-1} , \tag{2.61b}$$

$$c_{13} = \left\langle \frac{1}{c_{33}} \right\rangle^{-1} \left\langle \frac{c_{13}}{c_{33}} \right\rangle , \tag{2.61c}$$

$$c_{55} = \left\langle \frac{1}{c_{55}} \right\rangle^{-1} , \tag{2.61d}$$

and

$$c_{66} = \left\langle c_{66} \right\rangle , \tag{2.61e}$$

where the c_{ij} on the left-hand sides are the stiffnesses of the compound medium and those on the right-hand sides are the stiffnesses of the constituents. The sets (2.45) and (2.49) can be derived from the set (2.61) by specifying the stiffnesses of the constituents to be those of isotropic media. The sets (2.45) and (2.49) allowed conclusions on the range of the compound stiffnesses. No such conclusions are, at present, possible from set (2.61).

Finally, to use the Backus-type formulae from the above equations, all the layers must be stochastically stationary (possess the same average at all depths), the thickness of each layer and also the complete stack to be averaged must be much less than the incident wavelength. Practical application therefore takes the form of a sliding, frequency-dependent window, of a wavelength or so in size. Although such upscaling has been tested (for example, Esmersoy, Hsu and Schoenberg 1989), the best strategy for a continuously layered sequence has not yet been found. Indeed, the relationship between fine layering and seismic-scale anisotropy has not been widely demonstrated except in the laboratory, and has often brought inconclusive results in the field (Sams and Williamson 1994; Winterstein 1986). One possible reason for this is the scale of the fluctuations causing the anisotropy. This point is exemplified by Sams (1995) who showed additional anisotropy due to fluctuations existing at an extra fine scale (2.5 mm in electrical resistivity logs) below normal logging scale (10 cm to 10 m). Another reason is the role of multiple scattering, which is still an open question.

2.3.5 Fractures/cracks parallel to the bedding

Fractures/cracks parallel to the bedding can arise due to overpressure or unequal stresses. They also cause TIV anisotropy:

$$\mathbf{c}_{BPF} = \mathbf{c}_b - \begin{pmatrix} (\lambda_b + 2\mu_b)\zeta_b^2 e_n & \zeta_b\lambda_b e_n & \lambda_b e_n & 0 & 0 & 0 \\ \zeta_b\lambda_b e_n & (\lambda_b + 2\mu_b)\zeta_b^2 e_n & \lambda_b e_n & 0 & 0 & 0 \\ \lambda_b e_n & \lambda_b e_n & (\lambda_b + 2\mu_b)e_n & 0 & 0 & 0 \\ 0 & 0 & 0 & \mu_b e_t & 0 & 0 \\ 0 & 0 & 0 & 0 & \mu_b e_t & 0 \\ 0 & 0 & 0 & 0 & 0 & 0 \end{pmatrix}, \tag{2.62}$$

where the background (with stiffness matrix $\boldsymbol{c}_b$) is isotropic and $\zeta_b = \lambda_b/(\lambda_b+2\mu_b)$, as before. This leads to the Thomsen parameters

$$\varepsilon = 2\eta(1-\eta)e_n \ , \tag{2.63}$$

$$\delta = 2\eta(e_n - e_t) \ , \tag{2.64}$$

and

$$\gamma = \frac{e_t}{2} \ . \tag{2.65}$$

2.4 SHALES (TIV)

Shales make up a large proportion of most sedimentary basins, overlie hydrocarbon bearing reservoirs, and play a significant role in fluid flow. They are also known to contribute the strongest anisotropic response, and can be the cause significant errors in positioning events (Banik 1984) or can lead to misinterpretation of AVO (Chapter 7). Flat shale beds exhibit a predominantly TIV anisotropy that can vary in degree from 5 to 50 percent. Although it is tempting to use layering-related TIV to represent shales, to do so is incorrect. Shales are anisotropic due to the organization of their small-scale components (Figure 2.16), most noticeably due to the directional alignment of clay mineral platelets (Kaarsberg 1959; Jones and Wang 1981). Additional contributions from clay content, kerogen content, porosity and effective pressure make it difficult to predict the elastic properties of a shale system. All these anisotropic components can be regarded as "inherent" in the sense that they come from the internal structure of each individual shale and originate on the μm to mm scale, in contrast with sedimentary layering or laminations on a 1 m to 50 m scale.

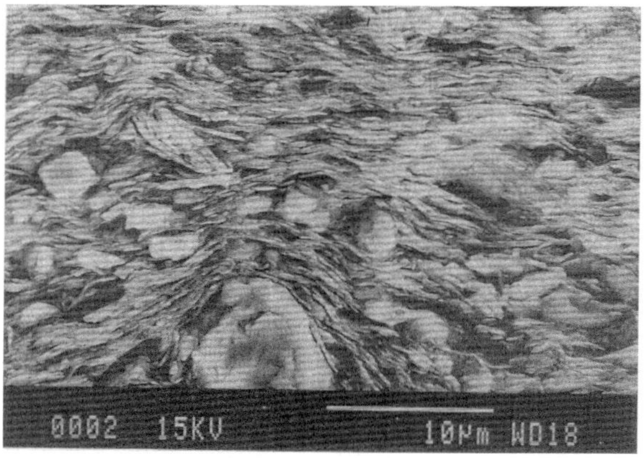

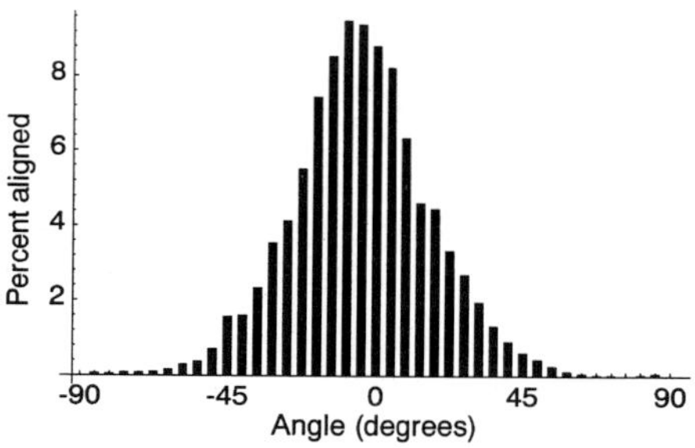

Figure 2.16. Scanning electron micrograph of a shale, extracted from Hornby et al. (1994). The platey particles are the clay minerals, whilst the larger, nearly spherical, particles are silt. The elastic properties of the composite can be computed by taking into account the clay and fluid phases, the orientation distribution of the minerals and also the unconnected silt particles. The orientation distribution of the platelets has been obtained from a digitized SEM image.

To connect the observed anisotropy to these microstructural elements remains a challenge, and current attempts to predict the strong anisotropy usually takes the form of a numerical recipe. It is particularly appropriate here to make use of the (macrofield) Thomsen parameters defined in the previous section. A specific feature of these rocks is that they appear to be mostly, if not entirely, positively anelliptic ($\varepsilon - \delta > 0$). These results will be discussed again in connection with the VSP methods in Chapter 7, and are shown for laboratory measurements in Section 2.4.3. The type of anellipticity is similar to that predicted for layering.

2.4.1 Numerical prediction using an SEM image

On the microscopic level the rocks are composed of connected distributions of fluid and flat platey clay mineral phases. The clay-fluid bundles are distributed sub-horizontally. Also included in the matrix are isolated inclusions of other minerals such as quartz and feldspar, pyrite and calcite cements.

Hornby et al. (1994) used the mineral composition and volume fraction (from laboratory data or logs), together with the shape and orientation measured from SEM images (see Figure 2.16) to construct the following recipe:

(a) calculate a bi-connected composite with equal concentrations of clay minerals and fluid;
(b) re-adjust the concentrations set in (a) to the desired amounts;
(c) account for the platelet orientations relative to the vertical axis;
(d) insert the quartz and other minerals.

(a) An effective TIV tensor **C** is calculated for an equal concentration of isotropic clay minerals (elastic properties are taken from the literature) and fluid-filled pores shaped as oblate spheroids (the Hudson inclusion in Figure 2.4) with their minor axis orientated in the vertical direction. The fourth-rank tensor **C** is used here because of the nature of the manipulations. With $\mathbf{C}_m$ and $\mathbf{C}_f$ representing, respectively, the properties of the clay minerals and the fluid, the so-called "self consistent" approximation gives

$$\overline{\mathbf{C}} = \left(\mathbf{C}_f\left(\mathbf{I}+\mathbf{K}_f\right)^{-1} + \mathbf{C}_m(\mathbf{I}+\mathbf{K}_m)^{-1}\right)\left(\left(\mathbf{I}+\mathbf{K}_f\right)^{-1} + (\mathbf{I}+\mathbf{K}_m)^{-1}\right)^{-1}, \tag{2.66}$$

where **I** is a fourth-rank identity matrix and

$$\mathbf{K}_f = \mathbf{G}\left(\mathbf{C}_f - \overline{\mathbf{C}}\right), \tag{2.67}$$

$$\mathbf{K}_m = \mathbf{G}\left(\mathbf{C}_m - \overline{\mathbf{C}}\right), \tag{2.68}$$

and the symmetric matrix **G** of tensor components is derived from definite integrals of Lin and Mura (1973)

$$G_{1111} = G_{2222} = \tfrac{\pi}{4}\int_0^1 \eta \frac{\left(1+3\eta+4\eta y(x)^2\right)\left(\eta+y(x)^2\right) - (1-\eta)^2 y(x)^2}{D}\mathrm{d}x, \tag{2.69}$$

$$G_{3333} = 2\pi\int_0^1 \eta \frac{y(x)^2\left(1+\eta y(x)^2\right)\left(1+y(x)^2\right)}{D}\mathrm{d}x \tag{2.70}$$

$$G_{1212} = G_{2121} = \tfrac{\pi}{4}\int_0^1 \frac{\left(1+\eta+4\eta y(x)^2\right)\left(\eta+y(x)^2\right) - 3(1-\eta)^2 y(x)^2}{D}\mathrm{d}x, \tag{2.71}$$

$$G_{1313} = G_{2323} = G_{3131} =$$

$$G_{3232} = \pi\int_0^1 y(x)^2 \frac{\left(1+\eta+2\eta^2 y(x)^2\right)\left(\eta+y(x)^2\right) - (1-\eta)y(x)^2 - \eta(1-\eta)\left(1+y(x)^2\right)}{D}\mathrm{d}x, \tag{2.72}$$

$$G_{1122} = G_{2211} = \tfrac{\pi}{4}\int_0^1 \frac{-\eta(1-\eta)\left(1+y(x)^2\right)}{D}\,\mathrm{d}x \;, \tag{2.73}$$

$$G_{1133} = G_{2233} = G_{3311} = G_{3322} = \pi\int_0^1 \frac{-\eta y(x)^2(1-\eta)\left(1+y(x)^2\right)}{D}\,\mathrm{d}x \;, \tag{2.74}$$

with

$$D = \eta\left(1+y(x)^2\right)\left(\left(1+\eta y(x)^2\right)\left(\eta+y(x)^2\right)-y(x)^2(1-\eta)^2\right) \tag{2.75}$$

and

$$y\left(x^2\right) = \frac{x^2}{a\left(1-x^2\right)} \;. \tag{2.76}$$

a in (2.76) is the aspect ratio of the inclusions and $\eta = \beta^2/\alpha^2$ as in previous Sections. In this formulation the aspect ratios a can be larger than the limit in the Hudson theory. The integrals can be evaluated numerically given the background medium and inclusion properties. Equation (2.66) is computed iteratively, by firstly setting $\overline{\mathbf{C}}$ to the clay matrix value $\mathbf{C}_m$, and then calculating the right-hand side. Once determined, the new $\overline{\mathbf{C}}$ is then fed back into the right-hand side, and the process continued. This is repeated until the output is close to the input. It should be noted that in application of the above procedure, it is necessary to estimate the clay properties. If tabulated values are not available (they are generally not well documented), it is possible to use data from laboratory experiments (Hornby et al. 1994).

(b) In the second stage of the shale calculation, the shale-fluid ratio from stage (a) is changed from 1:1 take the actual target porosity Φ into account. This is achieved by the "differential effective medium theory" (DEM). Here, an infinitesimal sub-volume of composite material is removed and replaced with a corresponding sub-volume of the constituent whose concentration is to be adjusted. The elastic properties of the composite are re-calculated and the process repeated. Each time the concentration (volume fraction) v of the mineral or fluid phase is altered in very small amounts of δv. For an increase in concentration of the fluid phase, this process can be written as

$$\overline{\mathbf{C}}(v+\delta v) = \overline{\mathbf{C}}(v) + \frac{\delta v}{1-v}\left(\mathbf{C}_f(v)-\overline{\mathbf{C}}(v)\right)\left(\mathbf{I}+\mathbf{G}\left(\mathbf{C}_f(v)-\overline{\mathbf{C}}(v)\right)\right)^{-1} \tag{2.77}$$

with a similar formula for the adjustments of the solid phase, if needed. This equation is applied repeatedly until the desired concentration is reached. Alternatively, the final result can be obtained by solving the differential equation for $\delta v \to 0$. The procedure is useful as it preserves the connectivity of the fluid and solid phases in the composite. The final result $\mathbf{C}_{cfb}$ can be regarded as the anisotropy for a single clay-fluid bundle.

(c) The result for a distribution of the clay-fluid bundles is obtained by a Reuss-Voigt-Hill average of the elastic properties derived from different orientations. The average is weighted according to the platlet orientation function $P(\chi)$ measured directly from the SEM image as a function of sub-horizontal dip χ (Figure 2.16). If this orientation function is not known, it can be estimated by means of a suitable orientation distribution (Sayers 1998). The rotation of each clay bundle can be performed by rotation of the fourth rank tensor $\mathbf{C}_{cfb}$ and its compliance counterpart. However, effective implementation can be achieved by applying a Bond transformation (see the Appendix) to the 6×6 stiffness matrix $\mathbf{c}_{cfb}$ and compliance matrix $\mathbf{s}_{cfb}$ derived from $\mathbf{C}_{cfb}$. Summing over all N dips and M azimuths produces the stiffness matrix

$$\bar{c}^{angle} = \sum_{i=1}^{N}\sum_{j=1}^{M} D(\chi_i) M(\chi_i,\phi_i) c_{cfb} M^T(\chi_j,\phi_j) \ . \tag{2.78}$$

The rotation matrices **M** for the Bond transformation are given in the Appendix. Because of circular symmetry it is sufficient to sum over only a few azimuths. A similar result $\bar{\mathbf{s}}_{angle}$ can be obtained for the compliance, from which the stiffness average is obtained:

$$\mathbf{c}_{total} = \frac{\bar{\mathbf{c}}_{angle} + \bar{\mathbf{s}}_{angle}^{-1}}{2} \ . \tag{2.79}$$

This average of the different TI bundles produces a TIV anisotropy, but reduces the overall anisotropy depending upon the distribution. In application of this procedure, Hornby et al. 1994) conclude that the primary controlling factor in the anisotropy is the orientation function, since this was critical for their final match to the laboratory data of Jones and Wang (1981). This platelet distribution limits the degree of anisotropy, with a wider distribution leading to a weaker anisotropy for a given volume fraction of clay.

(d) The final stage is to insert the isolated silt particles. This is achieved by re-visiting stage (b), and increasing the concentration of silt in small stages using its known elastic properties.

2.4.2 An approximate expression

Schoenberg et al. (1996) have suggested a simple three-parameter model as a first approximation for the elastic behavior of shales. The model visualizes shales as being composed of relatively hard, thin platelets aligned roughly along one direction in a soft matrix. The stiffness matrix takes the form

$$\mathbf{c} = \begin{pmatrix} \lambda+2\mu_h & \lambda & \lambda & 0 & 0 & 0 \\ \lambda & \lambda+2\mu_h & \lambda & 0 & 0 & 0 \\ \lambda & \lambda & \lambda+2\mu_v & 0 & 0 & 0 \\ 0 & 0 & 0 & \mu_v & 0 & 0 \\ 0 & 0 & 0 & 0 & \mu_v & 0 \\ 0 & 0 & 0 & 0 & 0 & \mu_h \end{pmatrix}, \tag{2.80}$$

where λ, μ_v, and μ_h are the three parameters of the model. This is a special case of a medium with rigid planar stiffeners outlined in the fracture case in Section 2.2.4, where we had $\mu_h - \mu_v = S_n = S_t > 0$. The model consists of an isotropic matrix in which horizontal rigid shale laminae are embedded. The matrix gives the Thomsen parameters

$$\varepsilon = \frac{S_n}{\lambda + 2\mu} , \tag{2.81}$$

$$\delta = 0 , \tag{2.82}$$

and

$$\gamma = \frac{S_t}{\mu} . \tag{2.83}$$

2.4.3 Empirical relationships

Seismic anisotropy in shales is complicated: besides the preferred orientation of clay particles, there is the additional dependence upon factors such as chemical softening (due to fluid-clay interaction and swelling), the strong anisotropy due to fine laminations of organic matter (kerogen), and compaction (Vernik and Liu 1997). Additional contributions can arise from micro-cracks parallel to the bedding caused by hydrocarbon generation in thermally mature shales (Vernik 1993). All of these factors depend upon the pressure, temperature and pressure of the rock. To understand the interplay between these factors, Vernik and Liu (1997) have compiled a comprehensive database of anisotropies for low- and high-porosity black shales (Figure 2.17).

Another recent avenue for anisotropy prediction comes from another large database of published measurements showing an empirical correlation between α/β and the Thomsen parameters, implying that knowledge of this alone is sufficient to estimate the anisotropy. The data for saturated shales at low and high porosities measured at the ultrasonic scale can be approximately represented by (Ryan-Grigor 1998):

$$\delta \approx \frac{\left(3.87\,\frac{\alpha}{\beta} - 4.54\right)^2 - \left(\left(\frac{\alpha}{\beta}\right)^2 - 1\right)^2}{2\left(\frac{\alpha}{\beta}\right)^2\left(\left(\frac{\alpha}{\beta}\right)^2 - 1\right)} , \tag{2.84}$$

$$\varepsilon = 0.2090\frac{\alpha}{\beta} - 0.2397 , \tag{2.85}$$

and

$$\gamma = 0.4014\frac{\alpha}{\beta} - 0.5576 . \tag{2.86}$$

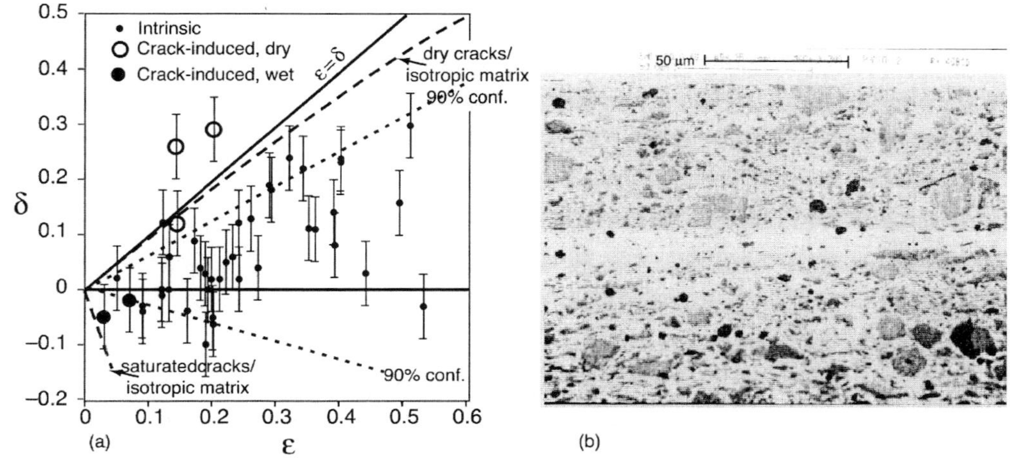

Figure 2.17. a: anisotropy measurements for black, kerogen rich, shales extracted from Vernik and Liu (1997). Laboratory measurements are taken at a high confining pressure of 70MPa and also a low confining pressure, with and without the crackinduced contribution. Predictions are also shown for gas and water-filled cracks. Thomsen's gamma values range from 0.02 to 0.59 for these shales.
b: SEM image of the shales displaying the distribution of kerogen (black) as lenticular microlayers dispersed among the clay (illite) matrix and also as sporadic layers (after Vernik and Nur 1992).

Shales become more anisotropic and anelliptic with increasing α/β. The large anisotropy of compacted shales is explained as having a high α/β. Shales with an α/β smaller than about 1.8 tend to have a near-zero to small positive values of δ and near zero values of anellipticity. As α/β increases, representing kerogen content or overpressure, δ becomes negative.

2.5 INTERSECTING SYSTEMS OF ALIGNED FEATURES

2.5.1 Intersecting fracture/crack sets

Natural fractures in rocks often have a number of intersecting sets of varying scale lengths (Figure 2.18), generally formed under different paleo-stress fields and strain conditions (Nelson 1985; Barton 1995). Depending upon the relative stresses present during the generation of each set, the fracture sets can have different inclinations (as in a conjugate pair of shear fractures) or strikes (different paleo-stress directions). Fracture statistics are likely to change with lithology. Depending upon their origin and subsequent history, some fractures can actively contribute to the seismic anisotropy, but others might not. In a similar way, it is expected that there should be several different crack sets in a rock.

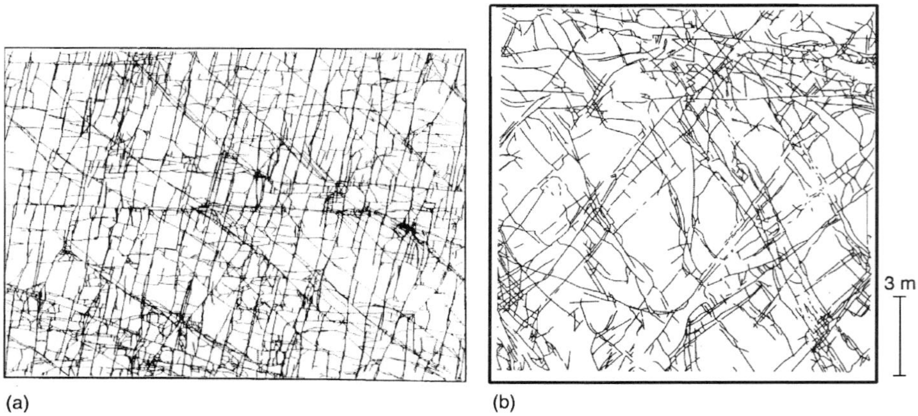

Figure 2.18. a: map of joints exposed on a 1m thick sandstone bedding plane from Whinney Hill Quarry, Lancashire. Diagram and results are extracted from Gillespie et al. (1993). Lithology is the Upper Carboniferous sandstones and shales. The joint system mapped comprises three principal sets, each of which appears to possess a regular spacing, or periodic distribution.

b: fracture trace map for a quarry floor in the Niagaran dolomite. Fractures are formed primarily as joints, defined by the absence of shear across them, with the longest fracture being 6.4m (Barton 1995).

The stiffness tensor for two intersecting fracture/crack sets can be formed by adding their individual compliances in (2.25) (Schoenberg and Sayers 1995), and then inverting to the stiffness domain. For weak anisotropy, this is equivalent to adding the first order stiffness terms

$$\mathbf{c}^{new} = \mathbf{c}_0 + \sum_{i=1}^{N} \mathbf{c}_i^{(1)} . \tag{2.87}$$

where $\mathbf{c}_0$ is the isotropic background matrix.

According to Hudson (1988), the accuracy can be improved without leaving the stiffness domain by calculating a second-order correction term to (2.87):

$$\mathbf{c}^{(2)} = \left(\sum \mathbf{c}_i^{(1)}\right)\mathbf{X}\left(\sum \mathbf{c}_i^{(1)}\right) , \tag{2.88}$$

where $\mathbf{X}$ is given by

$$\mathbf{X} = \begin{pmatrix} 58-17\eta & 1-\eta & 1-\eta & 0 & 0 & 0 \\ 1-\eta & 58-17\eta & 1-\eta & 0 & 0 & 0 \\ 1-\eta & 1-\eta & 58-17\eta & 0 & 0 & 0 \\ 0 & 0 & 0 & 60+15\eta & 0 & 0 \\ 0 & 0 & 0 & 0 & 60+15\eta & 0 \\ 0 & 0 & 0 & 0 & 0 & 60+15\eta \end{pmatrix} . \tag{2.89}$$

2.5.1.1 *Orthogonal sets*

With the help of (2.87), a vertical fracture set A is intersected by second vertical fracture set B which has been rotated in an anti-clockwise direction by 90° about the *Z*-axis. This process is aided by the use of the Bond transformation outlined in the Appendix. The new stiffness matrix is

$$\mathbf{c}^{ORTH} = \begin{pmatrix} \mathbf{A} & \mathbf{0} \\ \mathbf{0} & \mathbf{B} \end{pmatrix}, \tag{2.90}$$

where

$$\mathbf{A} = \begin{pmatrix} (\lambda+2\mu)(1-\zeta^2 e_n^B - e_n^A) & \lambda(1-e_n^B-e_n^A) & \lambda(1-e_n^A-\zeta e_n^B) \\ \lambda(1-e_n^B-e_n^A) & (\lambda+2\mu)(1-\zeta^2 e_n^A - e_n^B) & \lambda(1-e_n^B-\zeta e_n^A) \\ \lambda(1-e_n^A-\zeta e_n^B) & \lambda(1-e_n^B-\zeta e_n^A) & (\lambda+2\mu)(1-\zeta^2 e_n^B - \zeta^2 e_n^A) \end{pmatrix}. \tag{2.91}$$

and

$$\mathbf{B} = \begin{pmatrix} \mu(1-e_t^B) & 0 & 0 \\ 0 & \mu(1-e_t^A) & 0 \\ 0 & 0 & \mu(1-e_t^A-e_t^B) \end{pmatrix}. \tag{2.92}$$

If the two sets are identical in a background with $\alpha/\beta = 2$ – i.e., $\zeta = 1/2$ – then

$$\mathbf{c}^{ORTH} = \begin{pmatrix} (\lambda+2\mu)(1-\frac{5}{4}e_n) & \lambda(1-2e_n) & \lambda(1-\frac{3}{2}e_n) & 0 & 0 & 0 \\ \lambda(1-2e_n) & (\lambda+2\mu)(1-\frac{5}{4}e_n) & \lambda(1-\frac{3}{2}e_n) & 0 & 0 & 0 \\ \lambda(1-\frac{3}{2}e_n) & \lambda(1-\frac{3}{2}e_n) & (\lambda+2\mu)(1-\frac{1}{2}e_n) & 0 & 0 & 0 \\ 0 & 0 & 0 & \mu(1-e_t) & 0 & 0 \\ 0 & 0 & 0 & 0 & \mu(1-e_t) & 0 \\ 0 & 0 & 0 & 0 & 0 & \mu(1-2e_t) \end{pmatrix}. \tag{2.93}$$

This stiffness matrix has orthorhombic symmetry (see 2.3); the planes of symmetry coincide with the coordinate planes.

2.5.1.2 *Acute conjugate angle*

As in the previous section, there are two vertical fracture sets, but now they make an angle ϕ with each other. The stiffness matrix is determined from the two individual stiffness matrices as

$$\mathbf{c}^{MONO} = \mathbf{c}_0 + \mathbf{M}_Z(\phi)\mathbf{c}_A^{(1)}\mathbf{M}_Z^T(\phi) + \mathbf{c}_B^{(1)} . \tag{2.94}$$

This system has monoclinic symmetry with a stiffness matrix as in (2.4) with the individual entries

$$1-\frac{c_{11}}{\lambda+2\mu}=e_n^A+e_n^B\left(\cos^4\phi+\zeta^2\sin^4\phi+2\sin^2\phi\cos^2\phi\right)+e_t^B\sin^2 2\phi\ , \tag{2.95a}$$

$$1-\frac{c_{12}}{\lambda}=e_n^A+e_n^B\left(\cos^4\phi+\sin^4\phi+\left(1+\zeta^2\right)\sin^2\phi\cos^2\phi\right)-e_t^B\sin^2 2\phi\ , \tag{2.95b}$$

$$1-\frac{c_{13}}{\lambda}=e_n^A+e_n^B\left(\cos^2\phi+\zeta\sin^2\phi\right)\ , \tag{2.95c}$$

$$1-\frac{c_{22}}{\lambda+2\mu}=\zeta^2e_n^A+e_n^B\left(\zeta^2\cos^4\phi+\sin^4\phi+2\sin^2\phi\cos^2\phi\right)+e_t^B\sin^2 2\phi\ , \tag{2.95d}$$

$$1-\frac{c_{23}}{\lambda}=\zeta e_n^A+e_n^B\left(\zeta\cos^2\phi+\sin^2\phi\right)\ , \tag{2.95e}$$

$$1-\frac{c_{33}}{\lambda+2\mu}=\zeta^2\left(e_n^A+e_n^B\right)\ , \tag{2.95f}$$

$$\frac{c_{16}}{\lambda}=\sin\phi\cos\phi\left(\cos 2\phi\left(e_n^B+2e_t^B\right)+\sin^2\phi\zeta^2e_n^B-\cos^2\phi e_n^B\right)\ , \tag{2.95g}$$

$$\frac{c_{26}}{\lambda}=\sin\phi\cos\phi\left(-\cos 2\phi\left(e_n^B+2e_t^B\right)+\cos^2\phi\zeta^2e_n^B-\sin^2\phi e_n^B\right)\ , \tag{2.95h}$$

$$\frac{c_{36}}{\lambda}=\sin\phi\cos\phi\left(\zeta e_n^B-e_n^B\right)\ , \tag{2.95i}$$

$$1-\frac{c_{44}}{\mu}=\sin^2\phi\ \ e_t^B\ , \tag{2.95j}$$

$$\frac{c_{45}}{\mu}=-\sin\phi\cos\phi\ \ e_t^B\ , \tag{2.95k}$$

$$1-\frac{c_{55}}{\mu}=e_t^A+\cos^2\phi\ \ e_t^B\ , \tag{2.95l}$$

$$1-\frac{c_{66}}{\mu}=e_t^A+\sin^2 2\phi\left(\zeta^2-1\right)e_n^B+\cos^2 2\ \ \phi e_t^B\ . \tag{2.95m}$$

2.5.1.3 *Angular distributions of cracks*

This summation process can be continued further to simulate a distribution of fractures/cracks. In particular, crack systems under uni-axial compression or tension lead to a population that is randomly distributed around a prescribed *X*-axis, with the crack normals at some fixed angle θ to this axis. This gives a TIH medium with

$$1-\frac{c_{11}^{(1)}}{\lambda+2\mu}=e_n\left(\zeta+2\eta\cos^2\theta\right)^2+4\eta e_t\cos^2\theta\sin^2\theta \ , \tag{2.96a}$$

$$1-\frac{c_{33}^{(1)}}{\lambda+2\mu}=\frac{1}{2}\left(e_n\left(2\zeta^2+4\zeta\eta\sin^2\theta+3\eta^2\sin^4\theta\right)+e_t\eta^2 2\sin^2\theta\left(4-3\sin^2\theta\right)\right) , \tag{2.96b}$$

$$1-\frac{c_{12}^{(1)}}{\lambda}=\left(e_n\frac{\left(\zeta+\eta\sin^2\theta\right)\left(\zeta+2\eta\cos^2\theta\right)}{\zeta}-2e_t\frac{\eta}{\zeta}\sin^2\theta\cos^2\theta\right) , \tag{2.96c}$$

$$1-\frac{c_{66}^{(1)}}{\mu}=\frac{1}{2}\left(4e_n\sin^2\theta\cos^2\theta+e_t\left(\sin^2\theta+2\cos^2\theta-4\sin^2\theta\cos^2\theta\right)\right) , \tag{2.96d}$$

$$1-\frac{c_{44}}{\mu}=\frac{1}{2}\left(e_n\sin^4\theta+e_t\sin^2\theta\left(2-\sin^2\theta\right)\right) . \tag{2.96e}$$

For $\theta = 0$ this reduces to the case of aligned vertical cracks. For $\theta = 90°$, all the crack normals lie randomly oriented in a single vertical plane (they are coplanar). For this case we have $c_{44} = \mu(1 - e_t/2 - e_n/2) < c_{66} = \mu(1 - e_t/2)$. This particular distribution arises due to the closure of small cracks in the direction of the stress, and a corresponding stiffening of the material. Cracks aligned parallel to the stress are not affected. An orthorhombic stiffness matrix is possible if the stress is pure shear (Nur 1971). A shear stress causes some cracks to close, but some to open.

2.5.1.4 *More general crack distributions*

Further realism can be injected into the formulations above by including crack orientation, radius and aspect ratio distributions. One of the earliest papers treating such distributions was Nur (1971), in which an explanation for the effects observed in Nur and Simmons (1969) was attempted. Peacock and Hudson (1990) considered distributions in the context of an attenuation study, calculating the stiffnesses numerically. For orientation they used a "Fisher" probability-density function, which was symmetric about the mean orientation of the crack normals. Importantly, this still lead to a TIH medium, with an overall reduction in the difference between c_{44} and c_{66} for a fixed porosity. For aspect ratios with a log-normal distribution there is no appreciable difference from the results for a single aspect ratio. Another style of heterogeneity that must be considered is the discrete hierarchy of anisotropy mechanisms (Figure 2.19). Anisotropy from this system can be built up by combining the theories outlined above.

Figure 2.19. An example of sandstone architecture displaying alignments on several hierarchies of scale.
(a) cliff showing cross-bedded sandstones (Jurassic) from the Isle of Skye, Scotland.
(b) close-up of laminae in aeolian sandstone, courtesy of BG plc.
Diagrams are extracted from Pickup et al. (1994).

2.5.2 Intersecting layers and vertical fractures

Vertical fractures can be inserted into a TIV background given by $\mathbf{c}_b$. For strong TIV anisotropy the procedure yields (Hood and Schoenberg 1989)

$$\mathbf{c} = \mathbf{c}_b - \begin{pmatrix} c_{11}^b e_n & c_{12}^b e_n & c_{13}^b e_n & 0 & 0 & 0 \\ c_{12}^b e_n & \dfrac{c_{12}^{b\,2}}{c_{11}^b} e_n & c_{13}^b \left(\dfrac{c_{12}^b}{c_{11}^b} \right) e_n & 0 & 0 & 0 \\ c_{13}^b e_n & c_{13}^b \left(\dfrac{c_{12}^b}{c_{11}^b} \right) e_n & \dfrac{c_{13}^{b\,2}}{c_{11}^b} e_n & 0 & 0 & 0 \\ 0 & 0 & 0 & 0 & 0 & 0 \\ 0 & 0 & 0 & 0 & c_{55}^b e_t & 0 \\ 0 & 0 & 0 & 0 & 0 & c_{66}^b e_t \end{pmatrix} . \qquad (2.97)$$

For a very weakly anisotropic TIV background (2.44) helps to obtain

$$\mathbf{c}^{ORTH} = \mathbf{c}_0 - \begin{pmatrix} (\lambda_0+2\mu_0)(e_n-2\varepsilon) & \lambda_0\left(e_n-\frac{2\varepsilon}{\zeta}+4\gamma\frac{\eta}{\zeta}\right) & \lambda_0\left(e_n-\frac{2\delta}{\zeta}\right) & 0 & 0 & 0 \\ \lambda_0\left(e_n-\frac{2\varepsilon}{\zeta}+4\gamma\frac{\eta}{\zeta}\right) & (\lambda_0+2\mu_0)(\zeta^2 e_n-2\varepsilon) & \lambda_0\left(\zeta e_n-\frac{2\delta}{\zeta}\right) & 0 & 0 & 0 \\ \lambda_0\left(e_n-\frac{2\delta}{\zeta}\right) & \lambda_0\left(\zeta e_n-\frac{2\delta}{\zeta}\right) & (\lambda_0+2\mu_0)\zeta^2 e_n & 0 & 0 & 0 \\ 0 & 0 & 0 & 0 & 0 & 0 \\ 0 & 0 & 0 & 0 & \mu_0 e_t & 0 \\ 0 & 0 & 0 & 0 & 0 & \mu_0(e_t-2\gamma) \end{pmatrix}, \tag{2.98}$$

where $\mathbf{c}_0$ is the stiffness matrix of an isotropic background.

Such orthorhombic media have special wave properties, as will be shown in Chapter 3.

2.5.3 Bedding-parallel cracks in shales

This particular case was highlighted by the experimental studies of Vernik and Nur (1992) and Vernik (1993), who showed that intrinsic anisotropy is enhanced in thermally mature shales by bedding parallel micro-cracks caused by hydrocarbon generation and geopressuring From equation (2.64) and Figure 2.6 it can be seen that δ can be both positive or negative for bedding parallel cracks. δ is positive when the cracks are gas filled ($e_n > e_t$), and negative when the cracks are brine filled ($e_n < e_t$). Thus, the δ of the shales can be reduced or enhanced by a population of micro-cracks. With knowledge of the intrinsic shale anisotropy, the effect of this change on the wave behavior (Chapter 3) can be used as an indicator of the crack properties.

APPENDIX: THE BOND TRANSFORMATION

A.1 Rotation about the Z-axis

Clockwise rotation of a TIH matrix $\mathbf{c}$ about the Z-axis gives a new matrix $\mathbf{c}^A$. The rotation is achieved by the Bond transformation:

$$\mathbf{c}^A = \mathbf{M}_Z \mathbf{c} \mathbf{M}_Z^T \tag{A.1}$$

with the coordinate transformation matrix

$$\mathbf{M}_Z = \begin{pmatrix} \cos^2\phi & \sin^2\phi & 0 & 0 & 0 & 2\sin\phi\cos\phi \\ \sin^2\phi & \cos^2\phi & 0 & 0 & 0 & -2\sin\phi\cos\phi \\ 0 & 0 & 1 & 0 & 0 & 0 \\ 0 & 0 & 0 & \cos\phi & -\sin\phi & 0 \\ 0 & 0 & 0 & \sin\phi & \cos\phi & 0 \\ -\sin\phi\cos\phi & \sin\phi\cos\phi & 0 & 0 & 0 & \cos^2\phi-\sin^2\phi \end{pmatrix}. \tag{A.2}$$

This transformation gives the "monoclinic" stiffness matrix

$$\mathbf{c}^A = \begin{pmatrix} c_{11}^A & c_{12}^A & c_{13}^A & 0 & 0 & c_{16}^A \\ c_{12}^A & c_{22}^A & c_{23}^A & 0 & 0 & c_{26}^A \\ c_{13}^A & c_{23}^A & c_{33}^A & 0 & 0 & c_{36}^A \\ 0 & 0 & 0 & c_{44}^A & c_{45}^A & 0 \\ 0 & 0 & 0 & c_{45}^A & c_{55}^A & 0 \\ c_{16}^A & c_{26}^A & c_{36}^A & 0 & 0 & c_{66}^A \end{pmatrix} , \tag{A.3}$$

where

$$c_{11}^A = \cos^4\phi c_{11} + \sin^4\phi c_{33} + \frac{1}{2}\sin^2 2\phi(c_{12} + 2c_{66}) , \tag{A.4a}$$

$$c_{12}^A = \frac{1+\cos^2 2\phi}{2} c_{12} + \sin^2 2\phi \left(\frac{c_{11} + c_{33}}{4} - c_{66}\right) , \tag{A.4b}$$

$$c_{13}^A = \cos^2\phi c_{12} + \sin^2\phi(c_{33} - 2c_{44}) , \tag{A.4c}$$

$$c_{16}^A = \tfrac{1}{2}\sin 2\phi(\cos 2\phi(c_{12} + 2c_{66}) + \sin^2\phi c_{33} - \cos^2\phi c_{11}) , \tag{A.4d}$$

$$c_{22}^A = \sin^4\phi c_{11} + \cos^4\phi c_{33} + \frac{1}{2}\sin^2 2\phi(c_{12} + 2c_{66}) , \tag{A.4e}$$

$$c_{23}^A = \sin^2\phi c_{12} + \cos^2\phi(c_{33} - 2c_{44}) , \tag{A.4f}$$

$$c_{26}^A = \tfrac{1}{2}\sin 2\phi(-\cos 2\phi(c_{12} + 2c_{66}) + \cos^2\phi c_{33} - \sin^2\phi c_{11}) , \tag{A.4g}$$

$$c_{33}^A = c_{33} , \tag{A.4h}$$

$$c_{36}^A = \tfrac{1}{2}\sin 2\phi(c_{33} - 2c_{44} - c_{12}) , \tag{A.4i}$$

$$c_{44}^A = \cos^2\phi c_{44} + \sin^2\phi c_{66} , \tag{A.4j}$$

$$c_{45}^A = \tfrac{1}{2}\sin 2\phi(c_{44} - c_{66}) , \tag{A.4k}$$

$$c_{55}^A = \sin^2\phi c_{44} + \cos^2\phi c_{66} , \tag{A.4l}$$

$$c_{66}^{A} = \sin^2 2\phi \; \frac{c_{11} + c_{33} - 2c_{12}}{4} + \cos^2 2\phi c_{66} \;. \tag{A.4m}$$

A.2 Rotation clockwise about the *Y*-axis

Rotation of a TIH matrix **c** about the Y-axis by an angle θ gives a new stiffness matrix $\mathbf{c}^B$. The Bond transformation is

$$\mathbf{c}^B = \mathbf{M}_Y \mathbf{c} \mathbf{M}_Y^T \;,$$

where

$$\mathbf{M}_Y = \begin{pmatrix} \cos^2\theta & 0 & \sin^2\theta & 0 & -\sin 2\theta & 0 \\ 0 & 1 & 0 & 0 & 0 & 0 \\ \sin^2\theta & 0 & \cos^2\theta & 0 & \sin 2\theta & 0 \\ 0 & 0 & 0 & \cos\theta & 0 & \sin\theta \\ \frac{1}{2}\sin 2\theta & 0 & -\frac{1}{2}\sin 2\theta & 0 & \cos 2\theta & 0 \\ 0 & 0 & 0 & -\sin\theta & 0 & \cos\theta \end{pmatrix} . \tag{A.5}$$

and hence

$$\mathbf{c}^B = \begin{pmatrix} c_{11}^B & c_{12}^B & c_{13}^B & 0 & c_{15}^B & 0 \\ c_{12}^B & c_{22}^B & c_{23}^B & 0 & c_{25}^B & 0 \\ c_{13}^B & c_{23}^B & c_{33}^B & 0 & c_{35}^B & 0 \\ 0 & 0 & 0 & c_{44}^B & 0 & c_{46}^B \\ c_{15}^B & c_{25}^B & c_{35}^B & 0 & c_{55}^B & 0 \\ 0 & 0 & 0 & c_{46}^B & 0 & c_{66}^B \end{pmatrix} , \tag{A.6}$$

where

$$c_{11}^B = \cos^4\theta c_{11} + \sin^4\theta c_{33} + \sin^2 2\theta\left(\tfrac{1}{2}c_{12} + c_{66}\right) , \tag{A.7a}$$

$$c_{12}^B = \cos^2\theta c_{12} + \sin^2\theta\left(c_{33} - 2c_{44}\right) , \tag{A.7b}$$

$$c_{13}^B = \frac{\sin^2 2\theta}{4}\left(c_{11} + c_{33} - 4c_{66}\right) + \frac{1+\cos^2 2\theta}{2} c_{12} \;, \tag{A.7c}$$

$$c_{15}^B = \tfrac{1}{2}\sin 2\theta\left(\cos^2\theta c_{11} - \sin^2\theta c_{33} - \cos 2\theta\left(c_{12} + 2c_{66}\right)\right) , \tag{A.7d}$$

$$c_{22}^B = c_{33} \;, \tag{A.7e}$$

$$c_{23}^B = \sin^2\theta c_{12} + \cos^2\theta\left(c_{33} - 2c_{44}\right) , \tag{A.7f}$$

$$c_{25}^B = \tfrac{1}{2}\sin 2\theta\left(c_{12} - \left(c_{33} - 2c_{44}\right)\right) \ , \tag{A.7g}$$

$$c_{33}^B = \sin^4\theta c_{11} + \cos^4\theta c_{33} + \sin^2 2\theta\left(\tfrac{1}{2}c_{12} + c_{66}\right) \ , \tag{A.7h}$$

$$c_{35}^B = \tfrac{1}{2}\sin 2\theta\left(\sin^2\theta c_{11} - \cos^2\theta c_{33} + \cos 2\theta\left(c_{12} + 2c_{66}\right)\right) \ , \tag{A.7i}$$

$$c_{44}^B = \sin^2\theta c_{66} + \cos^2\theta c_{44} \ , \tag{A.7j}$$

$$c_{46}^B = \tfrac{1}{2}\sin 2\theta\left(c_{44} - c_{66}\right) \ , \tag{A.7k}$$

$$c_{55}^B = \tfrac{1}{4}\sin^2 2\theta\left(c_{11} - 2c_{12} + c_{33}\right) + \cos^2 2\theta c_{66} \ , \tag{A.7l}$$

and

$$c_{66}^B = \cos^2\theta c_{66} + \sin^2\theta c_{44} \ . \tag{A.7m}$$

For weak anisotropy caused by thin liquid-filled cracks,

$$\left|c_{12} - \left(c_{33} - 2c_{44}\right)\right| << \left|c_{11} - c_{33}\right| << \left|c_{44} - c_{66}\right| < 0.3c_{66} \ ; \tag{A.8}$$

the stiffnesses in the above can be approximated by

$$\mathbf{c}^B = \begin{pmatrix} c_{11}^B & c_{12}^B & c_{13}^B & 0 & c_{15}^B & 0 \\ c_{12}^B & c_{22}^B & c_{23}^B & 0 & 0 & 0 \\ c_{13}^B & c_{23}^B & c_{33}^B & 0 & c_{35}^B & 0 \\ 0 & 0 & 0 & c_{44}^B & 0 & c_{46}^B \\ c_{15}^B & 0 & c_{35}^B & 0 & c_{55}^B & 0 \\ 0 & 0 & 0 & c_{46}^B & 0 & c_{66}^B \end{pmatrix} , \tag{A.9}$$

where

$$c_{11}^B = c_{11} - \sin^2 2\theta\left(c_{44} - c_{66}\right) \ , \tag{A.9a}$$

$$c_{12}^B = c_{12} \ , \tag{A.9b}$$

$$c_{13}^B = c_{12} + \sin^2 2\theta\left(c_{44} - c_{66}\right) \ , \tag{A.9c}$$

$$c_{15}^{B} = \sin 2\theta \cos 2\theta \left(c_{44} - c_{66}\right) , \tag{A.9d}$$

$$c_{22}^{B} = c_{33} , \tag{A.9e}$$

$$c_{23}^{B} = \left(c_{33} - 2c_{44}\right) , \tag{A.9f}$$

$$c_{33}^{B} = c_{33} - \sin^2 2\theta \left(c_{44} - c_{66}\right) , \tag{A.9g}$$

$$c_{35}^{B} = -\sin 2\theta \cos 2\theta \left(c_{44} - c_{66}\right) , \tag{A.9h}$$

$$c_{44}^{B} = c_{44} - \sin^2 \theta \left(c_{44} - c_{66}\right) , \tag{A.9i}$$

$$c_{46}^{B} = -\tfrac{1}{2} \sin 2\theta \left(c_{44} - c_{66}\right) , \tag{A.9j}$$

$$c_{55}^{B} = c_{66} + \sin^2 2\theta \left(c_{44} - c_{66}\right) , \tag{A.9k}$$

$$c_{66}^{B} = c_{66} + \sin^2 \theta \left(c_{44} - c_{66}\right) . \tag{A.9l}$$

A.3 Rotation clockwise about the *Y* and *Z* axes

Rotating about the *Y*-axis by θ and then about the *Z*-axis by ϕ gives a "triclinic" set of elastic stiffnesses $\mathbf{c}^C$:

$$\mathbf{c}^C = \begin{pmatrix} c_{11}^C & c_{12}^C & c_{13}^C & c_{14}^C & c_{15}^C & c_{16}^C \\ c_{12}^C & c_{22}^C & c_{23}^C & c_{24}^C & c_{25}^C & c_{26}^C \\ c_{13}^C & c_{23}^C & c_{33}^C & c_{34}^C & c_{35}^C & c_{36}^C \\ c_{14}^C & c_{24}^C & c_{34}^C & c_{44}^C & c_{45}^C & c_{46}^C \\ c_{15}^C & c_{25}^C & c_{35}^C & c_{45}^C & c_{55}^C & c_{56}^C \\ c_{16}^C & c_{26}^C & c_{36}^C & c_{46}^C & c_{56}^C & c_{66}^C \end{pmatrix} , \tag{A.10}$$

With the c_{ij} from Section A.2 this gives

$$c_{11}^{C} = \cos^4 \phi c_{11}^{B} + \sin^4 \phi c_{22}^{B} + \tfrac{1}{2} \sin^2 2\phi \left(c_{12}^{B} + 2c_{66}^{B}\right) , \tag{A.11a}$$

$$c_{12}^{C} = \sin^2 2\phi \left(\frac{c_{11}^{B} + c_{22}^{B}}{4} - c_{66}^{B} \right) + \frac{1 + \cos^2 2\phi}{2} c_{12}^{B} , \tag{A.11b}$$

$$c_{13}^{C} = \cos^2\phi c_{13}^{B} + \sin^2\phi c_{23}^{B} \ , \tag{A.11c}$$

$$c_{14}^{C} = \sin\phi\cos^2\phi\left(2c_{46}^{B} - c_{15}^{B}\right) \ , \tag{A.11d}$$

$$c_{15}^{C} = \cos\phi\left(\cos^2\phi c_{15}^{B} + 2\sin^2\phi c_{46}^{B}\right) \ , \tag{A.11e}$$

$$c_{16}^{C} = \tfrac{1}{2}\sin 2\phi\left(\cos 2\phi\left(c_{12}^{B} + 2c_{66}^{B}\right) + \sin^2\phi c_{22}^{B} - \cos^2\phi c_{11}^{B}\right) \ , \tag{A.11f}$$

$$c_{22}^{C} = \sin^4\phi c_{11}^{B} + \cos^4\phi c_{22}^{B} + \tfrac{1}{2}\sin^2 2\phi\left(c_{12}^{B} + 2c_{66}^{B}\right) \ , \tag{A.11g}$$

$$c_{23}^{C} = \sin^2\phi c_{13}^{B} + \cos^2\phi c_{23}^{B} \ , \tag{A.11h}$$

$$c_{24}^{C} = -\sin 2\phi\left(\cos\phi\left(c_{46}^{B} - \sin^3\phi c_{15}^{B}\right)\right) \ , \tag{A.11i}$$

$$c_{25}^{C} = -\sin 2\phi\left(\sin\phi c_{46}^{B} - \tfrac{1}{2}\sin\phi c_{15}^{B}\right) \ , \tag{A.11j}$$

$$c_{26}^{C} = \tfrac{1}{2}\sin 2\phi\left(-\cos 2\phi\left(c_{12}^{B} + 2c_{66}^{B}\right) + \sin^2\phi c_{11}^{B} - \cos^2\phi c_{22}^{B}\right) \ , \tag{A.11k}$$

$$c_{33}^{C} = c_{33}^{B} \ , \tag{A.11l}$$

$$c_{34}^{C} = -\sin\phi c_{35}^{B} \ , \tag{A.11m}$$

$$c_{35}^{C} = \cos\phi c_{35}^{B} \ , \tag{A.11n}$$

$$c_{36}^{C} = \tfrac{1}{2}\sin 2\phi\left(c_{23}^{B} - c_{12}^{B}\right) \ , \tag{A.11o}$$

$$c_{44}^{C} = \cos^2\phi c_{44}^{B} + \sin^2\phi c_{55}^{B} \ , \tag{A.11p}$$

$$c_{45}^{C} = \tfrac{1}{2}\sin 2\phi\left(c_{44}^{B} - c_{55}^{B}\right) \ , \tag{A.11q}$$

$$c_{46}^{C} = \cos\phi\left(\sin^2\phi c_{15}^{B} + \cos 2\phi c_{46}^{B}\right) ,$$

(A.11r)

$$c_{55}^{C} = \sin^2\phi c_{44}^{B} + \cos^2\phi c_{55}^{B} ,$$

(A.11s)

$$c_{56}^{C} = \sin\phi\left(\cos^2\phi c_{15}^{B} + \cos 2\phi c_{46}^{B}\right) ,$$

(A.11t)

$$c_{66}^{C} = \tfrac{1}{4}\sin^2 2\phi \left(c_{11}^{B} + c_{22}^{B} - 2c_{12}^{B}\right) + \cos^2 2\phi c_{66}^{B} .$$

(A.11u)

CHAPTER 3 – FUNDAMENTALS OF SEISMIC ANISOTROPY ANALYSIS

In this chapter, the elastic stiffnesses defined by the theoretical models of Chapter 2 are translated into wave properties, from which the tools for lithology evaluation are shaped. From this process, vertical-incidence shear wave propagation through fracture-related TIH anisotropy emerges as being particularly important. Indeed, it is predicted that shear waves can provide an optimum estimate of the fracture parameter e_t (but not e_n, for which sub-horizontal P-wave propagation is also required) and orientation of the TIH anisotropy (and hence fracture strike). The growth in the multi-component VSP seismic technique described in Chapters 4, 5 and 6 began with the recognition of this simple and powerful interpretation. Oblique incidence using offset or multi-offset VSP introduces the possibility of sampling a rich range of anisotropic wave behavior which helps to determine the additional fracture indicator e_n, with extra shot line azimuths further aiding this task and making offshore measurement of e_t possible. For shales/sedimentary layering, the Thomsen parameters ε, δ and γ or their combinations can also be evaluated from a range of offsets using the moveout- and AVO techniques described in Chapters 7 and 8.

The Kelvin-Christoffel equation is the basis for all predictions of the anisotropic wave behavior. This equation arises from the conjunction of the fundamental equation of motion for an elastic medium and the anisotropic stress-strain relation, followed by substitution of a trial plane wave solution (Helbig 1994). This results in

$$\mathbf{Fu} = v^2\mathbf{u} \ , \tag{3.1}$$

from which the phase three velocities $v^{(i)}$ and permissible polarizations $\mathbf{u}^i$ for the plane waves propagating along each specific (wave normal) direction can be obtained by eigenanalysis of the 3×3 Kelvin-Christoffel matrix $\mathbf{F}$. The matrix is a function of the density-normalized elastic stiffnesses $a_{ij} = c_{ij}/\rho$. The eigenanalysis is, in principle, not difficult. Indeed, many standard mathematical packages have subroutines for this purpose. It is recognized, however, that an analytic solution provides insight into the importance of individual parameters or their combination and helps to construct general guidelines for interpretation. This approach requires expressions for the velocities derived from the eigenvalues of the characteristic equation

$$\left|\mathbf{F} - \mathbf{I}v^2\right| = 0 \ , \tag{3.2}$$

from which individual eigenvectors are obtained by solving

$$\mathbf{Fu}^i = v^{(i)2}\mathbf{u}^i \ . \tag{3.3}$$

For propagation in an arbitrary direction with inclination θ and azimuth ϕ, together with a complete set of 21 elastic stiffnesses, Helbig (1994) gives

$$\mathbf{F} = \begin{pmatrix} \begin{array}{l} a_{11}l_x^2 + a_{66}l_y^2 + a_{55}l_z^2 \\ +\ 2a_{56}l_yl_z \\ +\ 2a_{15}l_xl_z \\ +\ 2a_{16}l_xl_y \end{array} & \begin{array}{l} (a_{12}+a_{66})l_xl_y + \\ (a_{25}+a_{46})l_yl_z + \\ (a_{14}+a_{56})l_xl_z + \\ a_{16}l_x^2 + a_{26}l_y^2 + a_{45}l_z^2 \end{array} & \begin{array}{l} (a_{13}+a_{55})l_xl_z + \\ (a_{36}+a_{45})l_yl_z + \\ (a_{14}+a_{56})l_xl_y + \\ a_{15}l_x^2 + a_{46}l_y^2 + a_{35}l_z^2 \end{array} \\ \\ \begin{array}{l} (a_{12}+a_{66})l_xl_y + \\ (a_{25}+a_{46})l_yl_z + \\ (a_{14}+a_{56})l_xl_z + \\ a_{16}l_x^2 + a_{26}l_y^2 + a_{45}l_z^2 \end{array} & \begin{array}{l} a_{66}l_x^2 + a_{22}l_y^2 + a_{44}l_z^2 \\ +\ 2a_{24}l_yl_z \\ +\ 2a_{46}l_xl_z \\ +\ 2a_{26}l_xl_y \end{array} & \begin{array}{l} (a_{23}+a_{44})l_yl_z + \\ (a_{36}+a_{45})l_xl_z + \\ (a_{25}+a_{46})l_xl_y + \\ a_{56}l_x^2 + a_{24}l_y^2 + a_{34}l_z^2 \end{array} \\ \\ \begin{array}{l} (a_{13}+a_{55})l_xl_z + \\ (a_{36}+a_{45})l_yl_z + \\ (a_{14}+a_{56})l_xl_y + \\ a_{15}l_x^2 + a_{46}l_y^2 + a_{35}l_z^2 \end{array} & \begin{array}{l} (a_{23}+a_{44})l_yl_z + \\ (a_{36}+a_{45})l_xl_z + \\ (a_{25}+a_{46})l_xl_y + \\ a_{56}l_x^2 + a_{24}l_y^2 + a_{34}l_z^2 \end{array} & \begin{array}{l} a_{55}l_x^2 + a_{44}l_y^2 + a_{33}l_z^2 \\ +\ 2a_{34}l_yl_z \\ +\ 2a_{35}l_xl_z \\ +\ 2a_{45}l_xl_y \end{array} \end{pmatrix} . \tag{3.4}$$

The unit vector $(l_x, l_y, l_z)^T$ points in the direction of propagation, with its components given by

$$l_x = \cos\phi \sin\theta \ , \tag{3.5}$$

$$l_y = \sin\phi \sin\theta \ , \tag{3.6}$$

and

$$l_z = \cos\theta \ . \tag{3.7}$$

Despite the apparent complexity, the structure of the matrix already allows some fundamental statements to be made about the wave propagation. Firstly, the dimension of the matrix **F** is three, thus there are at most three distinct solutions (waves) for each specific phase direction (there may be more than three arrivals from a point source in a fixed direction, as will be discussed later). Secondly, **F** is symmetric, and thus the eigenvectors (polarization vectors) are mutually orthogonal and the eigenvalues (squared velocities) are always real and positive. Finally, the vector components appear as squared quantities in (3.4), indicating that diametrically opposite directions have the same properties.

For TIV and TIH anisotropy, most of the entries above are zero, and analytic expressions of practical value are just possible provided the anisotropy is not too strong.

3.1 SHALES AND SEDIMENTARY LAYERING – TIV

The TIV model is a good starting point as it represents the first step beyond isotropy. The symmetry axis is vertical, perpendicular to the horizontal beds. For this model, both the material properties and the wave behavior are independent of azimuth (i.e., they are azimuthally isotropic), and any vertical plane is a plane of symmetry. The specific Kelvin-Christoffel matrix is obtained from (3.4)

$$\mathbf{F} = \begin{pmatrix} a_{11}l_x^2 + a_{66}l_y^2 + a_{55}l_z^2 & (a_{12}+a_{66})l_x l_y & (a_{13}+a_{55})l_x l_z \\ (a_{12}+a_{66})l_x l_y & a_{66}l_x^2 + a_{11}l_y^2 + a_{55}l_z^2 & (a_{13}+a_{44})l_y l_z \\ (a_{13}+a_{55})l_x l_z & (a_{13}+a_{55})l_y l_z & a_{55}l_x^2 + a_{55}l_y^2 + a_{33}l_z^2 \end{pmatrix}. \tag{3.8}$$

The requisite dispersion relation can be obtained from the characteristic equation of (3.2)

$$\mathbf{F} - \mathbf{I}v^2 = \begin{vmatrix} a_{11}l_x^2 + a_{66}l_y^2 + a_{55}l_z^2 - v^2 & (a_{12}+a_{66})l_x l_y & (a_{13}+a_{55})l_x l_z \\ (a_{12}+a_{66})l_x l_y & a_{66}l_x^2 + a_{11}l_y^2 + a_{55}l_z^2 - v^2 & (a_{13}+a_{55})l_y l_z \\ (a_{13}+a_{55})l_x l_z & (a_{13}+a_{55})l_y l_z & a_{55}l_x^2 + a_{55}l_y^2 + a_{33}l_z^2 - v^2 \end{vmatrix} = 0. \tag{3.9}$$

Equation (3.9) can be considerably simplified by following the recipe for determinant manipulations given by Helbig (1994). This consists of firstly multiplying the first, second and third *columns* by l_x, l_y, and l_z, respectively. This is followed by respective division of the first, second and third *rows* by l_x, l_y, and l_z. Next, the second row is subtracted from the first, and then finally the first column added to the second. This process isolates zeros on the top row of the determinant and the relation becomes

$$\begin{vmatrix} a_{66}l_h^2 + a_{55}l_z^2 - v^2 & 0 & 0 \\ (a_{12}+a_{66})l_x l_y & a_{11}l_h^2 + a_{55}l_z^2 - v^2 & (a_{13}+a_{55})l_z^2 \\ (a_{13}+a_{55})l_x l_z & (a_{13}+a_{55})l_h^2 & a_{55}l_h^2 + a_{33}l_z^2 - v^2 \end{vmatrix} = 0 , \tag{3.10}$$

yielding

$$\left(a_{11}l_h^2 + a_{55}l_z^2 - v^2\right)\left(a_{55}l_h^2 + a_{33}l_z^2 - v^2\right) - \left(a_{13}+a_{55}\right)^2 l_h^2 l_z^2 = 0 , \tag{3.11}$$

and a separate (decoupled) equation

$$a_{66}l_h^2 + a_{55}l_z^2 - v^2 = 0 , \tag{3.12}$$

with $l_h^2 = l_x^2 + l_y^2$. Equation (3.12) gives vertical and horizontal velocities $\sqrt{a_{66}}$ and $\sqrt{a_{55}}$ (Figure 3.1). For (3.12), the eigenvector can be immediately obtained by feeding the eigenvalue $v_i^2 = a_{66}l_h^2 + a_{55}a_z^2$ back into (3.3), which after manipulation yields

$$\mathbf{u}_i = \begin{pmatrix} l_y \\ -l_x \\ 0 \end{pmatrix}. \tag{3.13}$$

This solution corresponds to a shear wave polarized transversely to the direction of propagation. As the plane of propagation also contains the in-line direction, this wave can be correctly called an SH wave. Equation (3.12) defines the phase velocity of this shear wave, which can be rewritten in terms of Thomsen's γ defined in (2.41) and the phase angle θ

$$v_\perp^2 = \beta_0^2\left(1 + 2\gamma \sin^2\theta\right) , \tag{3.14}$$

where the vertical velocity $\beta_0 = \sqrt{a_{55}}$. The decoupling of this SH solution and its dependence only on θ, the angle between the propagation direction and the symmetry axis, is a characteristic not just of TIV, but TI media in general. It will be seen later that it is of critical importance for fracture strike determination.

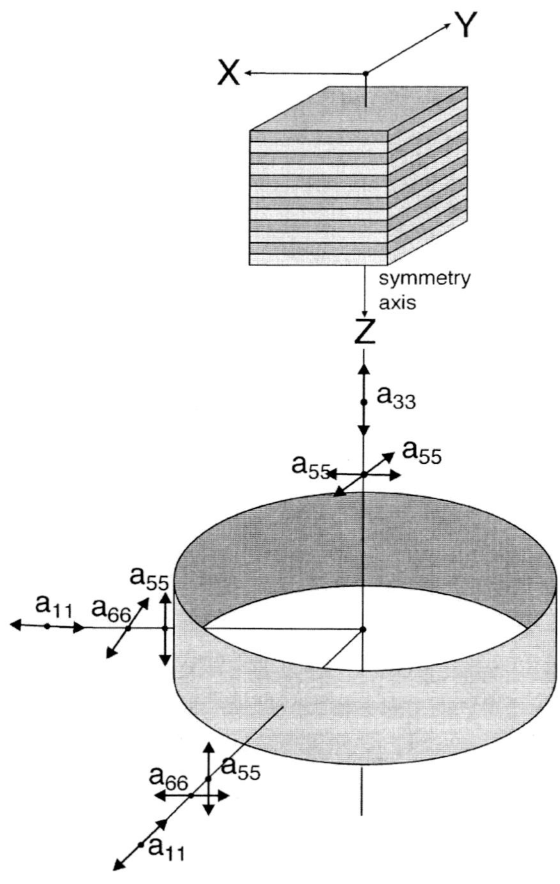

Figure 3.1. Eigenvalues determined from the Kelvin-Christoffel equation for directions along the *X*, *Y* and *Z*-axes in a TIV medium. For the anisotropy related to finely layered sequences and shales it is generally true that $a_{55} < a_{66}$. The double arrows indicate the corresponding eigenvectors and hence the polarization directions.

Equation (3.11), on the other hand, is more difficult to evaluate directly, except in specific directions. For propagation in the vertical direction, $l_x = l_y = 0$, and the solution for the faster (P-wave) is $v_i^2 = a_{33}$ and the other (qSV) is $v_i^2 = a_{55}$ (Figure 3.1). These vertical velocities correspond to α_0 and β_0, respectively, in Thomsen's expressions. For propagation in any horizontal direction (for which $l_z = 0$ and $l_x^2 + l_y^2 = 1$) solutions are $v_i^2 = a_{11}$ and $v_i^2 = a_{55}$, which become $\alpha_0^2(1 + 2\varepsilon)$ and β_0^2, where Thomsen's ε is defined in (2.40). Insight into a possible analytic solution of this quadratic in v_i^2 is provided by the re-arrangement

$$\left(a_{11}l_h^2 + a_{33}l_z^2 - v^2\right)\left(a_{55}l_h^2 + a_{55}l_z^2 - v^2\right) - \left(\left(a_{33} - a_{55}\right)\left(a_{11} - a_{55}\right) - \left(a_{13} + a_{55}\right)^2\right)l_h^2 l_z^2 = 0\ , \tag{3.15}$$

and hence

$$\left(a_{11}l_h^2 + a_{33}l_z^2 - v^2\right)\left(a_{55}l_h^2 + a_{55}l_z^2 - v^2\right) - \left(a_{33} - a_{55}\right)\left(a_{11} - a_{33} - \frac{\left(a_{33} - a_{55}\right)^2 - \left(a_{13} + a_{55}\right)^2}{a_{33} - a_{55}}\right)l_h^2 l_z^2 = 0\ , \tag{3.16}$$

After substitution of the velocity values for the known directions (including the definition of Thomsen's δ (see equation 2.42), and division by $\alpha_0^2\beta_0^2$, the equation becomes

$$\left((1 + 2\varepsilon)l_h^2 + l_z^2 - \frac{v^2}{\alpha_0^2}\right)\left(l_h^2 + l_z^2 - \frac{v^2}{\beta_0^2}\right) - 2\frac{1 - \eta_0}{\eta_0}(\varepsilon - \delta)l_h^2 l_z^2 = 0\ , \tag{3.17}$$

where, as in Chapter 2, $\eta_0 = \beta_0^2/\alpha_0^2$. For $\varepsilon = \delta$ (3.17) yields an elliptical and a circular slowness solutions. In fact, this particular condition is often referred to as "elliptical anisotropy". Elliptical anisotropy can be defined by other indicators, such as (Helbig 1994)

$$E^2 = \left(a_{11} - a_{55}\right)\left(a_{33} - a_{55}\right) - \left(a_{13} + a_{55}\right)^2 \tag{3.18}$$

and the normalized form of Schoenberg (1992)

$$\varepsilon_A = \frac{\left(a_{11} - a_{55}\right)\left(a_{33} - a_{55}\right) - \left(a_{13} + a_{55}\right)^2}{\left(a_{11} - a_{55}\right)\left(a_{33} - a_{55}\right)}\ , \tag{3.19}$$

all of which are zero for $\varepsilon = \delta$ for any strength of anisotropy. The exactly elliptical slowness solution corresponds to a *predominantly* longitudinal polarization (hence the term "quasi-P" with the abbreviation qP), while the circular solution is a quasi-shear (qS) wave. The latter is often called a qSV wave. Both waves have polarizations lying in the vertical (symmetry) plane, and the preferred nomenclature for the shear wave is $S_{\parallel}$. The phase velocities of the two velocities are given by

$$v_p^2 = \alpha_0^2\left(1 + 2\varepsilon\sin^2\theta\right) \tag{3.20}$$

and $v_{\parallel}^2 = \beta_0^2$ for the $S_{\parallel}$ waves.

For moderate deviations from elliptic anisotropy, solutions of equation (3.17) can be obtained as function of the first order perturbation $\varepsilon - \delta$. The range of validity of this perturbation solution defines "weak anisotropy". By absorbing this perturbation into each bracketed term, the results of Thomsen (1986) are recovered:

$$v_p^2 = \alpha_0^2\left(1 + 2\varepsilon\sin^2\theta - 2(\varepsilon - \delta)\sin^2\theta\cos^2\theta\right) = \alpha_0^2\left(1 + 2\delta\sin^2\theta\cos^2\theta + 2\varepsilon\sin^4\theta\right) , \tag{3.21}$$

and

$$v_\parallel^2 = \beta_0^2\left(1 + \frac{2(\varepsilon - \delta)\sin^2\theta\cos^2\theta}{\eta_0}\right) . \tag{3.22}$$

These three phase velocity dispersions are shown in Figure 3.2. A plot of anisotropy measurements from various origins is given in Figure 3.3. Most of the examples are weakly anelliptic. The majority of these materials are shales, for which the anellipticity is positive. Positive anellipticity ($\varepsilon > \delta$) means that qP waves in oblique directions are slower than for the elliptical case, and qSV waves are faster. This statement holds in general.

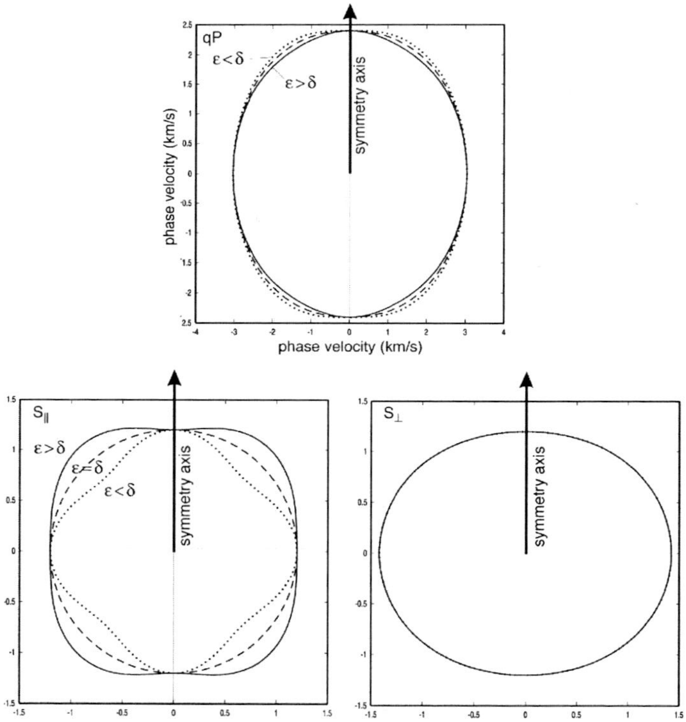

Figure 3.2. Phase velocity variations with direction of propagation in a vertical (symmetry) plane of a TIV medium. Three curves are shown for each wave type, corresponding to $\varepsilon > \delta$ (solid line), $\varepsilon = \delta$ (dashed line), and $\varepsilon < \delta$ (dotted line). The velocity surfaces for the $S_\parallel$ and P waves bulge in opposite directions.

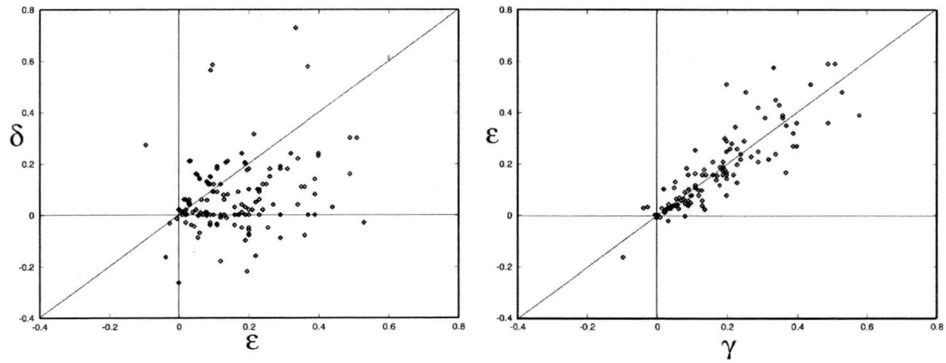

Figure 3.3. Cross-plots of the TI anisotropy measurements listed by Thomsen (1986), Vernik and Liu (1997) and Ryan-Grigor (1997). There are 161 values of ε and δ, and 117 ε.-γ pairs. Note, in particular, the divergence of the anisotropies from the ellipticity line $\varepsilon = \delta$ (top), with most media positively anelliptic ($\varepsilon > \delta$), There is also a strong correlation between γ and ε (bottom).

Exact formulae for the polarizations of these modes can be obtained from (3.3) without loss of generality by setting $l_x = 0$. It leaves only two non-zero off-diagonal terms in the Kelvin-Christoffel matrix in (3.8), which can be diagonalized by applying a rotation about the *Y*-axis. The operation effectively rotates the coordinate frame into the polarization directions, and is given exactly by the angle θ' with

$$\tan 2\theta' = \frac{\sin 2\theta(a_{13} + a_{55})}{(a_{33} - a_{55})\cos^2\theta - (a_{11} - a_{55})\sin^2\theta} . \tag{3.23}$$

The qP and $S_{\|}$ polarizations can be written in terms of their deviation Δ from longitudinal and normal directions:

$$\mathbf{u}_P = \begin{pmatrix} \sin(\theta + \Delta) \\ 0 \\ \cos(\theta + \Delta) \end{pmatrix}, \tag{3.24}$$

$$\mathbf{u}_{\|} = \begin{pmatrix} \cos(\theta + \Delta) \\ 0 \\ -\sin(\theta + \Delta) \end{pmatrix}. \tag{3.25}$$

(3.23) can be simplified by using the weak anisotropy approximation with the perturbation

$$\tan\Delta = \frac{\sin\theta\cos\theta}{1-\eta_0}\left(\delta + 2(\varepsilon - \delta)\sin^2\theta\right) . \tag{3.26}$$

There are three special characteristics of TI anisotropy that require careful consideration:

3.1.1 The line singularity

In most cases the phase velocity curves for the shear waves $S_{\perp}$ and $S_{\|}$ intersect (Figure 3.4). In three dimensions, these curves correspond to surfaces that are circularly symmetric about the symmetry (vertical) axis, and the point of intersection corresponds to a circular line of intersection. This particular feature is called a *line singularity*. Such line singularities are a peculiarity of TI media, and are not found in any other symmetry system (Crampin and Yedlin 1981). In directions defined by this singularity there is effectively isotropic propagation, albeit for a single incidence. More importantly, the fast and slow shear waves switch positions across it, so that the first arrival has a different polarization. The major influence of this intersection is interpretational, and it does not effect the overall wave propagation. This feature is of particular relevance in fracture-related TIH media and will be discussed further below. The incidence angle θ_{ls}, at which this singularity occurs can be determined from equations (3.14) and (3.22):

$$\cos\theta_{ls} = \sqrt{\frac{\eta_0\gamma}{\varepsilon-\delta}} \quad . \tag{3.27}$$

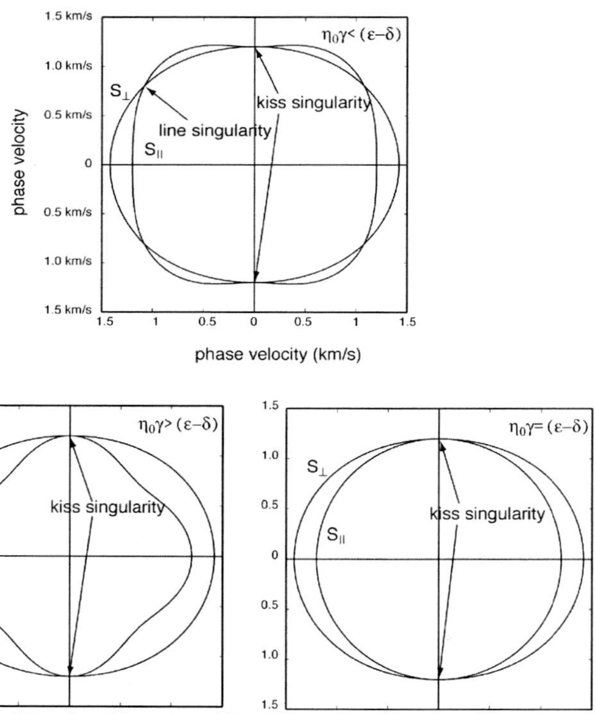

Figure 3.4. Effect of the TIV parameters on the position and existence of the line singularity at the intersection of the two shear velocity surfaces (a vertical section of which is shown above). Intersection occurs for only one out of the three scenarios above. The surfaces touch at a kiss singularity. Every singularity has its symmetrical "counterparts" (at least one). Symmetry axes generally correspond to kiss singularities.

Line singularities exist if (and only if) $\eta_0\gamma < \varepsilon - \delta$. Figure 3.5 shows that almost all materials in the database from Figure 3.3 satisfy this condition. For the rare case of elliptical anisotropy, the line singularity clearly has no meaning.

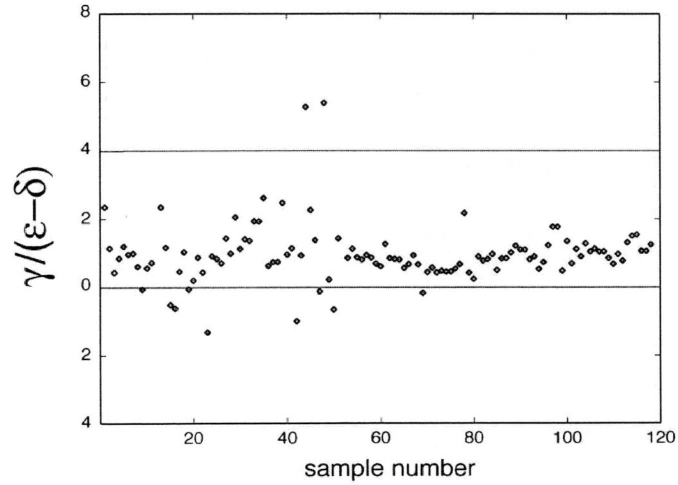

Figure 3.5. Plot based on the list of measurements in Figure 3.3, demonstrating that for most TI media there should be a line singularity. The horizontal line corresponds to the upper limit of existence, i.e., $\alpha/\beta = 2$.

3.1.2 Group velocity versus phase velocity

The difference between group- and phase velocity has only a minor impact on the near-offset, multi-component VSP method outlined in Chapters 4, 5 and 6, but a significant impact on the moveout and AVO analyses of Chapters 7 and 8. Indeed, obtaining a functional form for the group velocity forms the basis for deriving the moveout expressions. The first break at the receiver detects energy, with the arrival time depending on group velocity, not on the phase velocity obtained from the Kelvin-Christoffel equation. Figure 3.6 shows a snapshot of the energy radiated from a point source (an adequate representation of a VSP survey). In a homogeneous anisotropic medium the wave front is non-circular. Thus, although energy continues to move out radially along rays emanating from the source, the direction of propagation of each local planar portion of the wave front is different. This is because this segment of constant phase travels along the direction of the wave-front normal (Figure 3.6). Thus the group (ray) and phase angles are different, and thus group and phase velocities are different. The phase velocities determined from the Kelvin-Christoffel equation are not the appropriate velocities for interpretation based on travel-times in field surveys.

The exception to this occurs when acquisition is along particular symmetry directions, where the two velocities are equal. For weak anisotropy, Thomsen (1986) writes the group angle θ_{gr} in terms of the phase angle θ for qP waves

$$\tan\theta_P^{gr} = \tan\theta\left(1 + 2\delta + 4(\varepsilon - \delta)\sin^2\theta\right), \tag{3.28}$$

for $S_{\|}$ waves

$$\tan\theta_{\|}^{gr} = \tan\theta\left(1 + \frac{2}{\eta_0}(\varepsilon - \delta)\cos 2\theta\right), \tag{3.29}$$

and for $S_\perp$ waves

$$\tan\theta_\perp^{gr} = \tan\theta(1+2\gamma) \ . \tag{3.30}$$

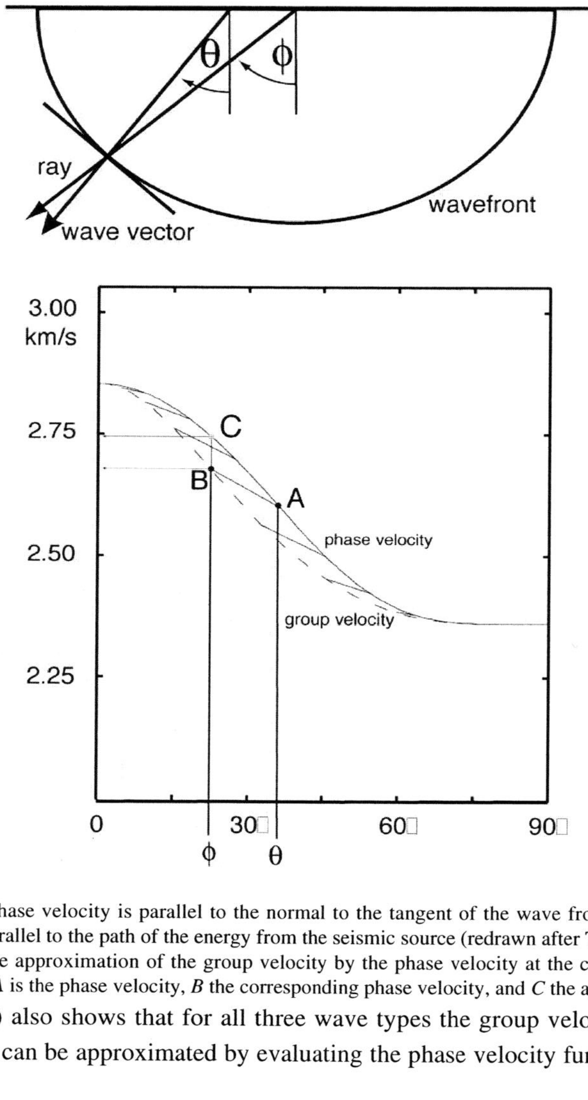

Figure 3.6. Top: the phase velocity is parallel to the normal to the tangent of the wave front, while the group velocity is parallel to the path of the energy from the seismic source (redrawn after Thomsen 1986).
Bottom: the approximation of the group velocity by the phase velocity at the corresponding group angle. Point *A* is the phase velocity, *B* the corresponding phase velocity, and *C* the approximation.

Thomsen (1986) also shows that for all three wave types the group velocity evaluated along the group (ray) angle can be approximated by evaluating the phase velocity function by replacing θ by θ_{gr} (Figure 3.6).

3.1.3 Cusps

The group velocity surface is generally rather complicated, as it is derived from the phase surface by differentiation (Helbig 1994). It follows that the greatest deviation between the two surfaces occurs when the phase velocity surface has high curvature. These can also lead to a rapid variation in the polarization directions for the shear waves, and for strong anisotropy to the appearance of cuspidal features (Musgrave 1970).

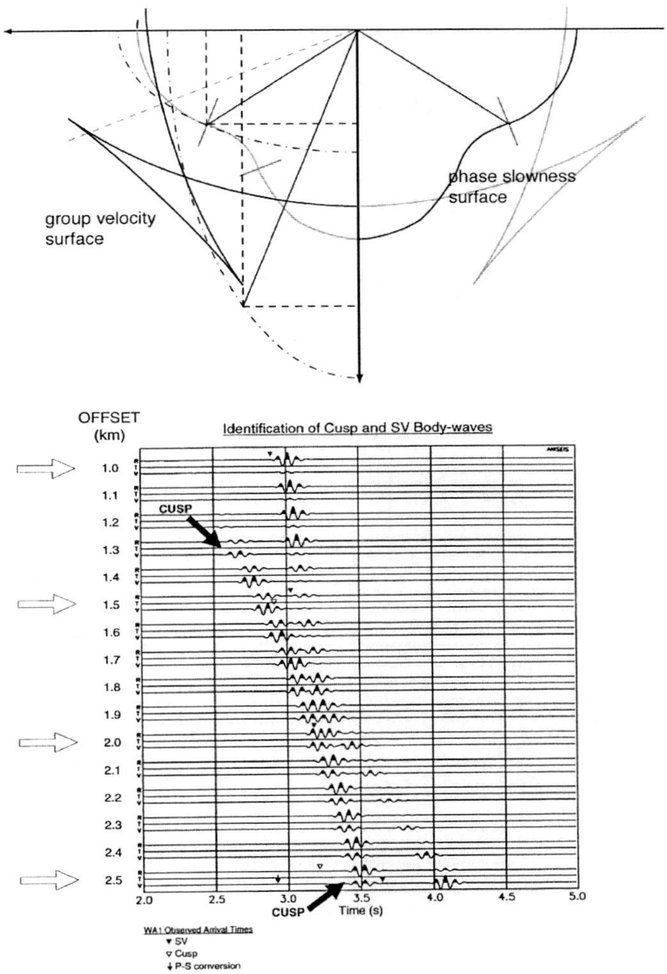

Figure 3.7. Top: a cusp on the group velocity surface is formed by a concavity on the phase slowness surface for the $S_{\parallel}$ wave.
Bottom: Three-component traces from modeling of walkaway VSP data shot over Maikop clays (Slater, 1997).

An illustration of the geometrical relationship between the phase and group surfaces is given by Helbig (1994). Cusps exist only in the qSV group velocity surface, and are caused by elements of inflection in the phase slowness surface (Figure 3.7). Cusps result in the possibility of two additional shear wave arrivals. These additions arise from the different branches of the group surface, and can result in signals arriving earlier than expected. They are important in velocity analysis and in the interpretation of effects in AVO studies. For $\varepsilon - \delta > 0$ cusps can occur in the vicinity of

$\theta = 45°$, for $\varepsilon - \delta < 0$ they can occur in the vicinity of $\theta = 0°$ and $\theta = 90°$. A cusp consists of three separate branches of the wave surface, two normal branches and one reversed branch. Between the cusps there is a range of directions where triplication occurs. Thus for each ray direction in the triplication range, there are three possible $S_{\parallel}$ arrivals. This behavior is illustrated in the simulated VSP data of Figures 3.7 and 3.8. The intensity varies along the branches, being highest close to the cusp tips. The wave on the reverse branch is phase-shifted by $\pi/2$ from the wave on the normal branches. Such features have been observed in producing fields where there is a particularly thick sequence of clays or shales. An example is the walkaway VSP study of Slater (1996) in the Juravskoe field, North Caucasus. Another cuspidal study is the fixed offset VSP over the Fiqa shales of the Natih field, Oman (Potters et al. 1999, Hake et al. 1998)).

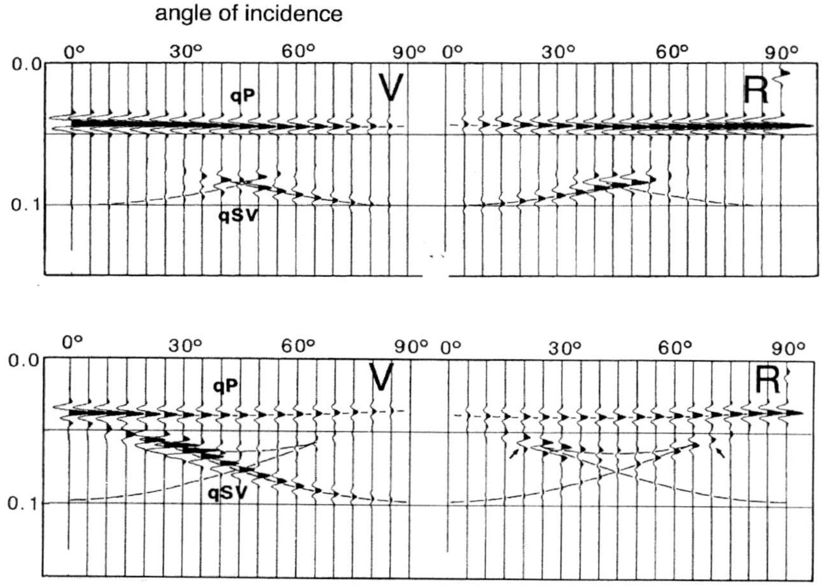

Figure 3.8. Numerical modeling of the predicted vertical and radial component traces from a circular profile by Kerner et al. (1989). The upper plot is for weak TIV ($\varepsilon = -0.04$, $\delta = 0.15$) and the lower for strong TIV (($\varepsilon = 0.10$, $\delta = 0.20$). Note the energy distribution along the different branches.

3.2 VERTICAL CRACKS/FRACTURES – TIH

The wave propagation behavior described in the last section is also applicable to crack/fracture-related TIH media. Indeed, the only difference is the orientation of the symmetry axis – for fracture strike along the Y-axis the symmetry axis is now the horizontal X-axis. The changes from TIV are purely geometrical, and the anisotropic phenomena remain the same. Thus, the symmetry plane that contains the direction of propagation determines the polarizations. The $S_{\perp}$ wave is polarized perpendicular to this symmetry plane (and therefore along the fracture strike), and $S_{\parallel}$ wave is polarized in the plane (Figure 3.9).

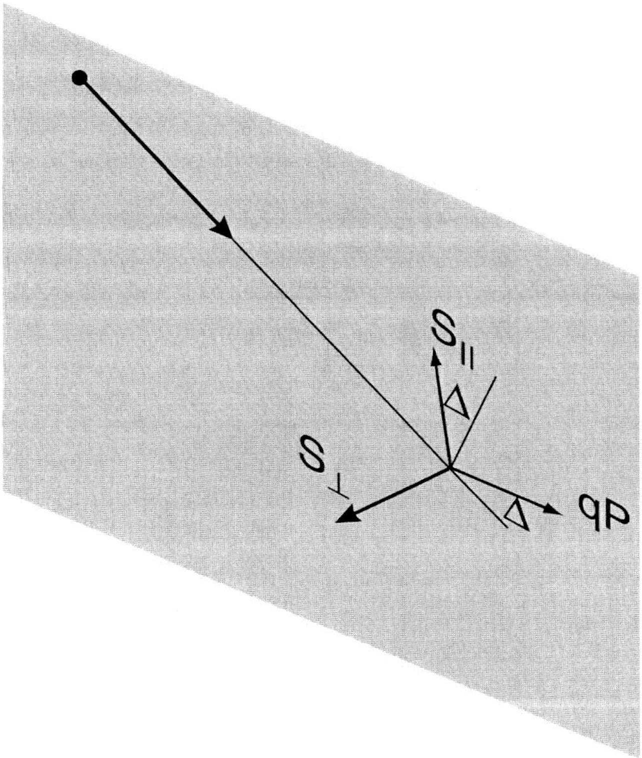

Figure 3.9. In a TI medium, each propagation direction lies in a symmetry plane. One of the shear wave polarizations is always perpendicular to this plane, while the other lies in the plane but generally at a slight angle to the normal to the propagation direction. This is always guaranteed to be the case.

The TIH phase velocity surfaces are obtained by rotating those for the TIV about the Y-axis by 90°. The effect of this rotation can be seen in Figure 3.10, which shows the expected results for three mutually orthogonal directions. Thus phase velocities can be calculated from Section 3.1 by replacing the angle θ in the TIV formulation by the angle χ subtended by the propagation direction and the symmetry axis. Another difference is that the anisotropy is now governed by the fracture indices e_t and e_n and thus has a different character.

3.2.1 Vertical incidence

This situation is particularly relevant to the near-offset multi-component VSP technique described in Chapters 4, 5 and 6. Near-vertical wave propagation in TIH media is equivalent to propagation in horizontal direction in TIV media: in both cases it is orthogonal to the symmetry axis. Propagation is considered to be in the X-Z plane as this is the only vertical symmetry plane. It is concluded from the previous section that shear waves are polarized parallel and perpendicular to this symmetry plane (and to the fractures). This correspondence can be algebraically confirmed by computations using equation (3.4). With a clockwise rotation of the fractures by ϕ about the vertical axis, the Bond transformation gives the arrangement of elastic stiffnesses

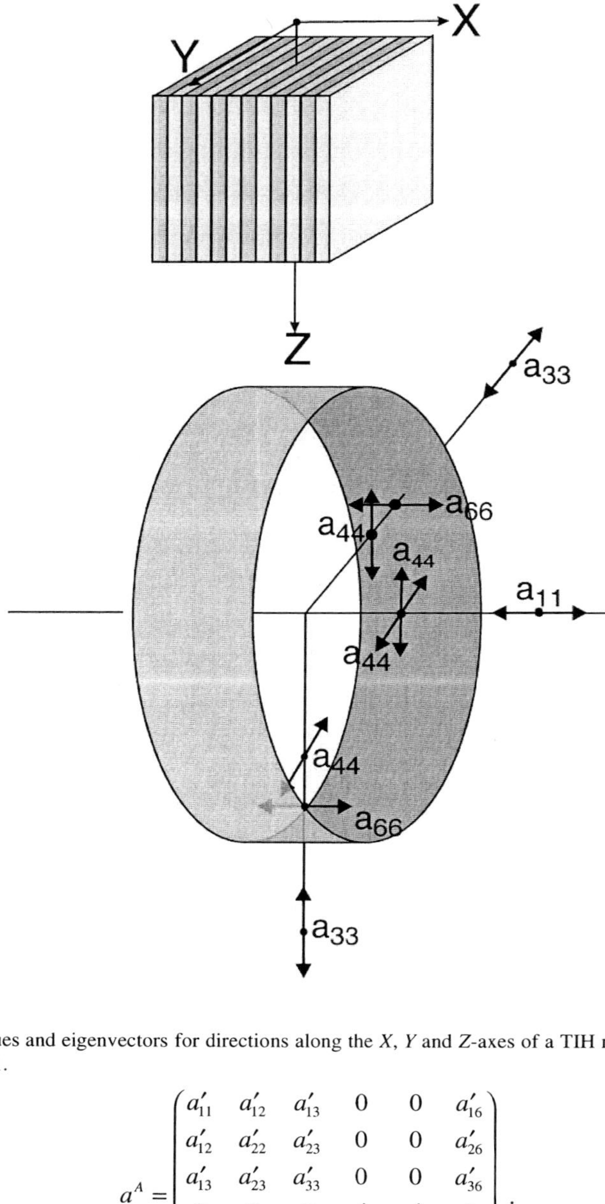

Figure 3.10. Eigenvalues and eigenvectors for directions along the *X*, *Y* and *Z*-axes of a TIH medium. Compare with Figure 3.1.

$$a^A = \begin{pmatrix} a'_{11} & a'_{12} & a'_{13} & 0 & 0 & a'_{16} \\ a'_{12} & a'_{22} & a'_{23} & 0 & 0 & a'_{26} \\ a'_{13} & a'_{23} & a'_{33} & 0 & 0 & a'_{36} \\ 0 & 0 & 0 & a'_{44} & a'_{45} & 0 \\ 0 & 0 & 0 & a'_{45} & a'_{55} & 0 \\ a'_{16} & a'_{26} & a'_{36} & 0 & 0 & a'_{66} \end{pmatrix}. \qquad (3.31)$$

where each a'_{ij} is a function of the a_{ij} and ϕ (Chapter 2, Appendix A.1). These are also equivalent to fractures striking at an azimuth of $\psi = \phi + 90°$ with respect to the X-axis. Substituting this – apparently monoclinic – arrangement into (3.4) gives

$$\mathbf{F}(\phi) = \begin{pmatrix} a_{44}\sin^2\phi + a_{66}\cos^2\phi & (a_{44} - a_{66})\sin\phi\cos\phi & 0 \\ (a_{44} - a_{66})\sin\phi\cos\phi & a_{44}\cos^2\phi + a_{66}\sin^2\phi & 0 \\ 0 & 0 & a_{33} \end{pmatrix} . \tag{3.32}$$

In this case decomposition is immediately possible, giving

$$\mathbf{F}(\phi) = \mathbf{C}_Z(\phi) \begin{pmatrix} a_{66} & 0 & 0 \\ 0 & a_{44} & 0 \\ 0 & 0 & a_{33} \end{pmatrix} \mathbf{C}_Z^T(\phi) , \tag{3.33}$$

where

$$\mathbf{C}_Z(\phi) = \begin{pmatrix} \cos\phi & \sin\phi & 0 \\ -\sin\phi & \cos\phi & 0 \\ 0 & 0 & 1 \end{pmatrix} , \tag{3.34}$$

with the eigenvalues given by a_{66}, a_{44}, and a_{33}. In terms of the fracture indices e_n and e_t the S- and P-wave solutions in ascending order are

$$v_{\parallel}^2 = a_{66} = \beta_0^2(1 - e_t) , \tag{3.35}$$

$$v_{\perp}^2 = \beta_0^2 , \tag{3.36}$$

and

$$v_P^2 = \alpha_0^2(1 - \zeta^2 e_n) . \tag{3.37}$$

where α_0 and β_0 are the velocities for the isotropic background medium and ζ is defined as in Chapter 2.

The corresponding (orthogonal) eigenvectors are given by the solution $\mathbf{u}^i$ of

$$\begin{pmatrix} a_{66} & 0 & 0 \\ 0 & a_{44} & 0 \\ 0 & 0 & a_{33} \end{pmatrix} \mathbf{C}_Z^T(\phi)\mathbf{u}^i = v^{(i)2}\mathbf{C}_Z^T(\phi)\mathbf{u}^i , \tag{3.38}$$

which yields the polarization vectors

$$\mathbf{u}_{\parallel} = \begin{pmatrix} \cos\phi \\ -\sin\phi \\ 0 \end{pmatrix} , \quad \mathbf{u}_{\perp} = \begin{pmatrix} \sin\phi \\ \cos\phi \\ 0 \end{pmatrix} , \quad \mathbf{u}_P = \begin{pmatrix} 0 \\ 0 \\ 1 \end{pmatrix} . \tag{3.39}$$

Thus, shear waves are indeed polarized parallel and perpendicular to the symmetry plane. The polarization of the (faster) $S_{\perp}$ wave is perpendicular to the symmetry plane, and hence along the fracture strike. The polarization of the (slower) $S_{\|}$ wave is parallel to the symmetry plane, and the P-wave points along the *Z*-axis. Importantly, these polarizations for the vertical propagation are independent of the elastic stiffnesses (Figures 3.11 and 3.12).

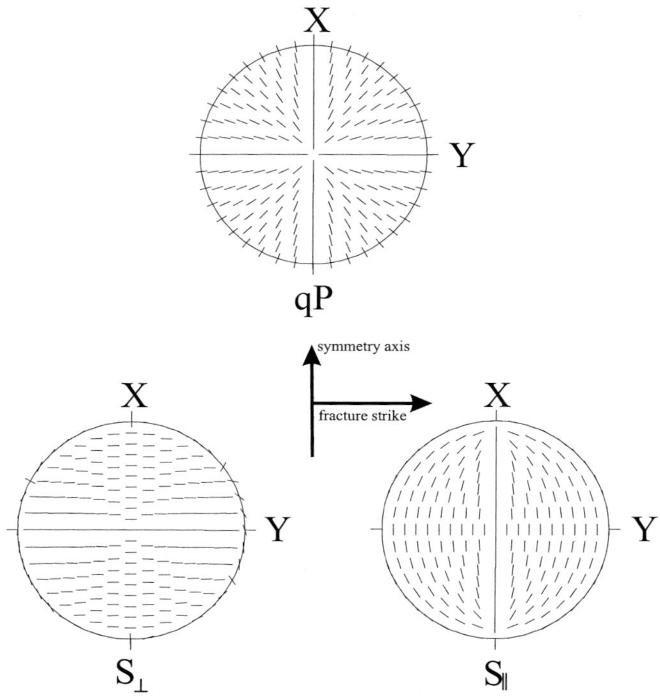

Figure 3.11. Horizontal components of the qP and S wave polarizations for a TIH medium, plotted on an equal area projection (see Appendix). The polarizations for each continuous mode are plotted separately, thus revealing the predicted behavior of alignment parallel to and mostly perpendicular to the fracture strike (*Y*-axis). The symmetry axis is the *X*-axis. Properties for vertical propagation are at the center of the plot, and horizontal directions around the perimeter.

The velocity expressions suggest that the travel times – $t_{\perp}$ for the $S_{\perp}$ wave and $t_{\|}$ for the $S_{\|}$ wave – are sufficient to determine the fracture index e_t:

$$e_t = \frac{t_{\perp}^2 - t_{\|}^2}{t_{\perp}^2} = \frac{v_{\|}^2 - v_{\perp}^2}{v_{\|}^2} = \frac{a_{44} - a_{66}}{a_{44}} \; . \tag{3.40}$$

In vertical shear-wave birefringence studies (Winterstein and Meadows 1991a, b) often the "velocity anisotropy" a_v is measured by

$$a_v = \frac{v_{\perp} - v_{\|}}{v_{\perp}} = 1 - \sqrt{1 - e_t} \; , \tag{3.41}$$

For moderate anisotropy this is *approximately* equal to $e_t/2$. The relation means that small values of a_v correspond to large values of relative compliance e_t (Figure 3.13).

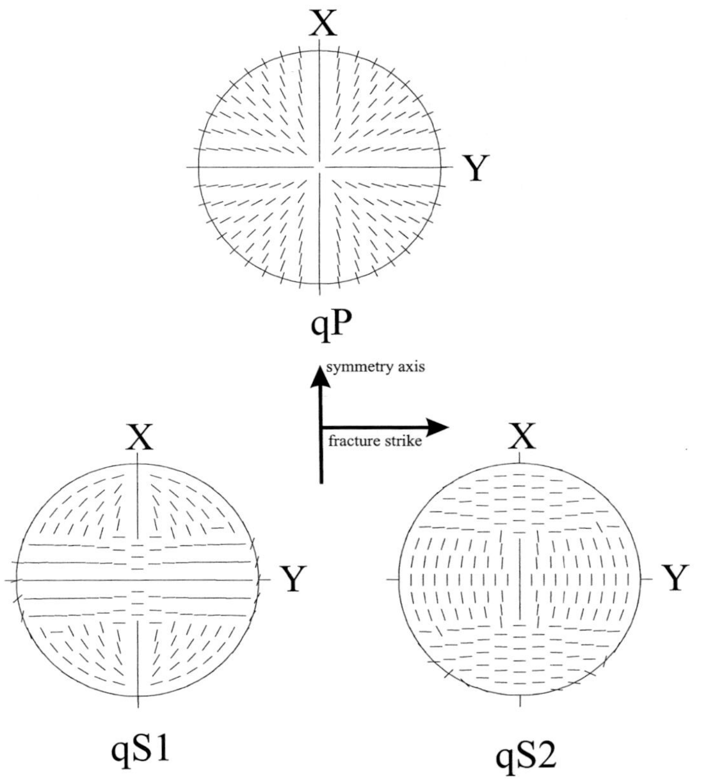

Figure 3.12. Horizontal polarizations for the qP- and S waves re-drawn from Figure 3.11, The only difference is a re-definition of the shear wave arrivals. The polarizations now represent the faster and the slower shear wave, respectively. The intersection between the two surfaces stands out as a prominent band of parallel polarizations. In the *X-Z* plane the line singularity occurs at 32°. This band restricts the available range of directions over which one can observe the correlation between the polarization and fracture strike.

An alternative measure of birefringence is the time-delay $\Delta\tau = t_{\parallel} - t_{\perp}$ (usually quoted in ms), from which $a_v = \Delta\tau/t_{\parallel}$ can be obtained. Chapter 5 gives examples of the determination of these quantities in different geographical locations.

The results above are powerful tools for crack or fracture-related anisotropy. However, some caution is required if the method is used as a universal indicator of fracture strike (Horne, MacBeth and Liu 1997). In particular, it was noted in Chapter 2 that for coplanar cracks and media whose pore space or grains could be modeled as prolate ellipsoids (long needle shapes with a circular cross section) $a_{44} < a_{66}$ is always true. This result does not alter the principal characteristics of the TIH medium, but it does mean that the $S_{\perp}$ wave is no longer the first arrival at normal incidence. Indeed, the polarization displays for this situation resemble those of Figure 3.12, but with the modes exchanged. Horne, MacBeth and Liu (1997) show that there are three materials in Thomsen's (1986) table that have this particular property: Calcareous sandstone, mudshale, and sandstone. This means that closer attention must be paid to *a priori* information

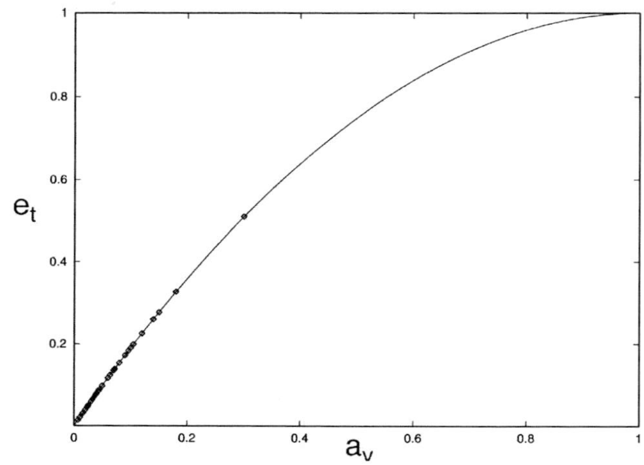

Figure 3.13. The relationship between shear wave velocity anisotropy a_v and the fracture index e_t. Note that the fractures are more compliant than the anisotropy measurements actually suggest, due to the square root function. The data points correspond to a range of vertical birefringence values extracted from Crampin (1994) for earthquake data, and some selected VSP and surface data.

3.2.2 Horizontal propagation

There is a simple and direct mapping of this case to the TIV case. The symmetry plane is horizontal, and thus the $S_\perp$ wave is polarized vertically (a true SV wave in this case), and the $S_\parallel$ wave is a true SH wave. The propagation azimuth ϕ with respect to the symmetry (X-) axis replaces the incidence angle θ. Alternatively, for fractures striking at an azimuth ψ to the propagation (X-) direction, $\theta = \phi = 90° - \psi$ (Figure 3.14). Thus all results of Section 3.1.1 are applicable upon making this change of angle, combined with conversion to fracture-specific parameters. Appropriate expressions for the Thomsen parameters can be obtained by rotating the fractures by 90° degrees about the Y-axis to simulate a TIV medium, and then using the expressions of (2.63), (2.64) and (2.65). The anellipticity $\varepsilon - \delta$ for these bedding-parallel fractures set can be positive or negative.

Substitution into the phase velocity expressions of (3.14), (3.21) and (3.22) produces the final fracture-specific velocities

$$v_\perp^2 = \beta_v^2\left(1 - e_t \sin^2\psi\right) , \tag{3.42}$$

$$v_\parallel^2 = \beta_v^2\left(1 - \left(\eta e_n - e_t\right)\sin^2 2\psi\right) , \tag{3.43}$$

and

$$v_P^2 = \alpha_v^2\left(1 - \eta e_t \sin^2 2\psi - \left(\left(1 - 2\eta\cos^2\psi\right)^2 - \left(1 - 2\eta\right)^2\right)e_n\right) , \tag{3.44}$$

which are similar to the expressions obtained by Schoenberg and Douma (1988). Here, α_v and β_v are the vertical velocities appropriate for the fracture-related TIH medium.

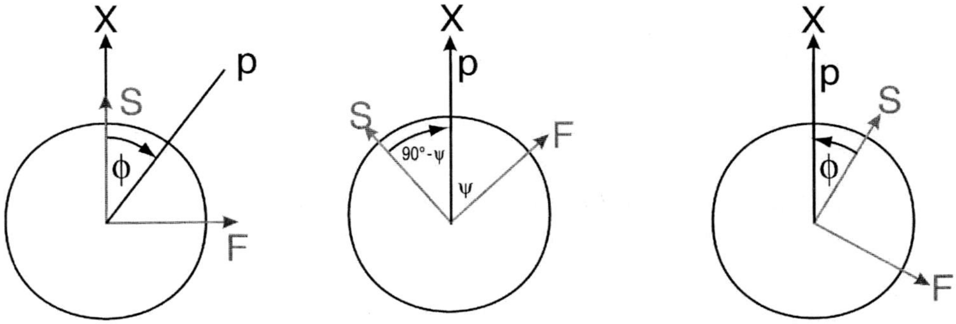

Figure 3.14. The different horizontal angles used in the expressions in Section 3.2. The propagation vector is denoted **p**, the symmetry axis S, and the fracture strike by F. Though the definition of the system changes, the relevant angle is in each case subtended by the symmetry axis and the direction of propagation. The rotation angle is ϕ, the fracture strike is ψ.

3.2.3 Propagation in arbitrary direction

The TIV results can also be generalized to an arbitrary propagation direction specified by an azimuth ϕ and an incidence θ with the propagation vector $\mathbf{p} = (\cos\phi \sin\theta, \sin\phi \sin\theta, \cos\theta)^T$. An identical scenario holds for oblique incidence in the XZ-plane for a fracture strike of $\psi = 90° - \phi$. The phase velocity of each wave type is governed only by the angle that the propagation direction makes with the symmetry axis. This angle χ is defined by $\cos\chi = \cos\phi \sin\theta = \sin\psi \sin\theta$. Generalization of equation (3.43), (3.44) and (3.45) gives

$$v_\perp^2 = \beta_v^2\left(1 - e_t \cos^2\chi\right) , \tag{3.45}$$

$$v_\parallel^2 = \beta_v^2\left(1 - (\eta e_n - e_t)\sin^2 2\chi\right) , \tag{3.46}$$

and

$$v_P^2 = \alpha_v^2\left(1 - \eta e_t \sin^2 2\chi - \left(\left(1 - 2\eta \sin^2\chi\right)^2 - (1 - 2\eta)^2\right)e_n\right) , \tag{3.47}$$

There is an infinite number of symmetry planes intersecting the symmetry axis (the horizontal X- axis). However, it is only the specific symmetry plane containing the propagation direction that determines the shear wave polarizations. The normal to this symmetry plane immediately determines the $S_\perp$ polarization in the components of the propagation vector. Substituting the fracture strike for the propagation azimuth gives

$$\mathbf{u}_\perp = \frac{1}{\sqrt{\cos^2\psi \sin^2\theta + \cos^2\theta}} \begin{pmatrix} 0 \\ -\cos\theta \\ \cos\psi \sin\theta \end{pmatrix} . \tag{3.48}$$

The horizontal projection of this vector remains orthogonal to the symmetry axis that is along the fracture strike. For a full suite of directions, this creates an array of parallel polarizations along the fracture strike (Figures 3.11 and 3.12). This particular attribute has been used by Crampin (1981) to interpret crack strike. The qP wave polarization deviates from the longitudinal direction by a small angle Δ that depends on the elastic stiffnesses, but still lies in the symmetry plane (Figure 3.9). This polarization is therefore a linear combination of the symmetry axis direction and the propagation direction. Calculation shows that

$$\mathbf{u}_P = \sin\Delta\begin{pmatrix}1\\0\\0\end{pmatrix} + (\cos\Delta - \sin\Delta\sin\psi\sin\theta)\begin{pmatrix}\sin\psi\sin\theta\\ \cos\psi\sin\theta\\ \cos\theta\end{pmatrix}, \tag{3.49}$$

where for normalization of the amplitude the vector should be divided by $1 - \sin^2\psi\sin^2\theta$. The second shear wave also has a polarization that lies in the symmetry plane, but is orthogonal to the qP wave

$$\mathbf{u}_{\|} = \cos\Delta\begin{pmatrix}1\\0\\0\end{pmatrix} - (\sin\Delta + \cos\Delta\sin\psi\sin\theta)\begin{pmatrix}\sin\psi\sin\theta\\ \cos\psi\sin\theta\\ \cos\theta\end{pmatrix} \tag{3.50}$$

with the same normalization factor as (3.49). When Δ is small, the horizontal component of the $S_{\|}$ wave is almost parallel to the symmetry axis at sub-vertical incidence, and tends to be perpendicular to the propagation vector when sub-horizontal (Figures 3.11 and 3.12). The magnitude of Δ can be derived from (3.24)

$$\tan\Delta = \frac{2\eta\sin\chi\cos\chi}{1-\eta}\left(e_n - e_t - 2(\eta e_n - e_t)\sin^2\chi\right). \tag{3.51}$$

3.2.3.1 *The line singularity*

The line singularity is particularly important for fracture analysis using seismic anisotropy, as it can limit the range of directions over which the polarization of the faster shear wave ($S_{\perp}$) can be interpreted as the fracture strike. The incidence angle at which the line singularity occurs now depends on the azimuth:

$$\sin\theta_{ls} = \frac{1}{2\sin\psi}\sqrt{1 - \frac{e_t}{4(\eta e_n - e_t)}}, \tag{3.52}$$

where $\eta e_n - e_t$ is the anellipticity.

The position of this intersection depends on fracture fill, crack geometry and the background properties ($\eta = \beta^2/\alpha^2$) and increases with azimuth. Figure 3.15 shows how the position varies with each of these parameters. The dependence on the ratio e_n/e_t means that there is no change for different porosities. For crack-related TIH the position of the line singularity strongly depends on the aspect ratio of the cracks (Figure 3.16) and the fill material (Figures 3.17 and 3.18).

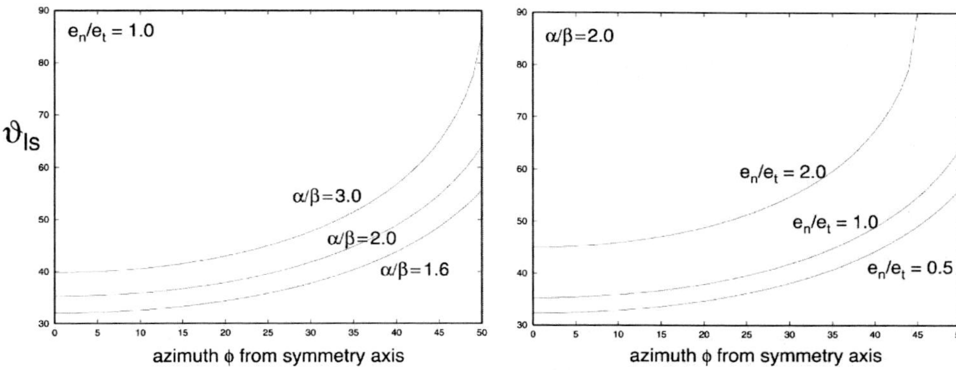

Figure 3.15. Prediction of the angle of incidence at which the line singularity occurs as a function of propagation azimuth against the symmetry axis (normal to the fracture strike). Curves are given for different values of α/β and e_n/e_t. Both parameters have similar influence.

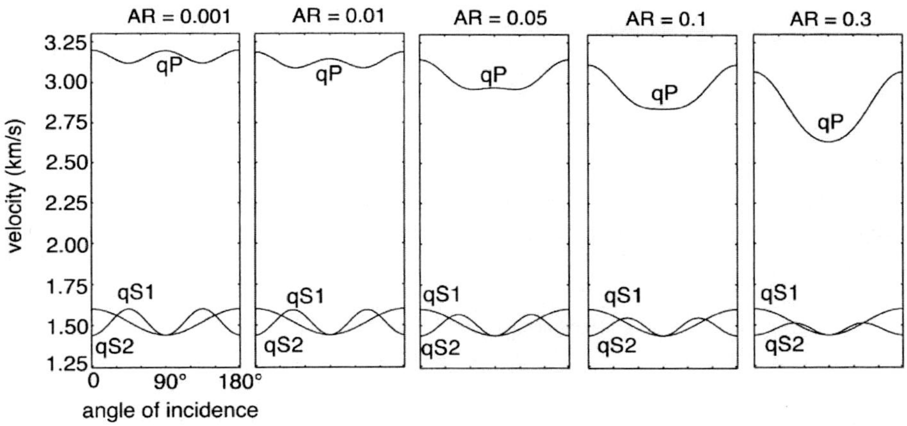

Figure 3.16. Phase velocity variations in the *X-Z* (fracture normal/symmetry) plane of a crack-related TIH medium. Note the change in the position of the line singularity with increasing crack aspect ratio (increasing e_n/e_t) and the large P-wave anisotropy.

For thin cracks filled with an incompressible liquid e_n is almost zero, thus the position of the line singularity depends only on the material properties; for $\alpha/\beta = 1.6$ it occurs at $\theta = 32°$. With increasing aspect ratio, with more gaseous fill, or with more communicative cracks, e_n/e_t increases and the position of the line singularity approaches the horizontal direction. A greater spread of values in a crack distribution can also effect the result. Angles beyond 32° degrees can occur for $\alpha/\beta > 2$.

The change in the position of the line singularity due to the change in aspect ratio is thought to be the cause of anisotropy changes observed prior to earthquake activity. A field example where a line singularity was detected is shown in Figures 3.19 and 3.20. Here, a simultaneous jump in polarizations combined with a reduction of the shear wave birefringence (in this case measured using time delays) is observed and matched. Detection of this line singularity has pinned down the inclination of the fractures in addition to details of their seismic response.

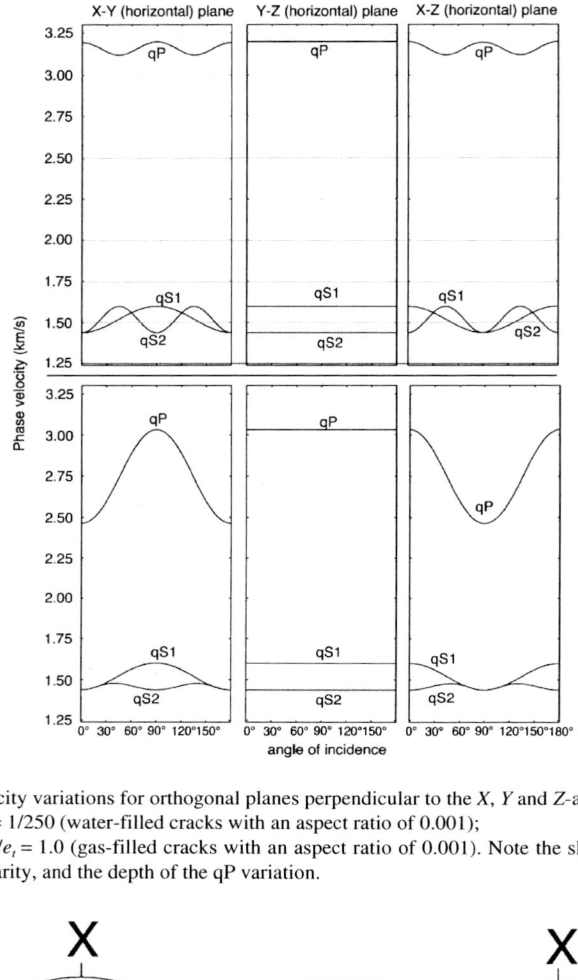

Figure 3.17. Phase velocity variations for orthogonal planes perpendicular to the *X*, *Y* and *Z*-axes of a TIH medium. Top: $e_n/e_t = 1/250$ (water-filled cracks with an aspect ratio of 0.001);
Bottom: $e_n/e_t = 1.0$ (gas-filled cracks with an aspect ratio of 0.001). Note the shift in the position of the line singularity, and the depth of the qP variation.

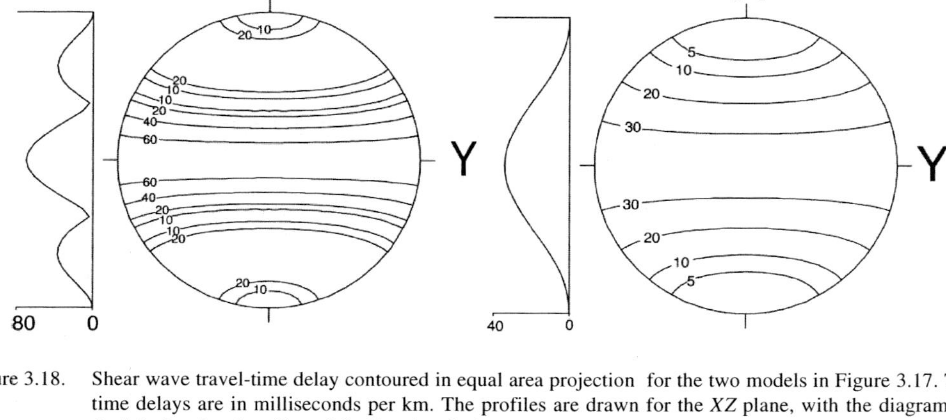

Figure 3.18. Shear wave travel-time delay contoured in equal area projection for the two models in Figure 3.17. The time delays are in milliseconds per km. The profiles are drawn for the *XZ* plane, with the diagram on the left showing the effect of the line singularity.

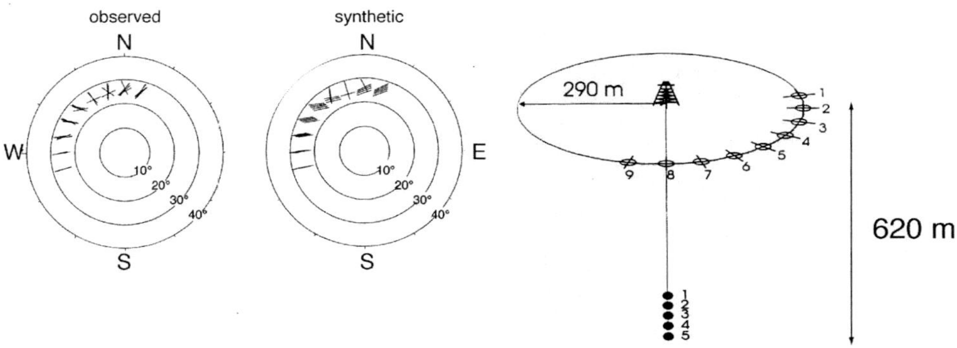

Figure 3.19. Acquisition geometry for an azimuthal VSP from which a line singularity was detected by Horne and MacBeth (1997). The lower diagram shows the observed shear wave polarizations with the rapid changes in polarization related to a sampling of directions either side of the shjear-wave singularity. These are compared with the results from synthetic seismograms computed for the best-fit solution. The final interpretation was an inclined fracture set with $e_t = 0.10$ and $e_n = 0.04$, and a dip of 10° SE. This agreed with *a priori* geological information for the site.

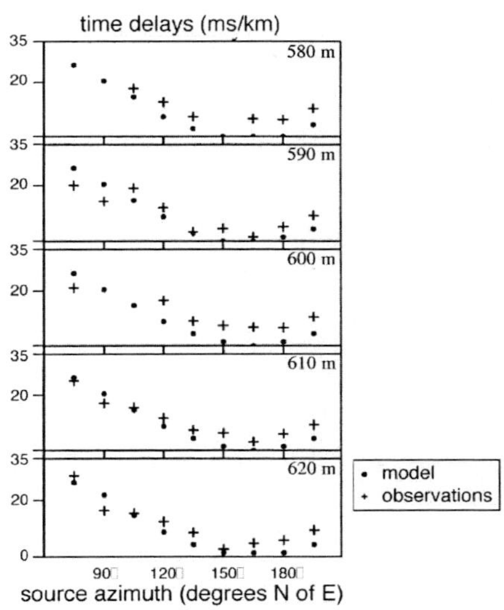

Figure 3.20. Observed and predicted time-delays for the shear wave birefringence variations in the azimuthal VSP of Figure 3.19. The predictions are based on the best fit model of inclined fractures. After Horne (1995).

3.2.4 Vertical incidence, inclined fracture set

Inclined fractures affect the shear wave interpretation by moving the position of the line singularity to a smaller incidence in the down dip direction, and larger incidence in the up dip direction (Figure 3.21). Compared to vertical fractures, the vertical birefringence is reduced and the $S_{\parallel}$ polarization is slightly rotated. For large inclinations, $S_{\parallel}$ is the faster shear wave; it does *not* lie along the fracture strike. It is important to identify this situation, but without *a priori* information or several different offsets this is difficult.

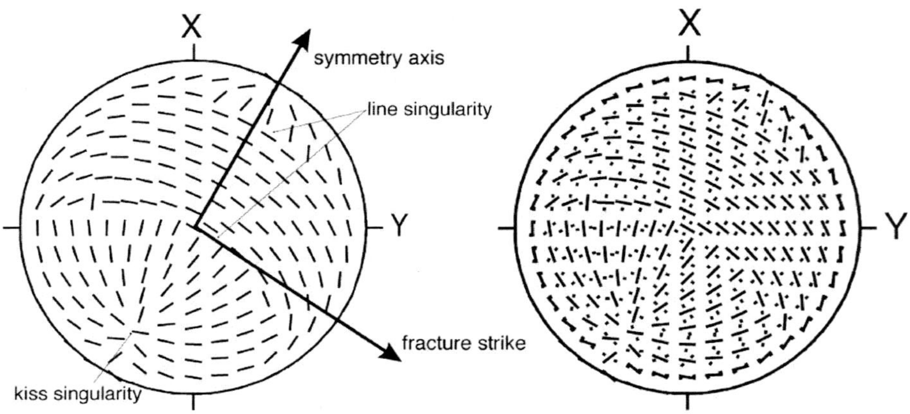

Figure 3.21. Left: the effect of arbitrary fracture strike and inclination is to move the band of parallel polarizations for the qS_1 wave away from the position of vertical incidence and towards the direction of inclination. Right: the horizontal polarization of the slower shear wave (broken bar) overlaid on the upper plot. The polarization is now slightly non-orthogonal to that of the faster shear wave.

Rotation of the TIH medium first about the *Y*-axis by Ω and then about the *Z*-axis by ϕ simulates fractures with an inclination of Ω and azimuth of $\psi = 90° - \phi$ with respect to the *X*-axis. A Bond transformation produces the relevant transformation of elastic stiffnesses to feed into (3.4). Vertical propagation considerably simplifies the matrix, leading to the equation

$$\begin{pmatrix} \cos^2\phi a'_{44} + \sin^2\phi a'_{55} & \sin\phi\cos\phi(a'_{44} - a'_{55}) & \cos\phi a'_{35} \\ \sin\phi\cos\phi(a'_{44} - a'_{55}) & \sin^2\phi a'_{44} + \cos^2\phi a'_{55} & -\sin\phi a'_{35} \\ \cos\phi a'_{35} & -\sin\phi a'_{35} & a'_{33} \end{pmatrix} \mathbf{u} = v^2\mathbf{u} , \tag{3.53}$$

where

$$a'_{44} = \frac{1}{2}(a_{44} + a_{66}) + \frac{1}{2}(a_{44} - a_{66})\cos 2\Omega , \tag{3.54}$$

$$a'_{55} = \frac{1}{8}(a_{11} - 2a_{12} + a_{66} + 4a_{55}) - \frac{1}{8}(a_{11} - 2a_{12} + a_{33} - 4a_{66})\cos 4\Omega , \tag{3.55}$$

and

$$a'_{35} = \frac{1}{4}(a_{11} + a_{33}) - \frac{1}{8}(a_{11} - 2a_{12} - a_{33} - 4a_{66})\sin 4\Omega . \tag{3.56}$$

Eigenanalysis can be accomplished by using the fact that the $S_\perp$ polarization is always independent of the inclination of the fractures and parallel to the fracture strike:

$$\mathbf{u}_\perp = \begin{pmatrix} -\sin\phi \\ \cos\phi \\ 0 \end{pmatrix} . \tag{3.57}$$

Rotating the eigensystem to align the *Y*-axis along this polarization simplifies (3.54) to

$$\mathbf{C}_z(\phi)\begin{pmatrix} a'_{44} & 0 & a'_{35} \\ 0 & a'_{55} & 0 \\ a'_{35} & 0 & a'_{33} \end{pmatrix}\mathbf{C}_Z^T(\phi)\mathbf{u} = v^2\mathbf{u} \ . \tag{3.58}$$

This result is expected, as this is now equivalent to oblique incidence in the fracture normal plane. A final rotation about the *Y*-axis gives

$$\mathbf{C}_Z(\phi)\mathbf{C}_Y(\Delta\theta)\begin{pmatrix} a'_{44} & 0 & 0 \\ 0 & a'_{55} & 0 \\ 0 & 0 & a'_{33} \end{pmatrix}\mathbf{C}_Y^T(\Delta\theta)\mathbf{C}_Z^T(\phi)\mathbf{u} = v^2\mathbf{u} \ , \tag{3.59}$$

and hence the polarizations

$$\mathbf{u}_P = \begin{pmatrix} \cos\phi & \sin\phi & 0 \\ -\sin\phi & \cos\phi & 0 \\ 0 & 0 & 1 \end{pmatrix}\begin{pmatrix} 0 \\ \sin\Delta\theta \\ \cos\Delta\theta \end{pmatrix} , \tag{3.60}$$

and

$$\mathbf{u}_{\|} = \begin{pmatrix} \cos\phi & \sin\phi & 0 \\ -\sin\phi & \cos\phi & 0 \\ 0 & 0 & 1 \end{pmatrix}\begin{pmatrix} 0 \\ \cos\Delta\theta \\ -\sin\Delta\theta \end{pmatrix} , \tag{3.61}$$

where for sub-vertical fractures

$$\Delta\theta \approx \frac{2\Omega}{1-\eta} \ . \tag{3.62}$$

The effect on the shear wave behavior is to change the horizontal polarization for $S_{\|}$ so that it is no longer orthogonal to the $S_{\perp}$. This creates an apparent non-orthogonality of the shear waves. Another effect is a reduction in the vertical birefringence calculated using the shear wave velocities. If an estimate of e_t is obtained as in (3.40), then

$$e'_t = \sin^2\chi e_t - \eta\sin^2 2\chi e_n \ , \tag{3.63}$$

where $\cos\chi = \sin\phi\,\sin\Omega$. For an inclination of 5° this amounts to $e_t/3$. A similar relation can be obtained for dipping shale/sedimentary layers, for which an apparent vertical birefringence emerges

$$e'_t \approx 2\gamma\sin^2\Delta\theta + \frac{\varepsilon - 2\delta}{2\eta_0}\sin^2 2\Delta\theta \ , \tag{3.64}$$

For the magnitude of anisotropies given by Thomsen (1986) with a typical $\varepsilon = 0.2$ and $\delta = 0.1$, a dip of 3° produces 4 % birefringence. These expressions are plotted in Chapter 5, and an assessment is made in the context of field measurement.

3.3 INTERSECTING SYSTEMS – MONOCLINIC AND ORTHORHOMBIC SYMMETRY

Depending upon their origin and subsequent history, intersecting systems of fractures can have a different seismic response and porosity. For intersecting vertical fractures the e_t values are generally the sum of the individual values. They result in an $S_\perp$ polarization that is the average weighted by the individual birefringence values. Sub-horizontal fractures or bedding can also have a significant effect on the wave field. Intersecting fractures create a comples new phenomenon: point singularities, that require some special attention.

3.3.1 Vertical incidence, two vertical fracture sets

The elastic stiffnesses for two fracture sets embedded in an isotropic matrix have been shown (Section 2.5.1) to be the linear sum of the individual matrices

$$\mathbf{a} = \mathbf{a}_0 + \mathbf{a}_A + \mathbf{a}_B \ , \tag{3.65}$$

provided that e_t and e_n are small. Here, $\mathbf{a}_0$ is the isotropic matrix, and $\mathbf{a}_A$ and $\mathbf{a}_B$ the individual fracture perturbations for fracture sets A and B (equation 2.94). This produces an orthorhombic arrangement when the fractures are orthogonal, but more generally a monoclinic distribution. In the latter case, the horizontal plane – containing the fracture normals – is the only symmetry plane. For TIH media rotated by different angles ϕ_A and ϕ_B from the X-axis, the Kelvin-Christoffel equation of (3.4) is

$$\left(\mathbf{F}(\mathbf{a}_0) + \mathbf{F}_A(\phi_A, \mathbf{a}_A) + \mathbf{F}_B(\phi_B, \mathbf{a}_B)\right)\mathbf{u} = v^2\mathbf{u} \ , \tag{3.66}$$

where $\mathbf{F}(\mathbf{a}_0) = \mathbf{F}_0$ is the contribution from the background isotropic matrix only, and

$$\mathbf{F}_A = \begin{pmatrix} \cos^2\phi a_{44}^A + \sin^2\phi a_{66}^A & \left(a_{44}^A - a_{66}^A\right)\sin\phi\cos\phi & 0 \\ \left(a_{44}^A - a_{66}^A\right)\sin\phi\cos\phi & \sin^2\phi a_{44}^A + \cos^2\phi a_{66}^A & 0 \\ 0 & 0 & a_{33}^A \end{pmatrix}, \tag{3.67}$$

with a similar term for $\mathbf{F}_B$. Written in terms of fracture indices and common background velocities α_0 and β_0, this is

$$\mathbf{F}_A = \begin{pmatrix} \cos^2\phi\beta_0^2 + \sin^2\phi\beta_0^2\left(1-e_t^A\right) & \beta_0^2 e_t^A \sin\phi\cos\phi & 0 \\ \beta_0^2 e_t^A \sin\phi\cos\phi & \sin^2\phi\beta_0^2 + \cos^2\phi\beta_0^2\left(1-e_t^A\right) & 0 \\ 0 & 0 & a_{33}^A \end{pmatrix}, \tag{3.68}$$

As both of the fracture-related matrices possess four zero entries, the diagonalization of the sum of $\mathbf{F}_0$, $\mathbf{F}_A$ and $\mathbf{F}_B$ can be achieved by a single rotation $\overline{\phi}$ about the vertical axis, where

$$\tan 2\overline{\phi} = \frac{e_t^A \sin 2\phi_A + e_t^B \sin 2\phi_B}{e_t^A \cos 2\phi_A + e_t^B \cos 2\phi_B} \ , \tag{3.69}$$

and the eigenvectors are $(\cos\bar{\phi},\ \sin\bar{\phi},\ 0)^T$, $(-\sin\bar{\phi},\ \cos\bar{\phi},\ 0)^T$ and $(0,\ 0,\ 1)^T$. A simplification of this is possible:

$$\bar{\phi} = \frac{e_t^A\phi_A + e_t^B\phi_B}{e_t^A + e_t^B}\ , \tag{3.70}$$

for $\phi_A < 45°$, $\phi_B < 70°$, or $\phi_A > 45°$ and $\phi_B > 20°$, with a corresponding accuracy of 5° for individual fracture indices up to 0.12. This agrees with the empirical result of Liu et al. (1993a) for a range of intersecting vertical fractures.

The eigenvalues are

$$v_\perp^2 = \frac{\beta_0^2}{2}\left(2 - e_t^A - e_t^B + \sqrt{e_t^{A2} + e_t^{B2} + 2e_t^Ae_t^B\cos 2(\phi_A - \phi_B)}\right)\ , \tag{3.71}$$

$$v_\parallel^2 = \frac{\beta_0^2}{2}\left(2 - e_t^A - e_t^B - \sqrt{e_t^{A2} + e_t^{B2} + 2e_t^Ae_t^B\cos 2(\phi_A - \phi_B)}\right)\ , \tag{3.72}$$

and

$$v_P^2 = \alpha_0^2\ , \tag{3.73}$$

where the $S_\perp$ and $S_\parallel$ waves are along the *Y* and *X* axes, respectively, as for vertical fractures. The vertical birefringence is

$$e_t' = \sqrt{\left(e_t^{A2} + e_t^{B2}\right)^2 - 4e_t^Ae_t^B\sin^2(\phi_A - \phi_B)}\ , \tag{3.74}$$

which for a small conjugate angle is approximately the sum of the two fracture indices. This means that the response is the same as for vertical fractures, and as in the previous section suggested, the analysis requires *a priori* information.

3.3.2 Two inclined fracture sets

If the fracture sets both have different inclinations and strikes, then equation (3.65) is again appropriate. Summation of the fracture perturbations and substitution into the Kelvin-Christoffel equation gives

$$\left(\mathbf{F}(\mathbf{a}_0) + \mathbf{F}(\phi_A, \mathbf{a}_A) + \mathbf{F}(\phi_B, \mathbf{a}_B)\right)\mathbf{u} = v^2\mathbf{u}\ , \tag{3.75}$$

After some algebra, eigenanalysis yields the horizontal polarization ϕ for the fast shear wave $S_\perp$ at vertical incidence

$$\tan 2\bar{\phi} = \frac{e_t^A\left(\cos^2 2\theta_A - \sin^2\theta_A\right)\sin 2\phi_A + e_t^B\left(\cos^2 2\theta_B - \sin^2\theta_B\right)\sin 2\phi_B}{e_t^A\left(\cos^2 2\theta_A - \sin^2\theta_A\right)\cos 2\phi_A + e_t^B\left(\cos^2 2\theta_B - \sin^2\theta_B\right)\cos 2\phi_B}\ . \tag{3.76}$$

This shows that the effect of the inclination is to provide weighting in the composite fracture strike. Calculations show that this averaging can change the shear wave polarization corresponding to a vertical fracture set by 5° to 10° in the presence of a weaker interfering sub-horizontal set. Figures 3.22 and 3.23 show the variations in horizontal polarization that can arise for various intersections and oblique ray paths. The intersections create a monoclinic medium, which has numerous point singularities – localized regions of disruption in the birefringence.

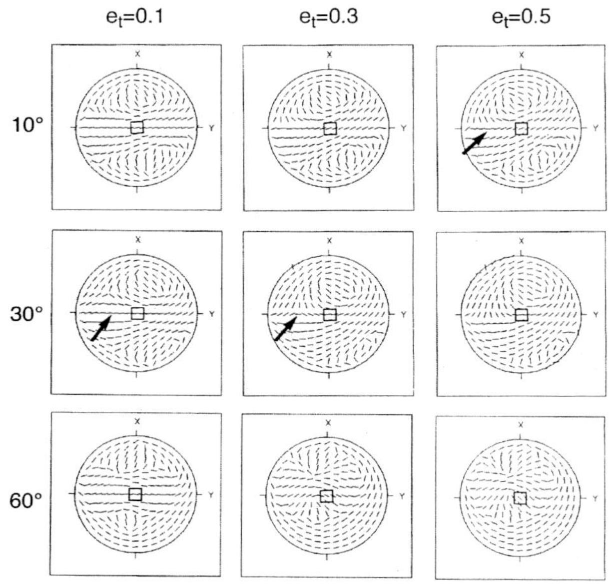

Figure 3.22. A selection of equal area projections displaying the horizontal polarization directions for the qS1 wave for two intersecting fracture sets. The strikes are 0° and 60°, respectively. The former is vertical, and the inclination of the latter assumes values of 10°, 30° and 60°. The parameter e_t for the vertical fracture set is fixed at 0.3, and for the second one assumes the values 0.1, 0.3 to 0.5. The square box indicates the directions sampled by the near-offset VSP (MacBeth 1996).

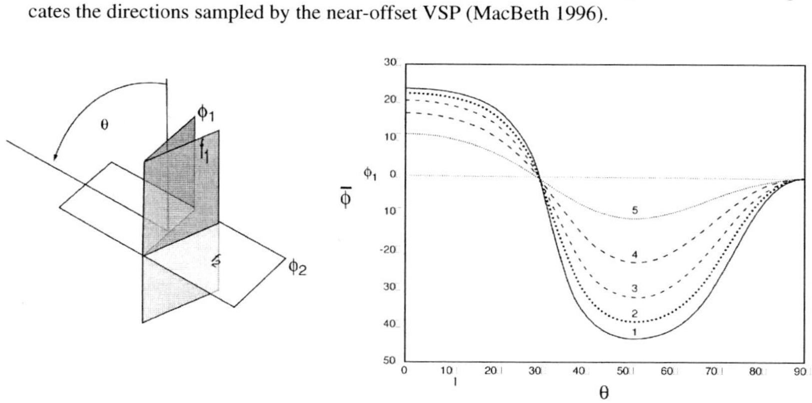

Figure 3.23. Variation of the "average" fracture strike $\bar{\phi}$ detected by the polarization of the faster shear wave, as a function of incidence angle ϑ. A vertical fracture set is combined with a second inclined fracture set, with strikes 0° and 30°, respectively. The fracture parameter e_t for vertical fracture set is 0.3, for the second it varies from 0.1 (curve 5) to 0.5 (curve1) (MacBeth 1996)..

3.3.3 Orthogonal fracture sets

An orthorhombic medium is obtained by adding the contributions from two orthogonal vertical fracture sets (equations 2.90 – 2.92). The anisotropy has three symmetry planes, one horizontal and two vertical containing the X and Y axes. The Kelvin-Christoffel matrix is

$$\mathbf{F} = \begin{pmatrix} a_{11}l_x^2 + a_{66}l_y^2 + a_{55}l_z^2 & (a_{12} + a_{66})l_x l_y & (a_{13} + a_{55})l_x l_z \\ (a_{12} + a_{66})l_x l_y & a_{66}l_x^2 + a_{22}l_y^2 + a_{44}l_z^2 & (a_{23} + a_{44})l_y l_z \\ (a_{13} + a_{55})l_x l_z & (a_{23} + a_{44})l_y l_z & a_{55}l_x^2 + a_{44}l_y^2 + a_{33}l_z^2 \end{pmatrix}. \tag{3.77}$$

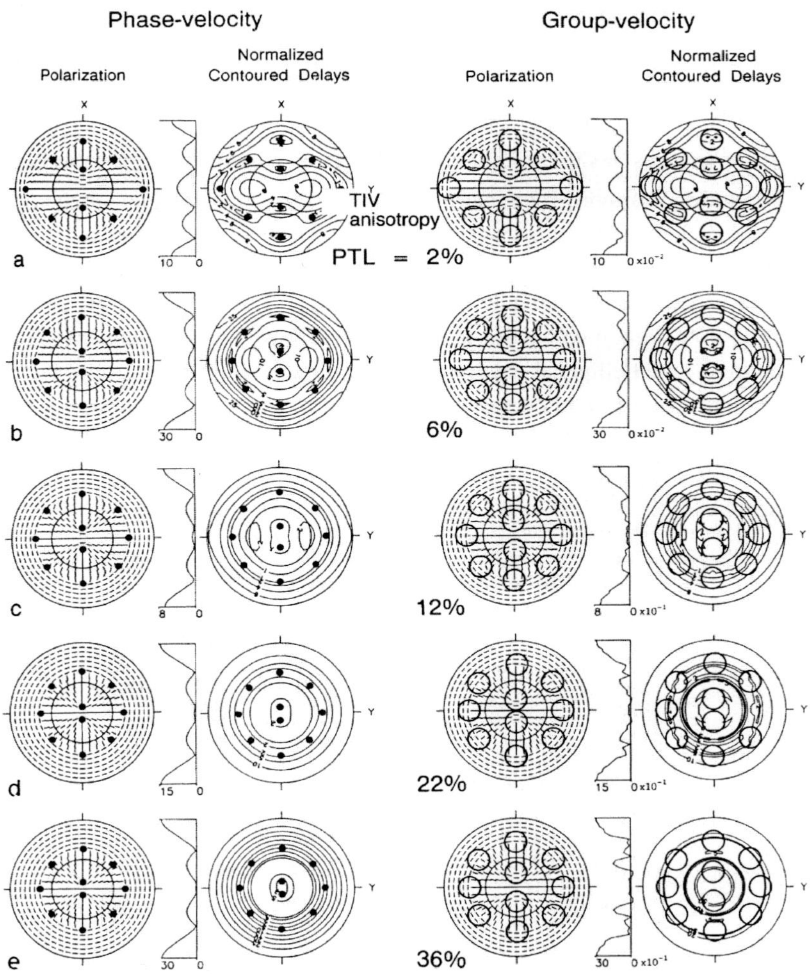

Figure 3.24. A selection of equal area projections of polarizations and time-delays (per km), for group and phase velocity directions, showing point singularities (solid circle). The crack-related anisotropy is fixed at 1 %. Note that with increasing TIV anisotropy the point singularities get closer to the vertical. This indicates the possibility of potential misinterpretation in near-offset VSP. After Wild and Crampin (1991).

For vertical incidence the shear wave velocities are readily obtained from expressions for the individual fracture-related TIH medium as the matrix is diagonal:

$$v_1^2 = \beta_0^2\left(1 - e_t^B\right) , \tag{3.78}$$

$$v_2^2 = \beta_0^2\left(1 - e_t^A\right) , \tag{3.79}$$

with the eigenvectors along the X- and Y-axes. The polarization of the faster shear wave depends upon which fracture set has the strongest seismic response. The P-wave velocity is

$$v_P^2 = \beta_0^2\left(1 - \zeta^2\left(e_n^A + e_n^B\right)\right) . \tag{3.80}$$

3.3.4 Vertical fractures in shales/sedimentary layering – orthorhombic symmetry

The wave properties of orthorhombic and monoclinic media involve point singularities. For reasons of symmetry, such features cannot exist in TI media (Duff 1960, Crampin and Yedlin 1981). A point singularity is a discrete point of contact between the two shear wave velocity surfaces. The rapid change of curvature associated with this pinch means that there can be cusps and other features of distortion in the group velocities. In this group velocity domain, such effects are not confined to a point but a small elliptical patch of disruption, for which the birefringence can be large and the polarizations exhibit rapid changes over small directions. The situation for orthorhombic materials in general can be exceedingly complicated as noted by Grechka and Obolentseva (1993), with up to twelve waves propagating along the ray direction (Fedorov 1968 quotes 150th order for the wave surface).

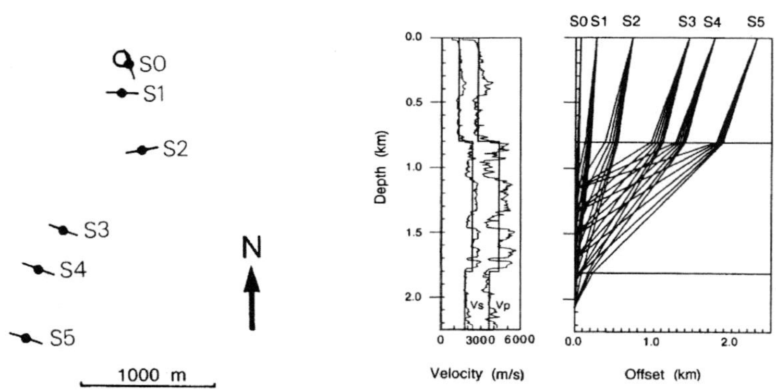

Figure 3.25. A multi-offset VSP shot in the Paris Basin was used by Bush and Crampin (1991) to obtain evidence of point singularities. The VSP employed a distribution of offsets at several different azimuths.

Point singularities occur in at least one symmetry plane , with their position depending upon the medium properties (Figure 3.24). A full gallery of possible variations has been displayed by Wild and Crampin (1991). Of particular interest is the case of weak fracturing in a strongly anisotropic shale, for which the singularities approach vertical incidence and could disrupt near-offset VSP measurements. Point singularities have been interpreted from multi-offset VSP by Bush and Crampin (1991), who show variations matched by synthetic seismograms and polarization diagrams (Figures 3.25 and 3.26).

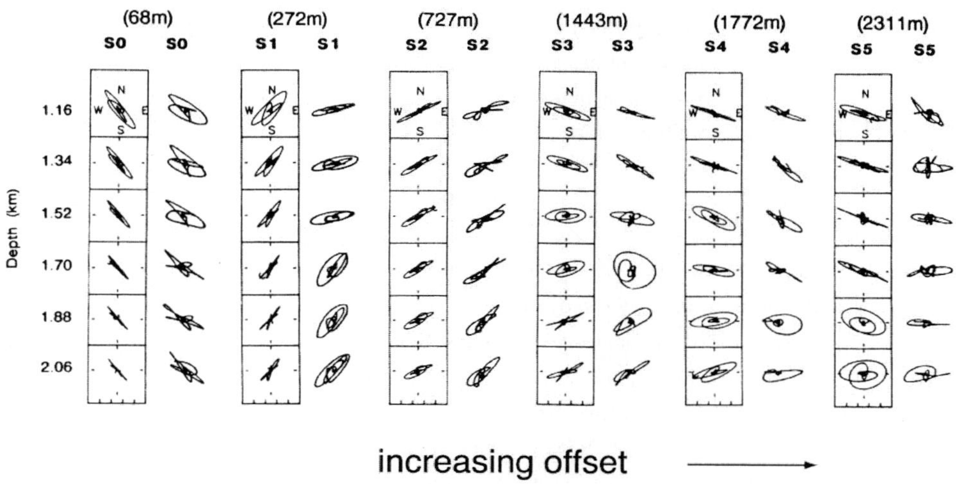

Figure 3.26. Match between horizontal particle motions for the field data from the acquisition in Figure 3.25, and synthetic seismograms (boxed) for an orthorhombic medium generated by cracking a TIV medium (Section 2.5.2). (Bush 1990).

APPENDIX: THE EQUAL-AREA PROJECTION

Figure A.1 shows the construction of an equal-area polar projection for a hemisphere of directions. Ray directions are projected onto a horizontal plane at the same azimuth at distance $2a$ $\sin(\theta/2)$ from the center O, where θ is the incidence angle of the ray (Aki and Richards 1980). The map resulting from this construction has the property that equal solid angles of directions are transformed to equal areas on the circular plane. The geographical analogue is preservation of the relative areas for pr ojection from the surface of a sphere.

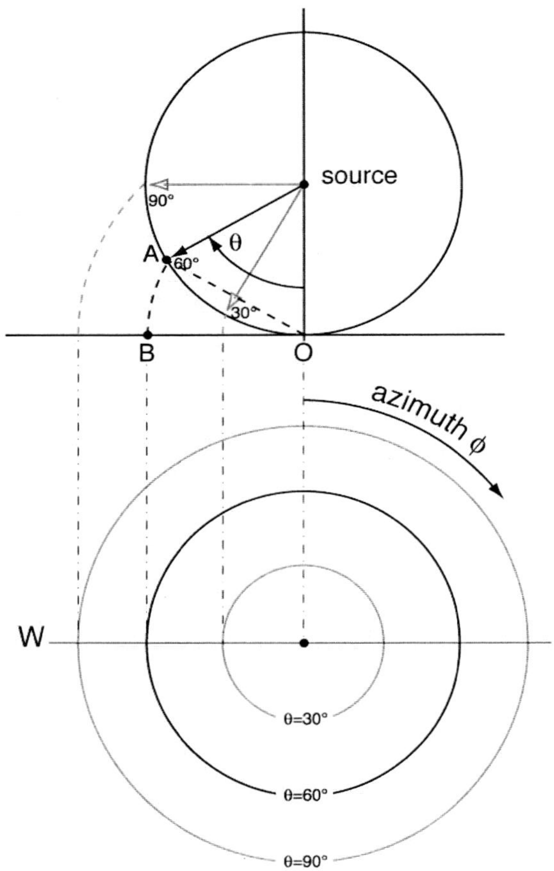

Figure A.1 The equal area projection used to visualize shear wave birefringence.

Equal area plots are used in Figures 3.11, 3.12, 3.18, 3.19, 3.21, 3.22 and 3.24. Polarizations and time-delays in the plots are calculated and plotted using the ANISEIS software (Taylor 1987).

CHAPTER 4 – PRE-REQUISITES FOR NEAR-OFFSET VSP ANALYSIS

To obtain high-quality anisotropy estimates from near-offset multi-component VSP requires correction for – or avoidance of – phenomena that prevent straightforward interpretation. It is an advantage of multi-component VSP that many of the effects can be easily detected or monitored to ensure limitation of their influence. Such effects arise due to the equipment, the specific style of acquisition, but also the nature of the earth. They conspire to prevent an accurate representation of the propagating wave field. The interaction of the source with the near-surface region, and the wave propagation through this zone provides the strongest footprint. Next in importance is the necessity for vector fidelity of the source and receiver, as this forms the axiom upon which all multi-component processing is founded. The third category includes inevitable errors due to misorientation of the acquisition components, together with unexpected polarity and coupling differences. Finally, the earth itself poses problems through the interfaces at which the seismic waves are generated, propagate through, and are recorded. In all instances, use can be made of the matrix expressions in the vector convolutional model (VCM) as an aid to the book-keeping of these individual effects throughout the processing and interpretation steps, and detecting inconsistencies.

4.1 THE VECTOR CONVOLUTIONAL MODEL

Although the underlying assumptions for application of the VCM are identical to those required for the operation of the scalar convolution (Sheriff 1983), the mathematical form is more restrictive. This is because it employs matrix algebra, and therefore requires the source, medium and receiver elements to be combined in a specific order. The following sub-sections assemble the basic building blocks for the application to multi-component VSP.

4.1.1 Isotropy

Consider an idealized near-offset VSP configuration, in which waves are emitted by point forces directed along the horizontal (*X* and *Y*) and vertical (*Z*) axes. The waves propagate vertically downwards by a distance *r* into three-component downhole receivers directed along the *X*-, *Y*- and *Z*-axes. For an unbounded, homogeneous and isotropic medium with density ρ and velocities V_P and V_S, the three-component displacements detected from each source are (Aki and Richards 1981)

$$s_X(t) = \frac{1}{4\pi\rho r}\left(\frac{1}{V_S^2} s\left(t - \frac{r}{V_S}\right) - s_N(t)\right)\begin{pmatrix}1\\0\\0\end{pmatrix}, \tag{4.1}$$

$$s_Y(t) = \frac{1}{4\pi\rho r}\left(\frac{1}{V_S^2} s\left(t - \frac{r}{V_S}\right) - s_N(t)\right)\begin{pmatrix}0\\1\\0\end{pmatrix},$$
(4.2)

and

$$s_Z(t) = \frac{1}{4\pi\rho r}\left(\frac{1}{V_P^2} s\left(t - \frac{r}{V_P}\right) + 2s_N(t)\right)\begin{pmatrix}0\\0\\1\end{pmatrix},$$
(4.3)

for a common source action $s(t)$. The first term on the right-hand-side of each equation corresponds to either a P or an S wave, and the second term $s_N(t)$ is the inevitable near-field term arising from the point sources:

$$s_N(t) = \frac{1}{r^2}\int_{r/V_P}^{r/V_S} \tau s(t-\tau)d\tau \ .$$
(4.4)

This particular arrival distorts the polarizations and waveforms of the waves of interest. This can be seen by considering an impulsive source action, for which it can be shown that $s_N(t)$ decays geometrically approximately as $1/r^2$, while the P and S waves decay as $1/r$. More specifically, the far-field terms dominates for distance r where

$$\frac{\omega r}{V_P} >> 1 \ .$$
(4.5)

Thus, for typical VSP frequencies in the vicinity of 35 Hz and $V_P = 3500$ m/s, this implies a source-receiver distance greater than 16 m. Consideration of the near-field terms is important for choosing the best set of traces for the near-surface deconvolution (see Section 4.2). However, for VSP investigations at exploration depths the near-field terms are not important. Neglecting the near-field terms, the nine traces generated by this arrangement can be grouped directly into a nine-component matrix

$$\mathbf{D}^{(9)}(t) = (\mathbf{s}_X(t), \ \mathbf{s}_Y(t), \ \mathbf{s}_Z(t)) = \begin{pmatrix} s\left(t-\frac{r}{V_S}\right) & 0 & 0 \\ 0 & s\left(t-\frac{r}{V_S}\right) & 0 \\ 0 & 0 & \frac{V_S^2}{V_P^2} s\left(t-\frac{r}{V_P}\right) \end{pmatrix},$$
(4.6)

or more concisely

$$\mathbf{D}^{(9)}(t) = \Lambda^{(9)}(t) * \mathbf{S}_0^{(9)}(t) \ .$$
(4.7)

Here, the isotropic wave propagator $\Lambda^9(t)$ is

$$\Lambda^{(9)}(t) = \begin{pmatrix} \lambda_S(t) & 0 & 0 \\ 0 & \lambda_S(t) & 0 \\ 0 & 0 & \lambda_P(t) \end{pmatrix}, \tag{4.8}$$

where $\lambda(t-r/V)*f(t) = f(t-r/V)$, and the source matrix is

$$\mathbf{S}^{(9)}(t) = \begin{pmatrix} s(t) & 0 & 0 \\ 0 & s(t) & 0 \\ 0 & 0 & s(t) \end{pmatrix}, \tag{4.9}$$

after an implicit rescaling of the vertical component traces. The factor of $1/(4\pi\rho\, r)$ is absorbed into the diagonal amplitudes. If the medium is a half-space and the source is applied to its free surface, the results can still be written in the same general form, though in view of the boundary condition at the free surface an amplitude adjustment is necessary.

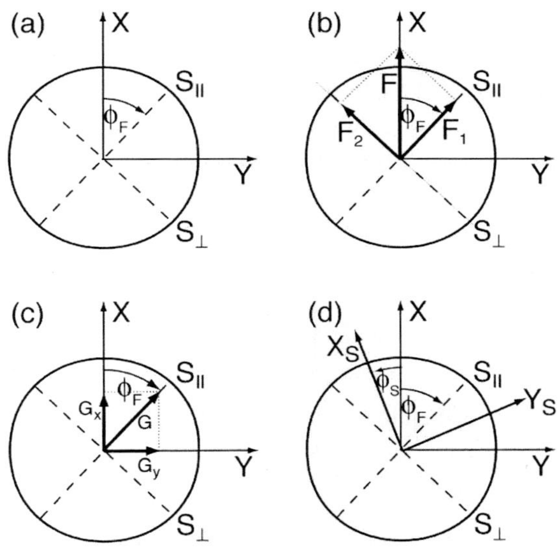

Figure 4.1. Plan view of horizontal plane displaying alignments relevant to multi-component VSP analysis for different idealized cases:

(a) Both source and receiver are aligned with the X and Y axes; the fracture strike makes an angle ϕ_F with the X-axis;

(b) The polarization from the X source (F) is decomposed into component vectors of magnitude F_1 and F_2 (where $F^2 = F_1^2 + F_2^2$) in the principal directions of the fracturing. The Y-source is treated similarly.

(c) After propagation, the amplitude G of the fast shear wave is projected back onto the receiver directions giving G_X and G_Y. The slow shear wave is treated similarly.

(d) The orthogonal source set X_S-Y_S is rotated by ϕ_S with respect to the acquisition axes.

4.1.2 TIH anisotropy

Consider the same VSP configuration as above, but this time the waves propagate first through a very thin isotropic layer (thickness r) and then into a zone of thickness R containing vertical fractures. The receiver is also in an isotropic medium. In the fractured medium, fast

$S_\perp$ and a slow $S_\parallel$ shear waves can propagate. Their polarizations are, respectively, parallel and perpendicular to the fracture strike ϕ_F, (Chapter 3), but not necessarily along the X and Y directions (Figure 4.1a). Assuming negligible transmission loss at the isotropic/anisotropic interface, vector decomposition of the shear wave polarization emitted by the X-directed force into components parallel and perpendicular to the fractures determines the partitioning into the $S_\perp$ and $S_\parallel$ waves (Figure 4.1b) and yields excitation amplitudes $\cos\phi_F\, s(t-r/V_S)$ and $-\sin\phi_F\, s(t-r/V_S)$ respectively. In a similar way, the force aligned along the Y-axis will give amplitudes $\sin\phi_F\, s(t-r/V_S)$ and $\cos\phi_F\, s(t-r/V_S)$. The result of the decompositions is

$$\mathbf{D}^{(9)}(t) = \begin{pmatrix} \cos\phi_F & -\sin\phi_F & 0 \\ \sin\phi_F & \cos\phi_F & 0 \\ 0 & 0 & 1 \end{pmatrix} \mathbf{S}^{(9)}(t) = \mathbf{C}_Z^T(\phi_F)\mathbf{S}_0^{(9)}(t) \ . \tag{4.10}$$

Note that the P waves remain undeviated and unchanged throughout this propagation process, and the isotropic zone shrinks to zero thickness ($r \to 0$). The $S_\parallel$, $S_\perp$ and P waves now propagate separately through the fractured medium, with the respective transit times $\tau_1 = R/V_\parallel$, $\tau_2 = R/V_\perp$, and $\tau_3 = R/V_P$. The wave amplitudes at the bottom of the fractured zone can be written

$$\mathbf{D}^{(9)}(t) = \left[\Lambda_M^{(9)}\mathbf{C}_Z^T(\phi_F)\right] * \mathbf{S}_0^{(9)}(t) \ . \tag{4.11}$$

where

$$\Lambda_M^{(9)}(t) = \begin{pmatrix} \lambda_1(t) & 0 & 0 \\ 0 & \lambda_2(t) & 0 \\ 0 & 0 & \lambda_3(t) \end{pmatrix} . \tag{4.12}$$

To calculate the recorded amplitudes, the two distinct shear wave polarizations must be projected onto the X and Y receiver axes (Figure 4.1c). This procedure can again be simulated by a rotation, but in the opposite direction to before

$$\mathbf{D}^{(9)}(t) = \mathbf{C}_Z^T(\phi_F)\Lambda_M^{(9)} * \left[\mathbf{C}_Z^T(\phi_F)\mathbf{S}_0^{(9)}(t)\right] \ . \tag{4.13}$$

In conclusion, it is observed that the resultant 9C matrix can be described in terms of two rotations about the vertical axis and a single time shift operator. The geometric operators take account of the difference in orientation between the acquisition coordinate frame and the fractures. The effect of this is to produce a mix of shear wave energy on all four components of the *XX*, *XY*, *YX* and *YY* sub-set of the 9C matrix, but to leave the P wave untouched. Indeed, this pattern can be clearly seen in the data of Figure 1.8. Equation (4.13), or its 4C subset, is the backbone of most of the processing and interpretation in multi-component VSP, and has been found to be satisfactory for most applications. It will be applied to fracture analysis in Chapter 5.

4.1.3 The requirement for vector fidelity of the source

Returning to the source, consider what may happen if the orthogonal pair of horizontal forces have some general orientation with respect to the acquisition axes *X* and *Y*. For example, they are rotated anti-clockwise by an angle ϕ_S (Figure 4.1d). According to the linear superposition rule, the resultant shear wave polarizations can still be decomposed into sub-vectors pointing in the *X* and *Y* directions. There is a purely geometric connection between this and the previous case:

$$\mathbf{S}_{new}^{(9)} = \mathbf{C}_Z^T(\phi_S)\mathbf{S}_0^{(9)} \quad . \tag{4.14}$$

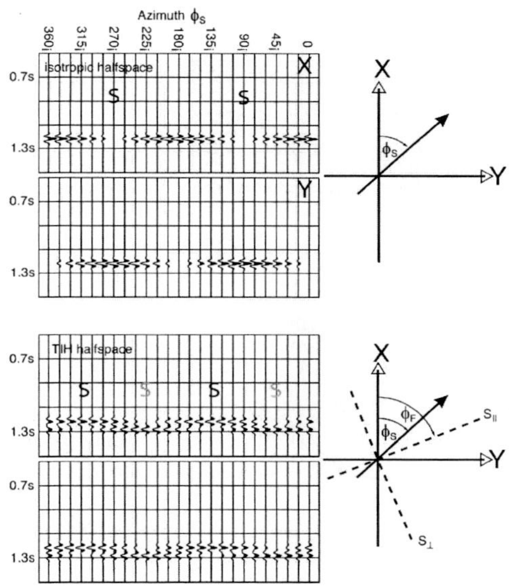

Figure 4.2. Two-component horizontal recordings for shear waves generated by a rotating horizontal force. Upper diagram: in the isotropic case zeros occur when the forces are aligned with either the *X*- or *Y* axis.
Lower diagram: in a TIH medium only the fast or slow waveforms appear when the forces are aligned with either principal direction of the medium. The effect provides an unambiguous determination of the fracture strike.

Thus, if a source is physically rotated about the vertical, specific anisotropy directions are reached where only the $S_{\|}$ or the $S_{\perp}$ waves are excited by each source of the orthogonal pair (Figure 4.2). Unfortunately, such rotation tests are rarely performed at a given location. More usually, a source pair deployed at a general orientation to the fracture strike must have the ability to predict the principal shear wave directions. A necessary condition is vector fidelity, that is, that a mathematical rotation operation on the data can simulate the physical act of rotation. This is critical to the success of the processing procedures based on (4.13). One way of testing vector fidelity is to perform a rotation test in the field, and to choose two orthogonal azimuths and then simulate what would be expected at the other azimuths (Figure 4.3). One requirement for this to succeed is balanced source wavelets and source coupling

(Winterstein and Meadows 1991a, Choi and Gangi 1991). Another requirement is an adequate understanding of the source mechanism.

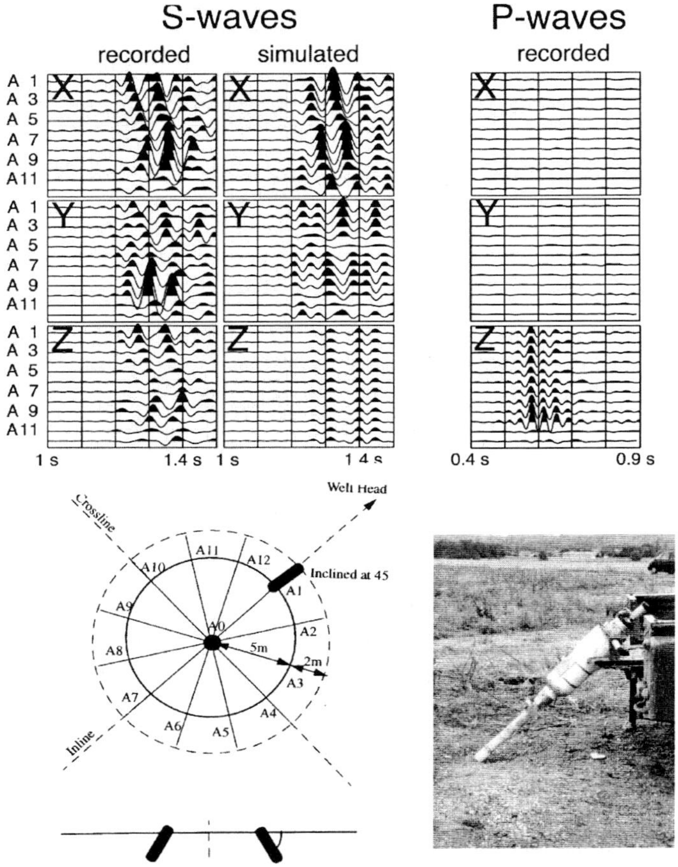

Figure 4.3. Rotation test for a prototype multi-component source consisting of a device for generating a controllable high-intensity shock wave in an angled plastic tube (MacBeth and Arnold 1998). The azimuth of the tubing was incremented over a full circle, with the shock gun shooting into a single downhole source. Vector fidelity is tested by selecting two orthogonal traces (marked by an asterisk), that are then used to predict the remaining two-component traces using mathematical rotation.

Many seismic sources cannot be adequately represented as point forces. Fortunately, the mathematics above can be generalized to admit a more arbitrary orientation for all three forces, and source mechanisms more exotic than just single forces (such as dipoles, couples, double couples or their combination). This then leads to the development of the source term in (4.13) as a representation for any three sources:

$$\mathbf{S}^{(9)}(t) \rightarrow \mathbf{S}^{(9)}_{gen}(t) = \begin{pmatrix} a_1(t) & a_2(t) & a_3(t) \\ b_1(t) & b_2(t) & b_3(t) \\ c_1(t) & c_2(t) & c_3(t) \end{pmatrix} * \mathbf{S}^{(9)}_0 = \mathbf{A}(t) * \mathbf{S}^{(9)}_0 \ , \tag{4.15}$$

where the scalar constants $a_i(t)$, $b_i(t)$, $c_i(t)$ include the radiation pattern of the composite sources, different source actions, in addition to differential or phase shift operators. Equation (4.15) expresses the fact that each of the three sources emits both a P wave and an S wave. With the continuing assumption of an isotropic source medium, the P waves are always polarized along the radial (vertical) direction, and the S waves polarization lies in the normal (horizontal) plane. In each case the waveform shape and relative proportion of the two waves change. Thus, wave propagation in the source medium yields the matrix of displacements

$$\mathbf{D}^{(9)}(t) = \Lambda^{(9)}(t) * \mathbf{S}^{(9)}_{gen}(t) \ . \tag{4.16}$$

Furthermore, when the P and S waves are incident upon a fractured zone, as in the previous case, they are treated exactly as before; equation (4.13) remains valid with the replacement of $\mathbf{S}^{(9)}(t)$ by $\mathbf{S}^{(9)}_{gen}(t)$. If a survey is shot with sources that require for their representation a mechanism more complicated than point forces, the implication is that knowledge of the scalars $a_i(t)$, $b_i(t)$, $c_i(t)$ provides a way of transforming that data to the ideal condition. Such rectification requires deconvolution applied as a post-multiplication of $\mathbf{D}^{(9)}(t)$.

For this operation it is vital that the wavelets and polarizations of the waves emitted from the source are predicted sufficiently well to allow the determination of $\mathbf{A}(t)$. It is also essential that the matrix $\mathbf{A}(t)$ has an inverse. This highlights a more general requirement on the vector fidelity of the source for multi-component acquisition – the sources must be capable of consistently generating different P and S wave vector amplitudes to allow the matrix to be inverted. Thus, an important test of multi-component capability for any new source is the ability to provide differently polarized shear waves, with a consistent predictable relationship between the physical orientation of the source and the polarizations. If this can be achieved with sources such as shaped charges, explosives or airguns in inclined bore holes, or even closely spaced vertical vibrators of opposite polarities, it opens up the possibility of more economic multi-component surveys. Section 4.2 demonstrates an extension of this idea to implement matrix inversion without knowledge of the source wavelets or orientation.

4.2 NEAR-SURFACE CORRECTION

The near surface presents one of the major challenges in the interpretation of multi-component seismic. In many cases it leaves a strong but unpredictable footprint on the recorded wave field, with a seismic character that is often distinctly different from that of the subsurface. Its multi-component signature can vary substantially over distances of a few meters laterally. One possible cause is variations in consolidation and porosity, that produce rapid spatial variations in velocity, attenuation, and anisotropy. Other heterogeneous features are scatterers such as inclined surfaces, near-surface bedding, and joints (Imhof et al. 1997); velocity gradients that focus excessive energy in specific directions (Ben-Menahem et al. 1991); and low velocities leading to multiples and local conversion (such as the Geysers example in Figure 1.2). There are many observations that suggest that the near surface is strongly anisotropic . The following table lists a few examples:

Table 1. Examples of near-surface anisotropy

Reference	Acquisition	Magnitude	Wavetype	Depth	Place/Lithology
1	VSP	5%, 12%	S	18 m–36 m	San Francisco Bay muds
2	VSP	several	S	≤ 80 m	till, sands and clay
3	sea floor	12%–15%	S	≤ 10 m	silty clay muds
4	crosshole/VSP	4%–9%	S	≤ 90 m	Tertiary shales
5	crosshole	≈ 30%	P	≤ 70 m	gneiss+schist
6	RVSP	≈ 10%	S	≤ 38 m	Conoco Borehole Test Facility Kay County, Oklahoma
7	VSP	23%	P	143 m	Dog Creek shale
		35%	S		
8	VSP	22%	P	58 m	Wills Point shale
		28%	S		
9	VSP	9%–14%	S	few 100 m	Cymric, Lost Hills, Railroad Gap
10	VSP	5%	S	47 m	soft sand-silt
11	VSP	≤ 10%	S	≤ 1920 m	soft, south-central Wyoming

Anisotropy values for the first few hundred meters. The magnitude of the maximum velocity anisotropy is shown, irrespective of whether it is TIH or TIV.

1 Lynn (1991);
2 Douma et al. (1990);
3 Berge et al. (1991);
4 Winterstein and Paulsson (1990);
5 Pratt et al. (1993);
6 Liu et al. (1993b);
7 Robertson and Corrigan (1983);
8 Robertson and Corrigan (1983);
9 Winterstein and Meadows (1991a, b);
10 Ghose and Takahashi (1992);
11 Kendall and Kendall (1996).

For a near-surface anisotropy of considerable magnitude, theory predicts a significant departure from the radiation patterns assumed in Section 4.2. Indeed, such anisotropy can generate S waves from explosions (Figure 4.4), unusually polarized P and S waves, and unexpected non-orthogonal radiation patterns from orthogonal sources (Slater 1997). In the former Soviet Union, the range of near-surface phenomena discussed above has been recognized under one generic category called "natural directivity" (Puzirev et al. 1985). One natural first-order approach for VSP is to "strip" the near-surface. Examples are shown in Chapter 6.

There is a second category of equipment-related coupling issues and non-linear vector signature issues that must be considered. For example, the source motion can be elliptic instead of linear (Cliet and Dubesset 1987 showed artifacts generated by the Marthor source – see Figure 1.7). Additional unknowns enter the multi-component data due to unexpected polarity changes or due to errors in the alignment of the source motions. Such errors occur when line-of-sight alignment is not possible or when inadequate surveying techniques are used. Queen (1987) observed errors of up to 30°. These additional unknowns add to the complexity of an already difficult problem. Such difficulties are particularly acute for reflection surveys where both source and receiver are on the free surface, and the near-surface complexities may not be easily removed from the section. VSP surveys have only one near-surface interaction and a geometry suitable for tackling this problem. Indeed, VSP techniques provide a method that can cope with most of the above effects. This method, a *multi-component downgoing deconvolution*, is the subject of the next sub-sections.

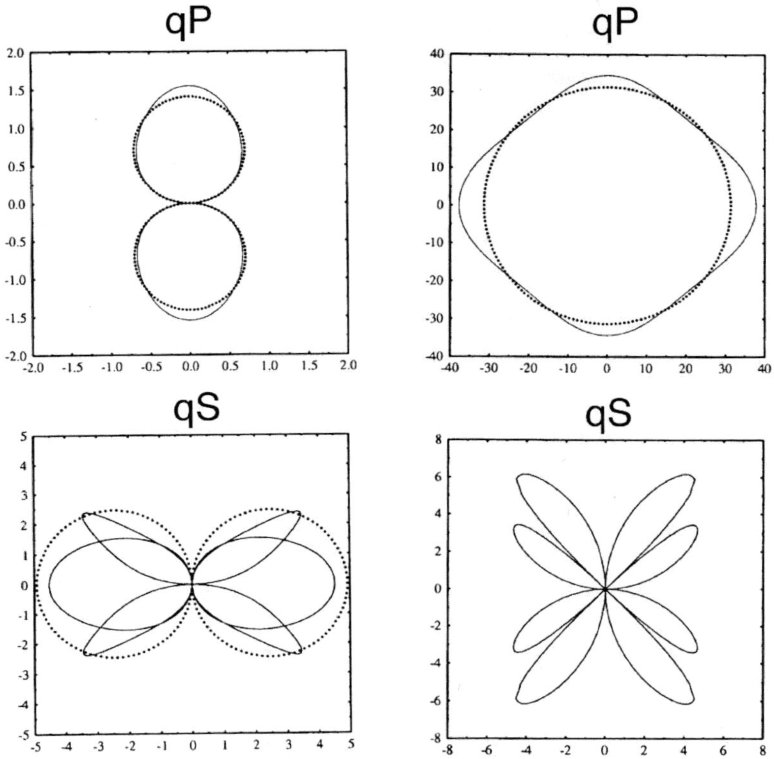

Figure 4.4. The effect of anisotropy on source radiation patterns, displayed as vertical sections. Left-hand column is for a vertical point force, the dotted lines correspond to the isotropic medium. Right-hand column is for an explosive source. The anisotropic medium has TIH symmetry in both cases. Diagrams are extracted from Gajewski (1993), who also provides a detailed mathematical description.

4.2.1 Basic principle

Near-surface and related effects can be eliminated by a technique similar to the usual scalar downgoing deconvolution (Smidt 1989). It is assumed that there is a datum level k in the subsurface where the displacements are independent of the wavenumber (Figure 4.5). Thus, the nine-component data matrix $\mathbf{D}_j(t)$ recorded at depth j may be factorized into the product of the downgoing wave field $\mathbf{D}_{k,S}(t)$ at the datum and an anisotropic propagator $\mathbf{D}_{j,k}(t)$ representing the anisotropic path between depths k and j:

$$\mathbf{D}_j(t) = \mathbf{D}_{j,k}(t) * \mathbf{D}_{k,S}(t) \ . \tag{4.17}$$

The object is to correct for $\mathbf{D}_{k,S}$, thus making (4.17) as similar to (4.13) as possible. The estimate of the subsurface operator $\mathbf{D}_{j,k}(t)$ can then be interpreted using the VCM

$$\hat{\mathbf{D}}_{j,k}(t) = \mathbf{C}(\phi_F)\Lambda_{j,k}(t)\mathbf{C}^T(\phi_F) * \mathbf{S}_0^{(9)}(t) \ . \tag{4.18}$$

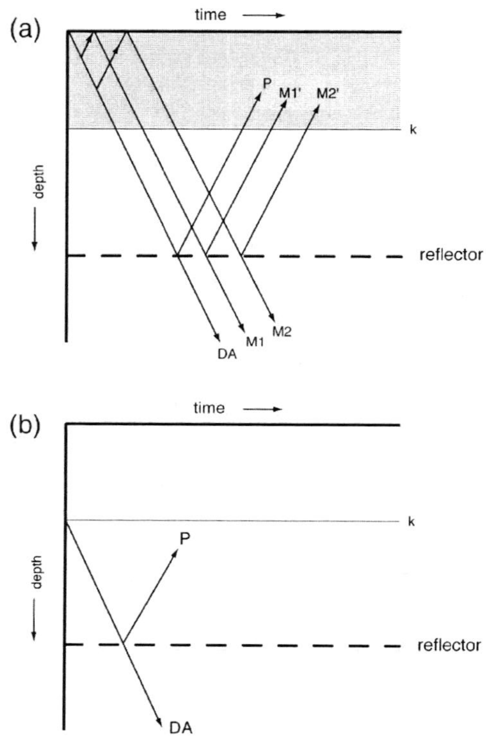

Figure 4.5. The central idea behind the near-surface correction is an estimate of the downgoing wave field from a group of receivers in the vicinity of datum level k. The estimate uses array processing techniques and determines the entire multi-component response.

The datum should determined in the far-field of the source and of near-surface scattering processes. In most commercial VSPs this is not a major problem. If the shallowest receiver position is chosen as the datum, the correction can be implemented by post-multiplying $\mathbf{D}_j(t)$ by an estimate of the inverse of $\mathbf{D}_{k,S}(t)$. The procedure is a form of layer stripping, and it may be visualized as placing the multi-component source set at level k. It should be emphasized that this particular operation is only possible because of the nine-component, near-offset nature of the data. It can only be accurately implemented if three-component recordings from each source are available.

4.2.2 Estimation of the downgoing wave field

4.2.2.1 Using multi-component data

Robust estimation of the downgoing wave field is performed in the frequency domain by taking a short vertical array of M equally spaced receivers around level k. It is assumed that the incident wave field consists of a finite number of (not necessarily linearly) polarized plane waves, each with a distinct slowness. The recordings are linked by a common transfer matrix $\Delta\mathbf{D}_k(\omega)$. This transfer matrix is a complex function, and represents changes in polarization state, amplitudes, and phase shifts due to propagation, but need not assume a special form. It must, however, relate to coherent wave propagation across the array and be mutually consistent for all receiver pairs such that each data matrix $\mathbf{D}_{k+j-1}(\omega)$, ($j$=1,$M$) in the array can now be defined according to

$$\mathbf{D}_{k+j-1}(\omega) = \Delta\mathbf{D}_k^{(j-1)}(\omega)\mathbf{D}_{k,S}(\omega) + \mathbf{N}_{k+j-1}(\omega) \ . \tag{4.19}$$

A consistency statistic can be used to define the choice of receivers (Zeng 1994). This consists of evaluating

$$\mathbf{U}_k(\omega) = (\Delta\mathbf{D}_k(\omega))^{-1}\Delta\mathbf{D}_{k-1}(\omega) = \mathbf{D}_k(\omega)\mathbf{D}_{k+1}^{-1}(\omega)\mathbf{D}_k(\omega)\mathbf{D}_{k-1}^{-1}(\omega) \ , \tag{4.20}$$

which should be equal to a unit matrix. A measure of closeness to this unit matrix is

$$\Gamma = \sqrt{\sum_j \frac{(u_{xx,j}-1)^2 + u_{xy,j}^2 + u_{yx,j}^2 + (u_{yy,j}-1)^2}{4n}} \ , \tag{4.21}$$

where the $u_{ab,j}$ are the components of the matrix $\mathbf{U}_k(\omega)$ at the frequency component j, and n is the number of frequencies in the band considered. Under ideal conditions Γ is zero. A low value of Γ is a necessary – but not a sufficient – condition for consistency. Without the noise term $\mathbf{N}_{k+j-1}(\omega)$, it would be unnecessary to use array processing as the first receiver level would immediately give the incident data matrix and hence the required inverse operator. With the noise term the result is achieved by (a) estimating the transfer matrix $\Delta\mathbf{D}_k(\omega)$, and (b) using this estimate to obtain $\mathbf{D}^{-1}_{k,S}(\omega)$.

(a) An implicit solution for the transfer function is possible by collecting (4.18) for each receiver into a single matrix equation

$$\mathbf{E}_{k+1}(\omega) = \Delta\mathbf{D}_k(\omega)\mathbf{E}_k(\omega) + \mathbf{N}_k(\omega) \ . \tag{4.22}$$

For three or fewer arrivals (only downgoing waves) $\mathbf{E}_k(\omega)$ consists of 3×3 (M–1) matrices, with the structure

$$\mathbf{E}_k(\omega) = (\mathbf{D}_k(\omega) \quad \mathbf{D}_{k-1}(\omega) \quad \dots \quad \mathbf{D}_{k+M-2}(\omega)) \ . \tag{4.23}$$

However, for more than three but fewer than six arrivals (when strong upgoing waves are present) data overlap results in a larger 6×3 (M–1) matrix

$$\mathbf{E}_k(\omega) = \begin{pmatrix} \mathbf{D}_k(\omega) & \mathbf{D}_{k+1}(\omega) & \dots & \mathbf{D}_{k+M-2}(\omega) \\ \mathbf{D}_{k+1}(\omega) & \mathbf{D}_{k+2}(\omega) & \dots & \mathbf{D}_{k+M-1}(\omega) \end{pmatrix} . \tag{4.24}$$

Equation (4.24) is over-determined for 9C data with three recording levels (3×6 data matrices), and even-determined for a single level and downgoing waves. In the assembly of these, the data matrix at each level must be normalized to the maximum amplitude for all traces. This means that the amplitude values for the transfer matrix relate only to multi-component effects and not to absolute amplitude decay. Under the assumption that the noise $\mathbf{N}_k(\omega)$ is uncorrelated with the seismic signal, the least-squares solution of (4.24) is

$$\hat{\Delta}\mathbf{M}_k(\omega) = \mathbf{E}_{k+1}(\omega)\mathbf{E}_k^H(\omega)\left(\mathbf{E}_k(\omega)\mathbf{E}_k^H(\omega) + e\mathbf{I}\right)^{-1} , \tag{4.25}$$

where the superscript H represents the “complex conjugate transpose”. The damping factor e introduces an *a priori* minimum length constraint into the estimates (Menke 1984). It is necessary to stabilize the inverse when for some frequencies in the band of interest amplitudes are near zero, but this must be used with care as it may bias the results. In practice, e can be adjusted by trial and error, and takes care of most difficulties. The auto-correlation matrix $\mathbf{E}_k(\omega)\mathbf{E}_k(\omega)^H$ becomes singular if similar information is contributed from each source, or if the number of waves is less than the dimension of the matrix (that is, the rank of the matrix is less than its order). The rank of this matrix can be tested by singular-value decomposition. With a strong upgoing wave field there may be some uncertainty over the number of distinct arrivals, and the eigenspectra have to be monitored.

In practice, the most satisfactory condition is when the number of independent and non-zero eigenvalues (arrivals) is equal to the matrix dimensions. This happens for strong down-going compressional and shear waves in most near-offset VSP. The procedure can be made more robust by extracting the upgoing wave field using independent application of an *f-k* filter or the median filter approach (Hardage 2000) for each individual component. This strategy is attractive because it reduces the number of arrivals, and hence the size and complexity of the matrix operations. It has the disadvantage that the signals are smoothed, and thus there may be edge effects and polarizations may be distorted. The estimate is also sensitive to variations in the source signature at each activation, and thus for consistency a set of levels on one sonde is preferred. In many practical applications a single trace set at one level is adequate.

(b) The transfer matrix $\hat{\Delta}\mathbf{D}_k(\omega)$ is now fed back into (4.19), giving the data matrix at each depth

$$\hat{\mathbf{D}}_{k,S}(\omega) = \left(\mathbf{Q}_k^H(\omega)\mathbf{Q}_k(\omega) + e\mathbf{I}\right)^{-1}\mathbf{Q}_k^H(\omega)\mathbf{P}_k(\omega) \tag{4.24}$$

or a direct estimate of $\mathbf{D}_{k,S}^{-1}(\omega)$:

$$\mathbf{D}_{k,S}^{-1}(\omega) = \left(\mathbf{P}_k^H(\omega)\mathbf{P}_k(\omega) + e\mathbf{I}\right)\mathbf{P}_k^H(\omega)\mathbf{Q}_k(\omega) , \tag{4.25}$$

where $\mathbf{P}_k(\omega)$ and $\mathbf{Q}_k(\omega)$ are the partitioned $3{\times}3(M{-}1)$ matrices

$$\mathbf{P}_k(\omega) = \left(\mathbf{D}_k(\omega) \quad \mathbf{D}_{k+1}(\omega) \quad \ldots \quad \mathbf{D}_{k+M-1}(\omega)\right)^T \tag{4.26}$$

and

$$\mathbf{Q}_k(\omega) = \left(\mathbf{I} \quad \hat{\Delta}\mathbf{D}_k(\omega) \quad \ldots \quad \hat{\Delta}\mathbf{D}_k^{M-1}(\omega)\right)^T . \tag{4.27}$$

For low-noise conditions, array processing is unnecessary and the equation reduces to $\hat{\mathbf{D}}_{k,S}^{-1}(\omega) = \mathbf{D}_k^{-1}(\omega)$. If the data matrices at each level are nearly identical due to low moveout across the levels, the propagator matrix is close to a unit matrix, and $\hat{\mathbf{D}}_{k,S}^{-1}(\omega)$ becomes the average of the data matrices in the chosen group. For realistic noisy conditions, the result lies somewhere between the two results. The design of the operator is now complete, and it can be used for post-multiplying each recorded nine-component data matrix. A frequency filter ensures a common bandwidth for the P and S wave data. The result should be a band-limited function with a zero-phase input wavelet.

4.2.2.2 *Using three-component data*

A similar – though less accurate – estimate for $\Delta\mathbf{D}_k(\omega)$ may be obtained by using vector displacements $\mathbf{d}_k(\omega)$ from a single source excitation (Cho and Spencer 1990; Zeng and MacBeth 1993a). By grouping together m of the three-component recordings $\mathbf{d}(t)$ an equation similar to (4.25) can be obtained, viz:

$$\left(\mathbf{d}_{k+1}(\omega) \quad \mathbf{d}_{k+2}(\omega) \quad \ldots \quad \mathbf{d}_{k+m}(\omega)\right) = \Delta\mathbf{D}_k(\omega)\left(\mathbf{d}_k(\omega) \quad \mathbf{d}_{k+1}(\omega) \quad \ldots \quad \mathbf{d}_{k+m-1}(\omega)\right) , \tag{4.28}$$

which may be inverted by using the equations above. Cho and Spencer (1992) show how this matrix can be extended to accommodate more events, although the recommendation for this particular VSP application is to use only the downgoing wave field in order to stabilize the inverse.

4.2.3 Nine-component field example

Figure 4.6 shows raw nine-component data from the CONOCO borehole test facility in Kay County, Oklahoma. These data are recorded between 900 m and 152 m at increments of 15 m in a vertical well. In-line (X) and cross-line (Y) horizontal vibrators are activated at offsets of 30 m, 36 m, and 39 m, and vertical (Z) vibrators at offsets of 41 m, 46 m and 50 m. The various offsets correspond to different pad locations, necessary when the baseplate digs into the ground. The data show evidence of an extended wave train which appears to be longer for the horizontal sources. This long wave train confuses deeper strong reflections and complicates interpretation. This may be caused by multiples in near-surface low-velocity zones. An alternative explanation could be differences between the true ground motion and measured motion leading to inaccurate correlation with the vibrator signal. There are also some unusual features: high compressional wave energy from the horizontal sources on the off-diagonal components $d_{ZX}(t)$ and $d_{ZY}(t)$ originating from above the shallowest recording level at 152 m, and a slightly lower but significant energy at shallow and deeper depths on the $d_{XZ}(t)$ and $d_{YZ}(t)$ components. The $d_{XZ}(t)$ and $d_{YZ}(t)$ components can be explained by the oblique incidence of the compressional waves at shallow depths. However, the increase of energy between 800 m and 900 m, where ray paths are predominantly vertical, must be associated with the medium. This feature is present throughout all processing stages.

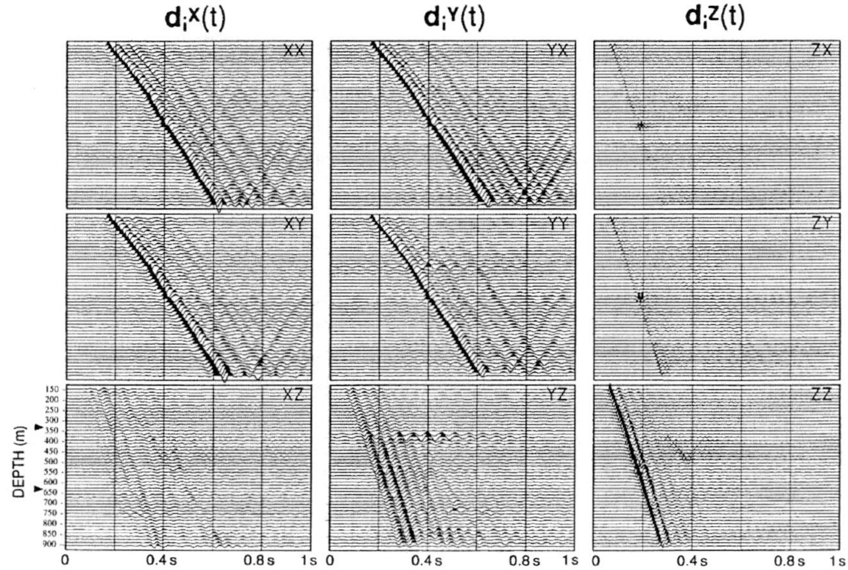

Figure 4.6. Nine-component VSP data from the Conoco Borehole Test Facility after correction for tool spin. The geophone axes and source motion are oriented cross-line (X), in-line (Y), and vertically down (Z) directions. The trace notation KJ refers to source K and receiver J. Traces are normalized to the maximum vector amplitude at each depth level (MacBeth et al. 1995).

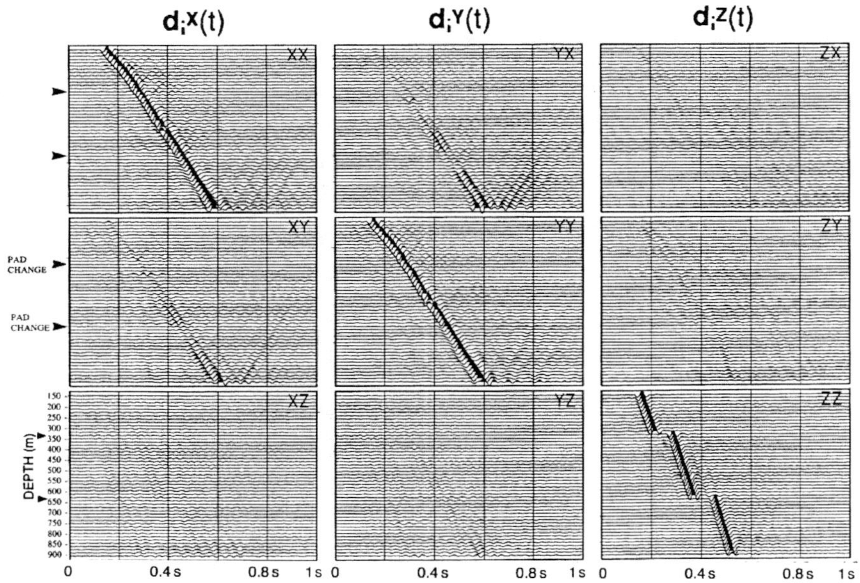

Figure 4.7. Nine-component data matrix of traces from Figure 4.6 corrected by a near-surface operators designed at the shallowest level, and then in the vicinity of traces 33 and 50. For display purposes the P-waves on the ZZ component have been re-aligned to the shear wave moveout after each successive deconvolution (MacBeth et al. 1995).

The downgoing wave field is determined using an *f-k* filter, and a near-surface correction is performed using an estimate of the inverse downgoing wave field evaluated from the shallow traces 4, 5 and 6, as near-field terms are strong for the first few levels. A nine-component operator $\mathbf{D}^{-1}_{k,S}(t)$ of 1.0 s duration is applied to the total wave field. The result is shown in Figures 4.7. Interestingly, it was necessary to re-apply the correction at each source pad change (340 m and 640 m) because the near-surface response is a function of activation.

In general, the process works effectively, especially for the shallowest levels, and the off-diagonal matrix elements $d_{ZX}(t)$, $d_{XZ}(t)$, $d_{YZ}(t)$, and $d_{ZY}(t)$ are now considerably reduced. The original waveforms at the shallowest levels have collapsed and the reflections are clearer and simpler. There is a gradual build-up of energy on the off-diagonal components indicative of an increase in delay between the two shear waves, and symptomatic of seismic anisotropy (Chapter 2, but see also Chapter 5). Since the procedure converts the first few traces to a unit diagonal, the onsets of the compressional wave coincide with the shear waves after each correction. A more sensitive measure of the effect of the operator can also be seen from cross plots of the particle motion (Figure 4.8). After the application of the near-surface correction, separation of the up and downgoing waves is possible (Figures 4.9 and 4.10). Figure 4.10 gives the result after adjustment for the asymmetry of the ray path to obtain two-way propagation at the datum level, by post-multiplying the data matrix by the contribution from the upgoing leg

$$\mathbf{D}^{up}_{k,j}(\omega) = \left(\mathbf{D}^{down}_{j,k}(\omega)\right)^T. \tag{4.29}$$

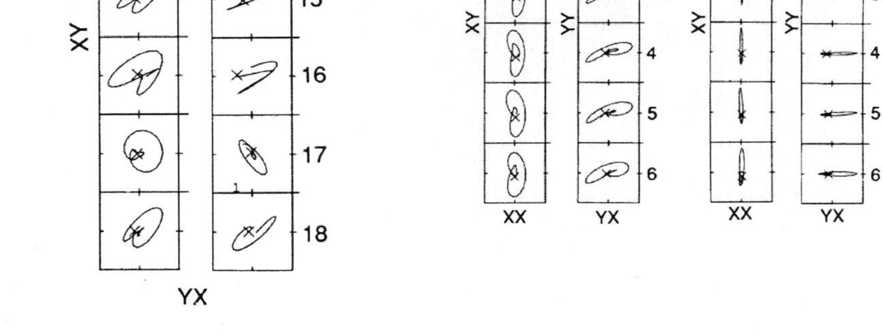

Figure 4.8. (a) Cross-diagonal symmetry check using plots of *XY* versus *YX* component, plotted before and after near-surface correction;

(b) Horizontal shear-wave motions for in-line (left) and cross-line (right) sources, plotted before and after near-surface correction.

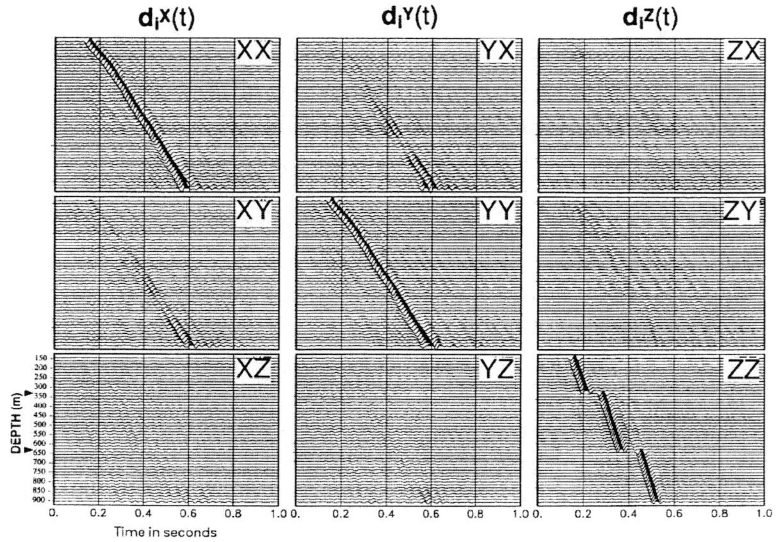

Figure 4.9. Downgoing wave field obtained by an *f-k* filter after near-surface correction (MacBeth et al. 1997).

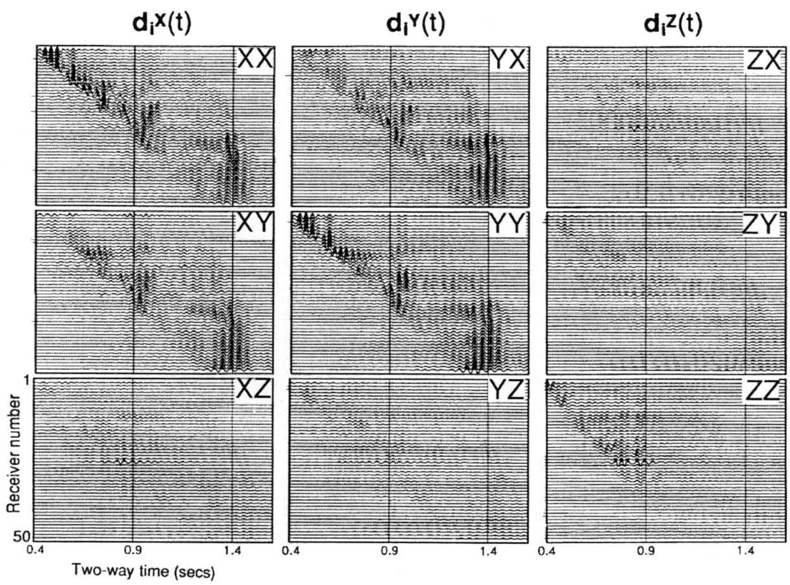

Figure 4.10. Upgoing wave field for the nine-component VSP after near-surface correction, *f-k* filtering and correction to a two-way response correction (MacBeth et al. 1997).

4.2.4 Four-component field example

Figure 4.11 shows the plan view of an acquisition used to record VSPs in three separate wells in the Romashkinofield, Russia (Cliet et al. 1991; Zeng and MacBeth 1996). The data are from well 98, which is deviated by up to 38°. For the VSP, two orthogonal pairs of motion are generated by VEIP electrodynamic directional pulse sources (Slater 1997). The orthogonal source set A (*N*122°*E*) and B (*N*212°*E*) were of similar strength but less powerful than the other orthogonal set C (*N*77°*E*) and D (*N*167°*E*) which are again matched in strength. The data are recorded by 90 receivers located between 440 m and 830 m for the four source orientations.

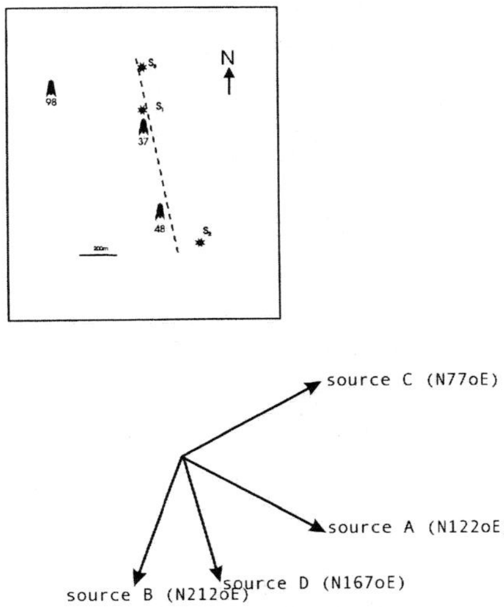

Figure 4.11. Acquisition geometry for the Romashkino VSP data. S_1, S_2 and S_3 are shear-wave source positions for wells 37, 48 and 98, respectively. The dashed line is a surface survey line. A four-component VSP is shot in well 98, with a VEIP electrodynamic truck aligned in four different orientations (Zeng and MacBeth 1996).

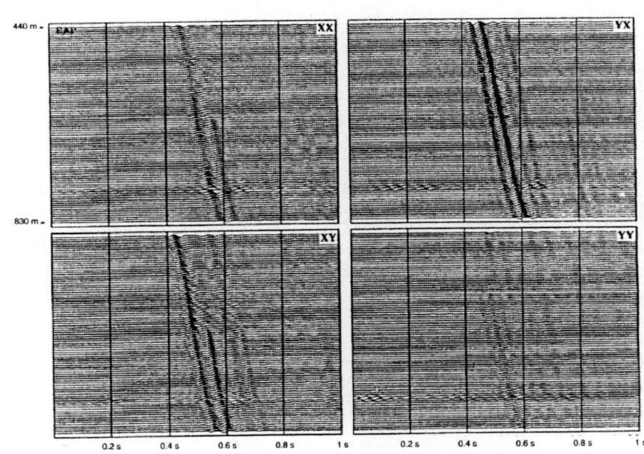

Figure 4.12. Data matrix of VSP recordings at well 98 for source combination A and B. The *XX* and *XY* traces are in-line (*X*) and cross-line (*Y*) recordings of the source A, the *YX* and *YY* traces are the corresponding recordings for source B (Zeng and MacBeth 1996).

Figure 4.12 shows the data matrix generated by the source combination A and B. The data are processed for the source pairs A–B and C–D separately. Application of anisotropy estimation techniques (Chapters 5 and 6) give shear wave polarization estimates for the two source pairs that are constant with depth but 30° apart (Figure 4.13). The time delays also differ, for one pair they increase from 16 ms to 28 ms, for the other they are 24 ms throughout. There are therefore clear differences between the source combinations, even though they are spatially very close. Figure 4.14 shows the traces after near-surface correction for source pair A and B, for which correction is required at two separate depths, as before. When the anisotropy estimation procedures are re-applied to these data matrices, major improvements are noticed (Figure 4.15). The estimates are nearly identical, and the results are independent of the sources. The interpretation value is considerably enhanced. The approach has also been applied to other wells in the same area with similar conclusions.

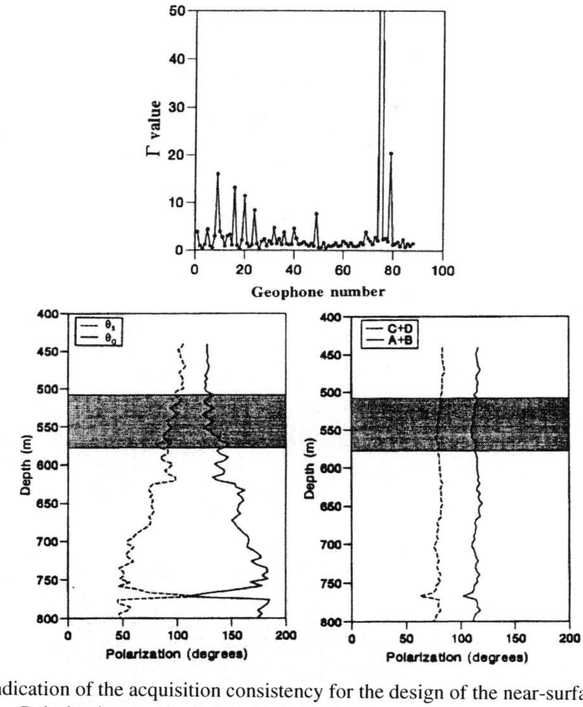

Figure 4.13. Top: indication of the acquisition consistency for the design of the near-surface operator. Bottom: Polarizations estimated by applying the DIT algorithm detailed in Chapter 6 to the source pairs A and B, and then C and D. The degree of similarity of the two lines in each plot is an indicator of the data quality for anisotropy estimation. The shaded area is the location of the reservoir zone (Zeng and MacBeth 1996).

4.2.5 Near-surface correction with mixed sources?

Near-surface correction has been demonstrated for multi-component data generated with specialized sources arranged in orthogonal orientations. In good-data areas, the multi-component framework can also be extended to deal with non-specialized sources – necessary in areas of restricted access, or when multi-component sources are not available. According to the VCM and (4.17), the downgoing wave field $\mathbf{D}_{k,S}(t)$ could instead be generated by a group of sources contributing different levels of P and S wave energy. Indeed, this downgoing wave field grouping could be a composite of three (or more) three-component recordings from a variety of source types. For more than three sources, stacking is recommended to enhance the signal-to-noise ratio and improve the matrix inverse. A synthetic seismogram test of this approach is shown in Figure 4.16, which favorably compares with the orthogonal source case (Figure 4.17). This suggests a role for a shallow far-field monitor placed in the borehole alongside the moveable sonde. Should such a technique be possible, it would considerably enhance the VSP data quality by providing a means of compensating for the multi-component variability of the source. Conceivably, this technique might also be extended to converted waves in offset VSP. Unfortunately, this is not feasible as shear wave amplitude changes too rapidly with depth. Tests showed that offsets giving angles of incidence 3°–6° are optimal; for such angles the converted shear wave amplitudes are not significant.

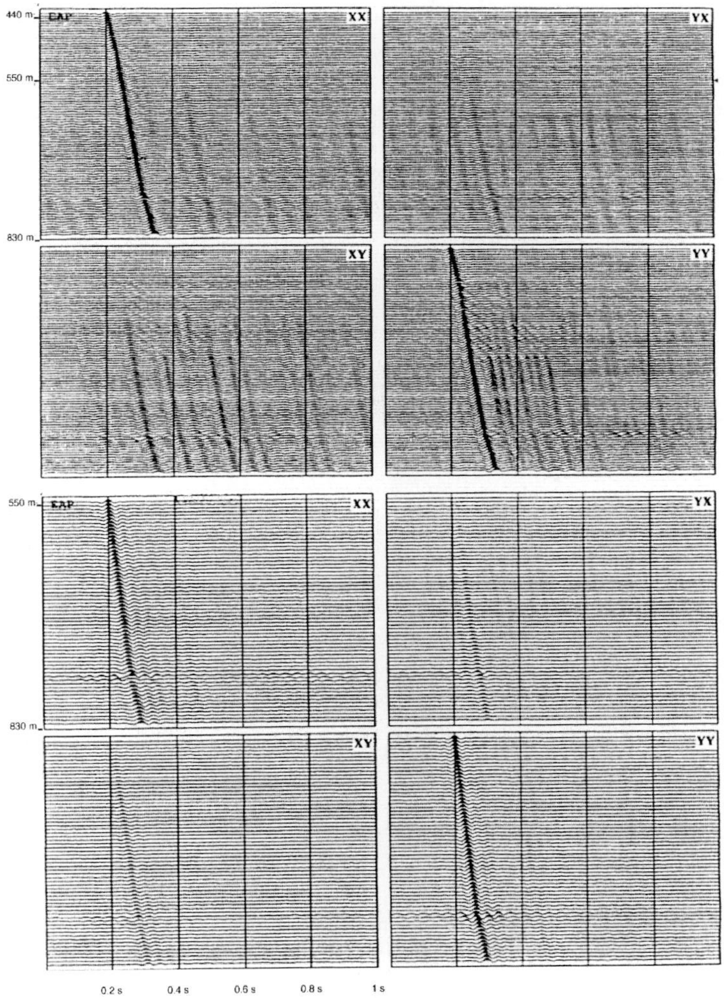

Figure 4.14. Data matrix of VSP recordings at well 98 for the source combination A and B, after near-surface correction based at depths of 445m (top) and 555m (bottom). The arrival time has been shifted by 200 ms for display purposes (Zeng and MacBeth 1996).

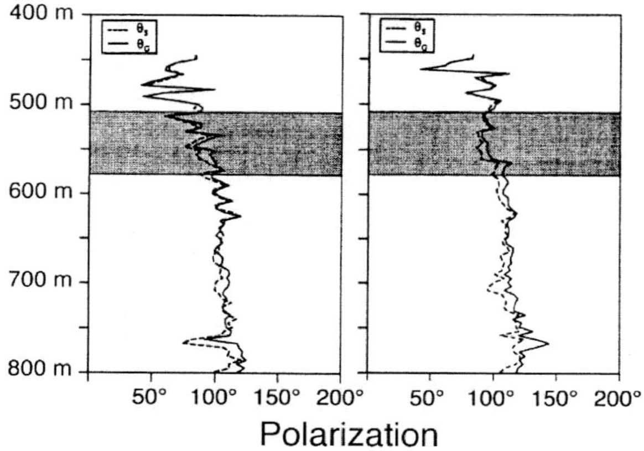

Figure 4.15. As in Figure 4.13, except after near-surface correction. Note the substantial improvement of data quality and consistency between the two source pairs (Zeng and MacBeth 1996).

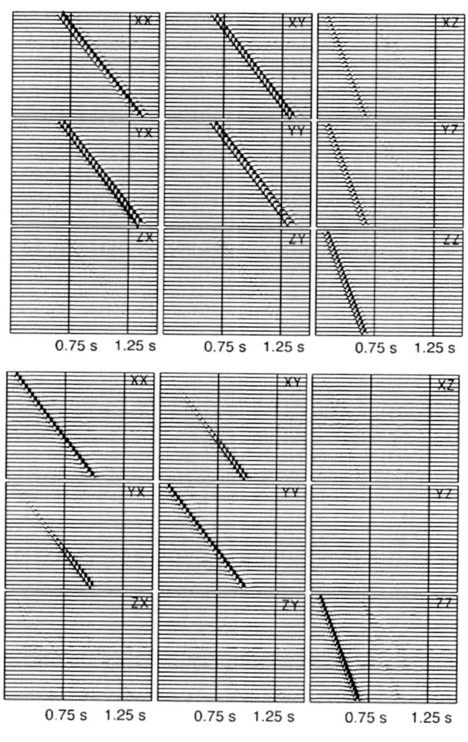

Figure 4.16. Nine-component VSP matrices from three-component synthetic seismograms generated using three different sources, all at an offset of 50 m. The model is a uniform TIH half-space overlain by a differently oriented TIH layer. The first source is a point force dipping at 30° in the in-line direction; the second is dipping at 40° and is directed 70° from the in-line direction; the final is a vertical force. A near-surface correction is applied to the data based on the shallowest traces. The upper diagram shows the data before correction, the bottom diagram after correction. The procedure yields the 9C matrix expected for this medium. It can be directly compared with the ideal result in Figure 4.17.

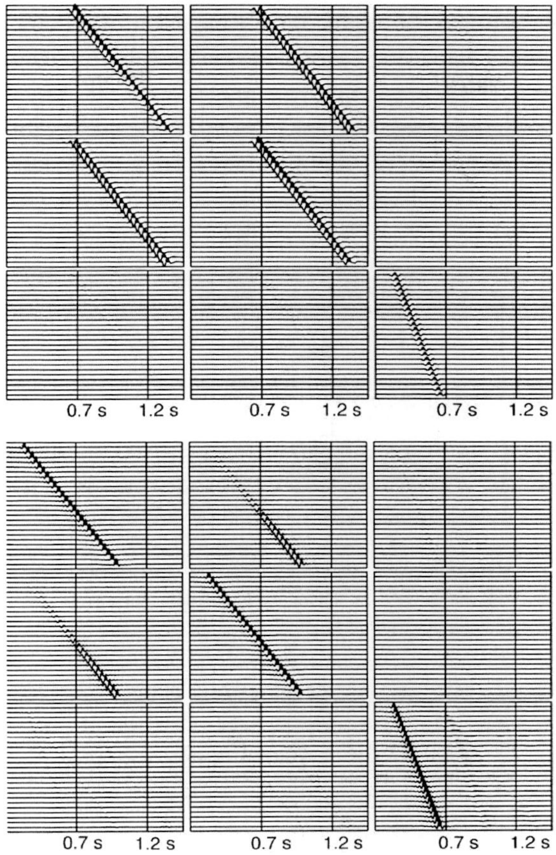

Figure 4.17. Nine-component VSP synthetics generated using orthogonal point forces in horizontal and vertical directions. The model is the same as Figure 4.16. Before (top) and after (bottom) near-surface correction.

4.3 CORRECTION FOR RECEIVER TOOL MIS-ORIENTATION

For anisotropy analysis with multi-component VSP it is preferable to acquire data with the source and geophone axes aligned along the global frame X-Y-Z, thus ensuring that (4.13) is applicable. This is not always possible, as an individual sonde may twist as it is unlocked, moved and locked at a new depth due to the torque acting on the cable. This effect leads to the introduction of a differential rotation, even between the individual recording elements on the tool string (Figure 4.18).

These mis-orientations are generally of random character and highly unpredictable. Additionally, some tools have the receivers fixed relative to the external package, with the borehole trajectory determining its orientation. The consequences of these mis-orientations for near-offset VSP is that the fast and slow shear waves are projected onto the sensor frame X_G-Y_G-Z_G (Figure 4.19) and not along the source motions. The effects on data matrix symmetry are shown in Figure 4.20, the implications of which are followed up in Chapter 5. The geometric connection between the desired three-component recordings in the global coordinates and the actual three-component recording $\mathbf{d}(t)$ in the sensor frame is

$$\mathbf{d}'(t) = \mathbf{G}\mathbf{d}(t) \ , \tag{4.30}$$

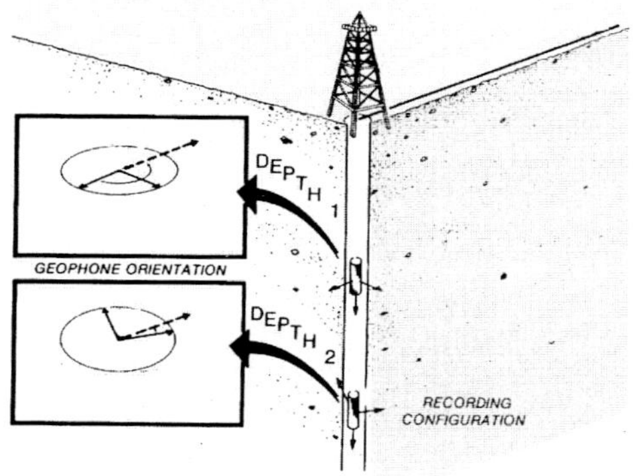

Figure 4.18. The elements of the downhole sensor package twist during movement between levels, thus making the recorded polarizations inconsistent.

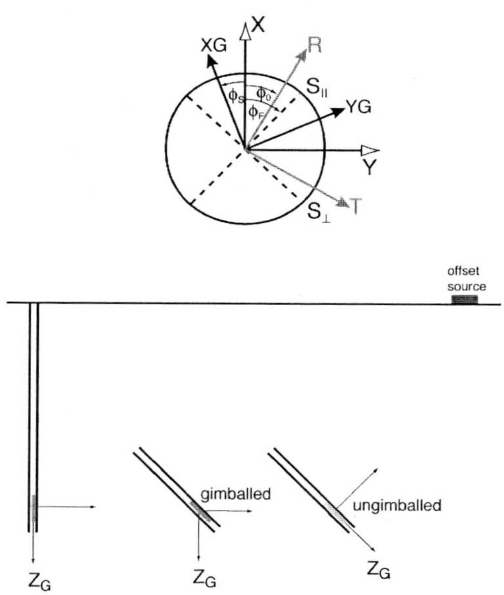

Figure 4.19. Tool twist affects the orientation of the horizontal receiver axes X_G and Y_G, so that they are no longer parallel to the acquisition-defined X-Y frame. The tool Z_G might also not be aligned with the Z axis. The correction procedure for the horizontal axes (or the horizontal projection of X_G and Y_G) first aligns along the radial (R) and transverse (T) directions as defined by the source-receiver vector using an offset source. If the azimuth ϕ_0 of this particular source is known, another rotation achieves the final alignment.

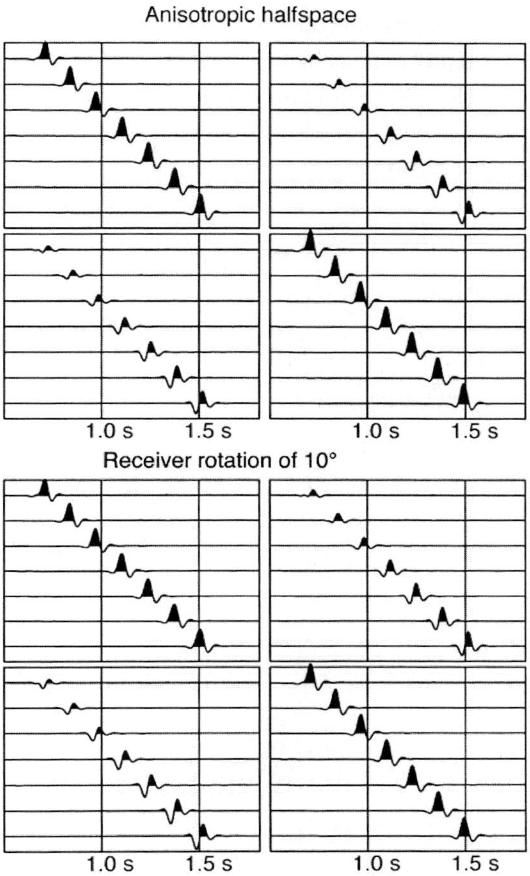

Figure 4.20. Four-component shear-wave VSP for a uniform anisotropic half-space.
Top: perfect acquisition alignment.
Bottom: 10° horizontal receiver rotation.
Note the asymmetry in the cross-diagonals, This property is not confined only to this situation (see Chapter 6).

where **G** contains projection angles relevant to a particular depth. This reorientation matrix can be determined relative to some fixed external reference field using instruments attached to the downhole package. Suitable devices with this capability are inclinometers, magnetometers or gyroscopes (Hardage 2000). The inclinometer measures the deviation of Z_G against the true vertical, but has the disadvantage of adding cost and weight to the total instrument configuration. The magnetometer (in-hole magnetic compass) gives X_G relative to true North, but works only in an uncased hole. Also, gyroscopes tend to drift, the geomagnetic field is distorted locally by the casing, and the orientation devices may give rise to unwanted resonances, which limit the seismic resolution. Additionally, the gyroscope can double the acquisition time. Thus, a widely accepted alternative for downhole orientation, which is considered both cost-effective and accurate, is to utilize the motion of the direct P-wave recorded from offset sources. Therefore, for near-offset VSP use is usually made of one or more offset sources also. An example of this is shown in Figure 4.21, where additional vertical vibrators are activated at offsets 213 m and 457 m at a common azimuth of 155°. The tests by Queen (1987) provide a quantitative evaluation of the expected error of this technique. Comparing results from an offset VSP designed with a number of source azimuths at a common offset with gyroscope data from the same experiment, they found a systematic difference of +3° and –16°.

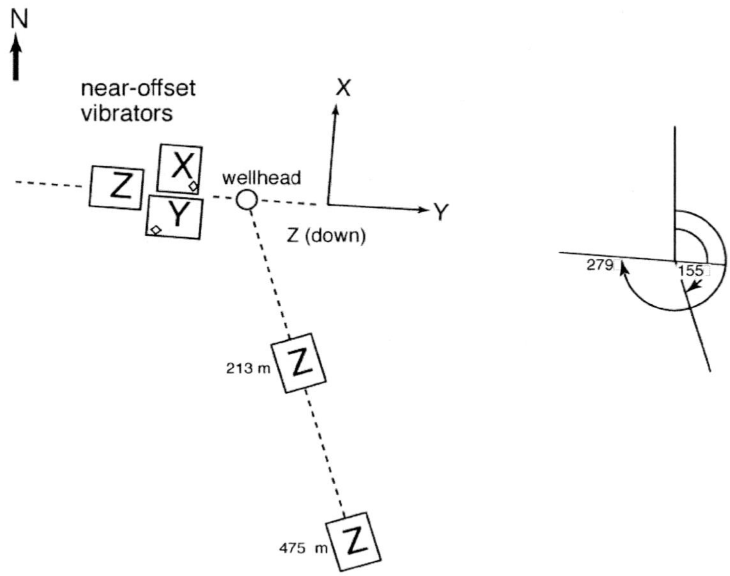

Figure 4.21. Acquisition geometry for the nine-component VSP shot at the CONOCO Borehole Test Facility, Kay County, Oklahoma. Rectangles marked *Z* indicate vibrators with vertical motion, those marked *X* and *Y* indicate vibrators with cross-line and inline horizontal motions, respectively. Far-offset *Z*-vibrators are used for reorientation of the sensors (MacBeth et al. 1995)

4.3.1 Misorientation correction using seismic data

This method relies upon the stability of the P-wave arrival to provide a measure of the tool orientation. Ideally, an arbitrarily oriented downhole receiver records a linear P-wave motion $\mathbf{d}^P(t)$ in the sensor frame. Under the assumption that the polarization lies in the sagittal plane (vertical plane containing the source and receiver), alignment along this motion automatically aligns the tool axes along the source-receiver directions and hence the radial (R) and transverse (T) directions of wave propagation. As this direction is known, the acquisition coordinate system can then be rotated into the desired coordinate system. Alignment in the vertical plane requires additional knowledge of the ray path direction through ray tracing. In practice, a window of samples around the P-wave motion is chosen, the size of which varies depending on the linearity of the motion (Figure 4.22). Usually the first quarter of the wavelet is selected for analysis. The success of this approach to date attests to the general robustness and stability of the P-wave, in spite of lateral variations in material properties and structural complications.

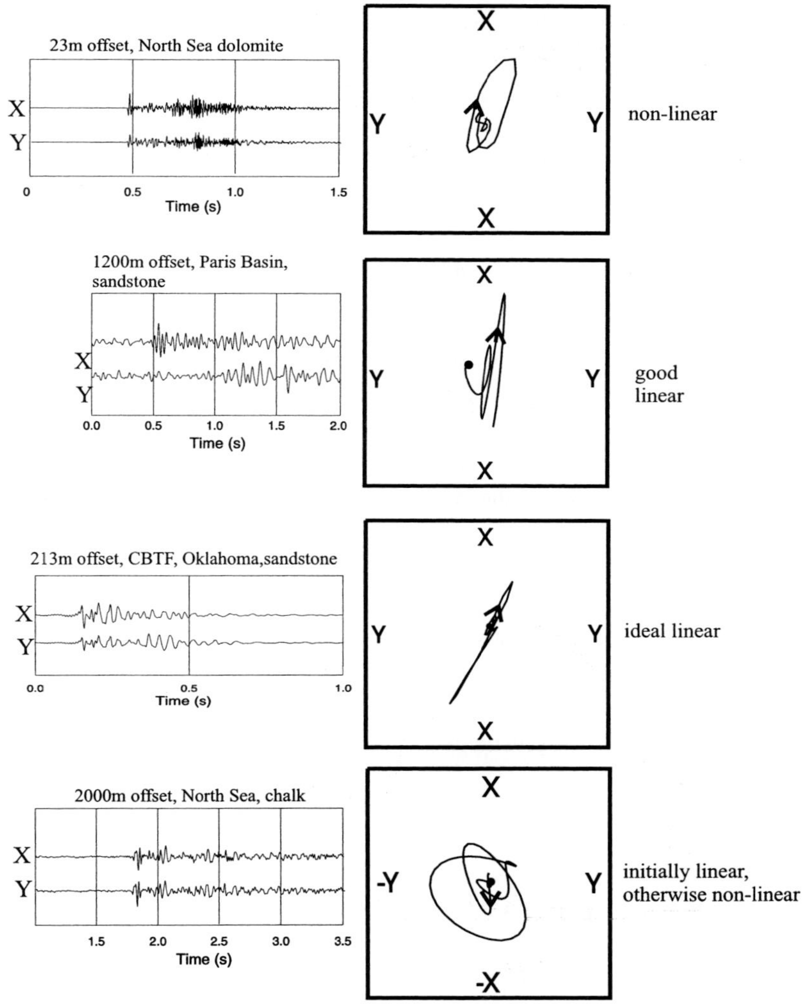

Figure 4.22. Examples of horizontal P-wave motions from different locations and offsets, displaying a range of behavior. The arrowhead indicates the direction of the first motion. Data courtesy of the British Geological Survey.

4.3.1.1 Rotation in the horizontal plane only

For a vertical borehole or gimballed receivers in a deviated well the Z_G-axis always remains along the true vertical (Z) (Figure 4.19). The – mutually orthogonal – horizontal components are arbitrarily oriented in the horizontal plane and only a single estimate of the rotation angle $\hat{\phi}_G$ about the vertical is required. The connection between the motion $\mathbf{d}''$ expressed in the global system and that expressed in the R-T coordinates is

$$\mathbf{d}''^P(t) = \begin{pmatrix} d_X''^P(t) \\ d_Y''^P(t) \\ d_Z''^P(t) \end{pmatrix} = \begin{pmatrix} \cos\hat{\phi}_G & \sin\hat{\phi}_G & 0 \\ -\sin\hat{\phi}_G & \cos\hat{\phi}_G & 0 \\ 0 & 0 & 1 \end{pmatrix} \begin{pmatrix} d_X^P(t) \\ d_Y^P(t) \\ d_Z^P(t) \end{pmatrix} . \tag{4.31}$$

Multiplication with this matrix should transform the three-component recordings as close to the assumed form

$$\mathbf{d}''^P(t) = \begin{pmatrix} a \\ 0 \\ b \end{pmatrix} \tag{4.32}$$

as possible. There are several ways to calculate this estimate, all of which converge for noise-free data and truly linear polarization. The following is a selection of recipes:

(a) A simple method is to calculate the particle motion alignment directly from the recordings $\mathbf{d}^P(t)$ and recover ϕ_G by inversion. This can be done by inspection of the particle motions, or automatically by first calculating the instantaneous horizontal polarization

$$\phi_I(t_i) = \tan^{-1}\left(\frac{d_Y^P(t_i)}{d_X^P(t_i)}\right) . \tag{4.33}$$

The instantaneous horizontal polarization is a measure of the position vector of an individual point i, and when interpreted as a function of t_i gives the predominant alignment over the time-window of interest. However, by itself it is a poor indicator of orientation and is strongly sensitive to noise. This sensitivity can be reduced by a weighting modulus $m^2(t)$ (Di Siena et al. 1984)

$$m^2(t_i) = d_X^{P2}(t_i) + d_Y^{P2}(t_i) \tag{4.34}$$

followed by binning of the resultant function according to angle. The histogram function is then consulted to yield the most probable value for ϕ_G.

(b) Another criterion used for estimation is the maximum energy $\hat{E}(\phi_G)$ on the transformed X (R)-component in (4.31) (Di Siena et al. 1984) in terms of the original components $d_X^P(t_i)$ and $d_Y^P(t_i)$ for an appropriate window $i = 1, N$ around the P-wave arrival. Thus

$$E(\phi_G) = \sum_i \left|\hat{d}_X''\right|^2 = \sum_i \left(d_X^P(t_i)\cos\phi_G + d_Y^P(t_i)\sin\phi_G\right)^2 , \tag{4.35}$$

for which the least-squares estimate

$$\hat{\phi}_G = \frac{1}{2}\tan^{-1}\left(\frac{2\sum_i d_X^P(t_i)d_Y^P(t_i)}{\sum_i d_X^P(t_i)d_X^P(t_i) - \sum_i d_Y^P(t_i)d_Y^P(t_i)}\right) \tag{4.36}$$

can be recovered. Another possibility is to minimize the Y (T)-component, which for noise-free data gives the same result (4.36).

(c) One of the most important procedures maximizes the energy projected from the three-component recording onto an estimate of the required direction (Esmersoy 1984). This approach is particularly useful if the signal contains noise uncorrelated with the individual components. Mathematically, the objective may be met by performing an eigenanalysis of the correlation (covariance) matrix **P**

$$\mathbf{P} = \sum_i \mathbf{d}^P(t_i)\left(\mathbf{d}^P(t_i)\right)^T \tag{4.37}$$

or

$$\mathbf{P} = \frac{1}{N}\begin{pmatrix} \sum_i \left(d_X^P(t_i)-\bar{d}_X^P\right)^2 & \sum_i \left(d_X^P(t_i)-\bar{d}_X^P\right)\left(d_Y^P(t_i)-\bar{d}_Y^P\right) & \sum_i \left(d_X^P(t_i)-\bar{d}_X^P\right)\left(d_Z^P(t_i)-\bar{d}_Z^P\right) \\ \sum_i \left(d_Y^P(t_i)-\bar{d}_Y^P\right)\left(d_X^P(t_i)-\bar{d}_X^P\right) & \sum_i \left(d_Y^P(t_i)-\bar{d}_Y^P\right)^2 & \sum_i \left(d_Y^P(t_i)-\bar{d}_Y^P\right)\left(d_Z^P(t_i)-\bar{d}_Z^P\right) \\ \sum_i \left(d_Z^P(t_i)-\bar{d}_Z^P\right)\left(d_X^P(t_i)-\bar{d}_X^P\right) & \sum_i \left(d_Z^P(t_i)-\bar{d}_Z^P\right)\left(d_Y^P(t_i)-\bar{d}_Y^P\right) & \sum_i \left(d_Z^P(t_i)-\bar{d}_Z^P\right)^2 \end{pmatrix}, \tag{4.38}$$

where $\bar{d}_X^P$, $\bar{d}_Y^P$ and $\bar{d}_Z^P$ represent averages of the three components defined by

$$\bar{d}_X^P = \frac{1}{N}\sum_i d_X^P(t_i)\ , \tag{4.39}$$

$$\bar{d}_Y^P = \frac{1}{N}\sum_i d_Y^P(t_i)\ , \tag{4.40}$$

$$\bar{d}_Z^P = \frac{1}{N}\sum_i d_Z^P(t_i)\ . \tag{4.41}$$

Eigenanalysis decomposes the matrix into the form

$$\mathbf{P} = \mathbf{U\Lambda U}^T\ , \tag{4.42}$$

where the matrix

$$\mathbf{U} = \begin{pmatrix} u_{11} & u_{12} & u_{13} \\ u_{21} & u_{22} & u_{23} \\ u_{31} & u_{32} & u_{33} \end{pmatrix} \tag{4.43}$$

is formed by columns of normalized eigenvectors, and the diagonal matrix

$$\Lambda = \begin{pmatrix} \lambda_1 & 0 & 0 \\ 0 & \lambda_2 & 0 \\ 0 & 0 & \lambda_3 \end{pmatrix} \tag{4.44}$$

contains the eigenvalue elements. As the matrix **P** is real and symmetric, the eigenvalues are real and the eigenvectors orthogonal (Kanasewich 1981). For a linearly polarized arrival, the dominant eigenvalue λ_1 corresponds to the linear polarization of the P-wave. It measures the centroid of the motion in the window.

The horizontal components of u_{11} and u_{12} give a direct estimate of the desired horizontal rotation angle

$$\hat{\phi}_P = \tan^{-1}\left(\frac{u_{12}}{u_{11}}\right) . \tag{4.45}$$

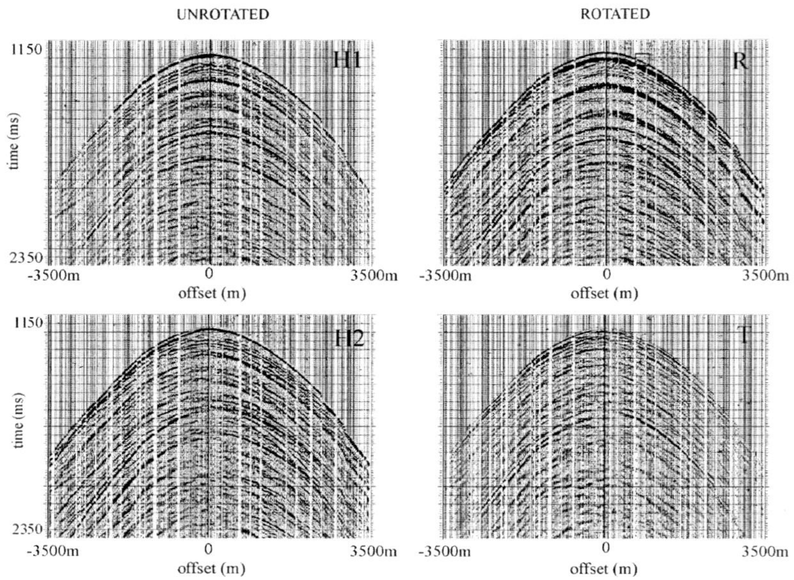

Figure 4.23 Result of the eigenanalysis rotation on the horizontal components H_1 and H_2 of three-component marine walkaway VSP data from the North Sea. The gimballed receiver tool was in a deviated portion of the well and lay to one side of the surface line of shots. Rotation gives the radial and transverse components and diminishes the initial part of the P-wave motion. The energy remaining on the transverse component is due to converted shear waves from the vicinity of the receiver and from the deeper subsurface. Data courtesy of the British Geological Survey and Norsk Hydro.

Two examples of application of this technique for different VSP scenarios are shown in Figures 4.23 and 2.24. This polarization analysis can also be applied in the frequency domain by replacing the covariance matrix by the spectral equivalent

$$S = \sum_j \mathbf{d}^P(\omega_j)(\mathbf{d}^P(\omega_j))^H , \tag{4.46}$$

where only the real parts need be analysed for a strictly linear polarization. This form has the particular advantage that it is possible to focus on specific frequency bands. A quantitative assessment of the overall degree of linearity L^P in the chosen time window may be obtained from the eigenvalues and used to judge the quality of the final result. Esmersoy (1984) recommends the ratio of linearly polarized power to residual power

$$L^P = \frac{2\lambda_1}{\lambda_2 + \lambda_3} - 1 \tag{4.47}$$

as a quality measure. L^P lies between 0 (circular or random polarization) and 1 (purely linear polarization).

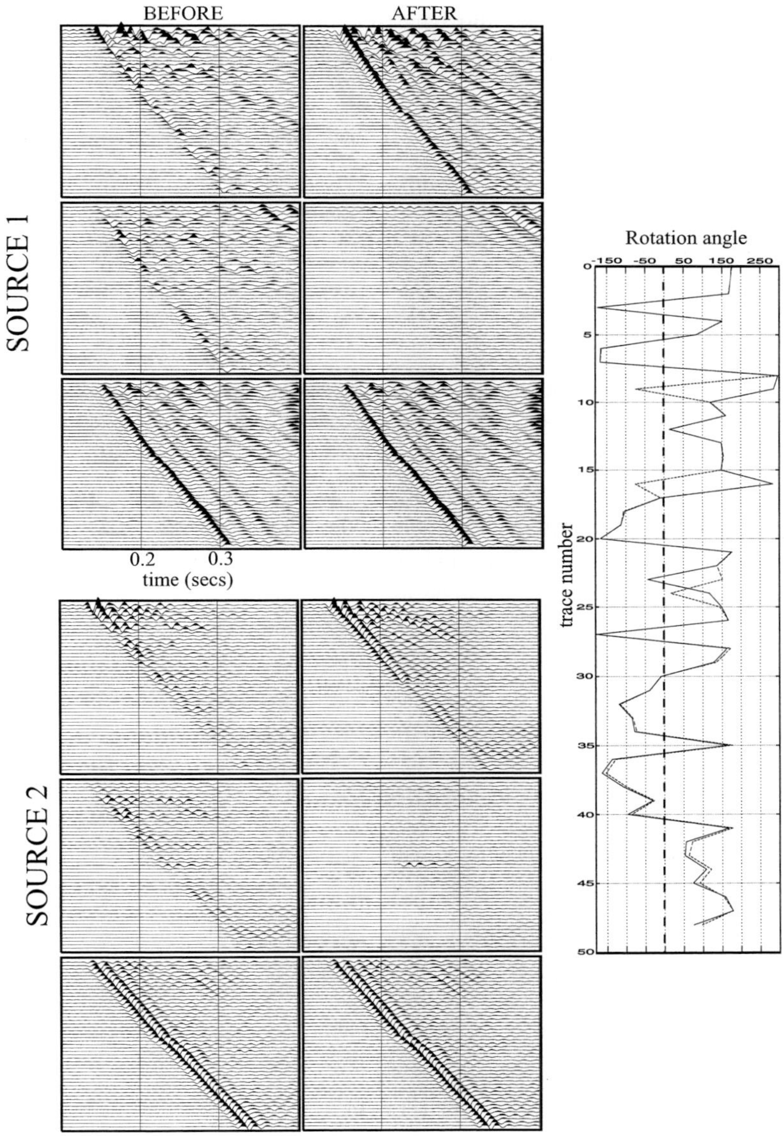

Figure 4.24. Three-component recordings from the two vertical sources in Figure 4.21 placed at a common azimuth. There is good agreement between the two sets of rotation estimates, with the occasional 180° shift leading to a polarity change. Data courtesy of the British Geological Survey.

4.3.1.2 Non-gimballed receiver tool in a deviated well, but single fixed axis

The receiver axis Z_G is determined by the well trajectory, while the orientation of the other two components is unknown. For this general case, equation (4.15), relating the incoming polarization $\mathbf{d}^P(t)$ to the polarization $\mathbf{d}''(t)$ that would be recorded had the receivers been in the *R-T* coordinates, becomes

$$\mathbf{d}''^P(t) = \begin{pmatrix} \cos\phi_H & \sin\phi_H & 0 \\ -\sin\phi_H & \cos\phi_H & 0 \\ 0 & 0 & 1 \end{pmatrix} \begin{pmatrix} \cos\theta_W & 0 & -\sin\theta_W \\ 0 & 1 & 0 \\ \sin\theta_W & 0 & \cos\theta_W \end{pmatrix} \begin{pmatrix} \cos\phi_G & \sin\phi_G & 0 \\ -\sin\phi_G & \cos\phi_G & 0 \\ 0 & 0 & 1 \end{pmatrix} \mathbf{d}^P(t) \tag{4.48}$$

or

$$\mathbf{d}''^P(t) = \mathbf{C}_Z(\phi_H)\mathbf{C}_Y(\theta_W)\mathbf{C}_Z(\phi_G)\mathbf{d}^P(t) \ , \tag{4.49}$$

with all angles positive for rotations in the anti-clockwise direction. Equations (4.48, 49) are based on the fact that any general 3D rotation may be written as three separate rotations about two axes. The angle θ_W defines the well deviation obtained from the well survey. The angle between the source-receiver direction and the horizontal projection of the well trajectory is ϕ_H. This angle might not be known due to source navigation/survey error and should be regarded as an unknown in this problem. The angle ϕ_G represents a rotation about the borehole axis. Following (4.35) above, a solution can be sought by maximizing the energy on the resulting *X*-component in terms of ϕ_H and ϕ_G. An alternative criterion is to minimize the energy on the *Y*-component. Analytic solutions are unwieldy, therefore a numerical search procedure is usually employed. A search using the Gauss-Newton method is proposed by Hokstad (1995), although he recommends calculating the gradient vectors analytically.

4.3.1.3 Non-gimballed receiver tool in a deviated well

When the orientation of the receiver components is not known with confidence, either due to the lack of a well survey or incorrect functioning of the sonde, (4.41) can be used in a optimization scheme to determine all three angles. Clearly this has a less robust solution than the procedure discussed in the previous section and may encounter many local minima. In this situation, Greenhalgh and Mason (1995) propose to achieve the reorientation by re-expressing the composite rotations by a single rotation of Ω about a single axis with direction $\mathbf{a} = (a_1, a_2, a_3)^T$ relative to the global *X-Y-Z* axes (Altmann1986). Thus (4.25) leads to

$$\mathbf{d}''^P(t) = \mathbf{B}\mathbf{d}^P(t) \ ,$$

where

$$\mathbf{B} = \mathbf{I}\cos\Omega + (1-\cos\Omega)\begin{pmatrix} a_1^2 & a_1a_2 & a_1a_3 \\ a_1a_2 & a_2^2 & a_2a_3 \\ a_1a_3 & a_2a_3 & a_3^2 \end{pmatrix} + \sin\Omega \begin{pmatrix} 0 & -a_3 & a_2 \\ a_3 & 0 & -a_1 \\ -a_2 & a_1 & 0 \end{pmatrix} , \tag{4.50}$$

with the parameters given by

$$\Omega = \cos^{-1}(\mathbf{u} \cdot \mathbf{v}) \tag{4.51}$$

and

$$\mathbf{a} = \begin{pmatrix} v_2 u_3 - v_3 u_2 \\ v_3 u_1 - v_1 u_3 \\ v_1 u_2 - v_2 u_1 \end{pmatrix} . \tag{4.52}$$

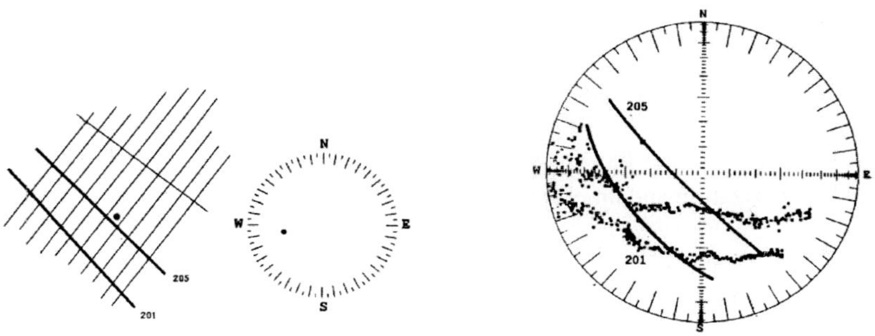

Figure 4.25. Figure extracted from Greenhalgh and Mason (1995) showing the shot lines and well location in the southern North Sea, highlighting lines 201 and 205, and an upper hemisphere equal-area projection of the apparent P-wave directions against the anticipated directions. A rotation of 37° about the vertical, followed by a tilt of 5° produces alignment.

These are determined from an estimate $\mathbf{u}$ of the polarization vector in the sensor frame (determined by eigenanalysis of the recorded motion $\mathbf{d}^P(t)$ as in (c) above – (u_{11}, u_{12}, u_{13}) and $\mathbf{v} = (v_1, v_2, v_3)^T$, which is the expected (ideal) polarization vector in the acquisition R-T-Z frame (see equation 4.13). Applications by Greenhalgh and Mason (1995) suggest the form

$$\mathbf{v} = \begin{pmatrix} \cos\overline{\theta} \\ 0 \\ \sin\overline{\theta} \end{pmatrix} , \tag{4.53}$$

where the angle of incidence $\overline{\theta}$ must be determined by ray tracing if complete reorientation is required. The approximation is sufficient if reorientation of only the X_G and Y_G axes is the primary objective (Figure 4.25). The accuracy of the procedure is improved if several sources are used, with each source position j yielding an individual estimate of the rotation axis $\mathbf{a}_j$ using the predominant motion $\mathbf{u}_j$ in the sensor frame but a constant anticipated result. A robust average of the rotation axes is obtained by calculating the dominant eigenvector of the matrix $\mathbf{QQ}^T$, where

$$\mathbf{Q} = (\mathbf{a}_1, \ \mathbf{a}_2, \ \ldots \ \mathbf{a}_M) . \tag{4.54}$$

An average of the required rotation angle is

$$\hat{\Omega} = \frac{1}{M}\sum_{j=1}^{M}\cos^{-1}\left(\mathbf{u}_j \cdot \mathbf{v}\right) . \qquad (4.55)$$

4.3.2 Vector fidelity of the receiver

Rotation is a valid procedure for reorientation if the mathematical expressions above adequately simulate the physical rotation of the sensing device; this situation is exactly analogous to that of the source. Vector fidelity is guaranteed provided the receiver-formation coupling is balanced, there is no cross-coupling, and the directional response of each sensor obeys projection geometry at all frequencies. This is exemplified by the laboratory experiments of Sondergeld and Rai (1992) who confirm fidelity using specially designed receiver mounts, both for isotropic (fused quartz) and anisotropy (McNabb shale) samples (Figure 4.26). In the field, however, problems could arise due to a change in the coil response with wave propagation direction. Such changes have been noted by Schulte and Edelmann (1988). They can lead to a swing in polarization and an elliptical motion.

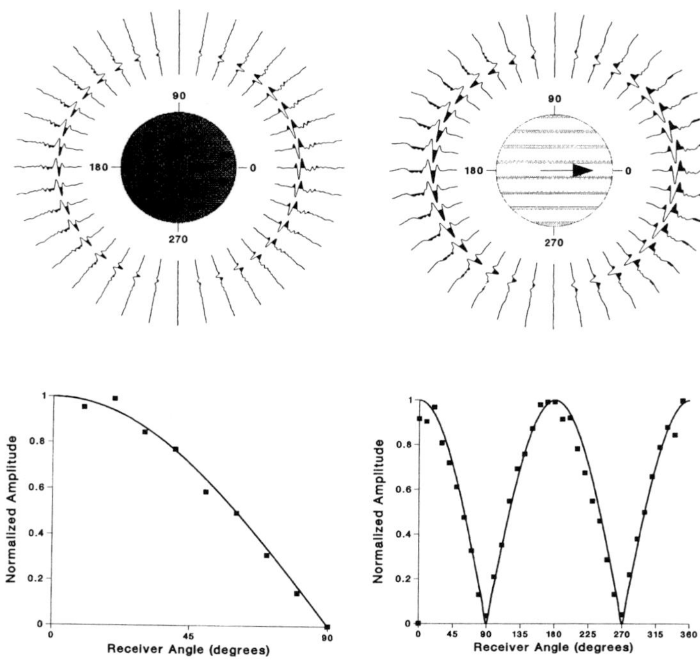

Figure 4.26. Laboratory test showing vector fidelity, after Sondergeld and Rai (1992). The receiver is capable of rotating relative to the material fabric while remaining in the same location. The source polarization remains fixed, and points from right to left in both cases. The left-hand side shows the resultant waveform and amplitude variations for an isotropic specimen, the right-hand side shows the resultant waveform for a shale. The propagation direction in both cases is perpendicular to the page, along the cylinder axis.

An example of vector infidelity is the tests of Ghose and Takahashi (1991). They constructed a downhole VSP experiment in which the lower half of the borehole tool containing the three-component geophones was free to rotate about the vertical (well) axis. The receiver tool was rotated in the borehole relative to the fixed offset and near-offset source positions. They found that the frequency response of the instrument was non-linear in azimuth and could not be predicted adequately by rotation (Figure 4.27). Furthermore, they discovered that the shear-wave polarization changed with the azimuth of the detector, and that the shear-wave anisotropy was affected.

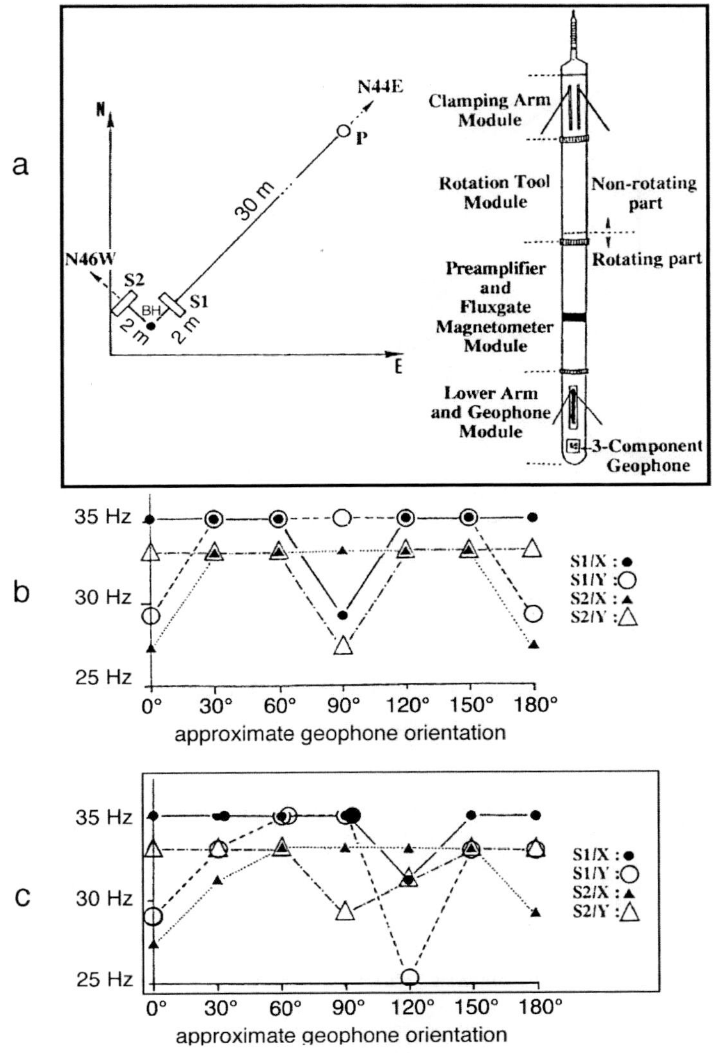

Figure 4.27. Test for vector fidelity using receivers. The lower part of the sensor package is free to rotate.
(a) field layout for the tests.
(b) variation of sensor's frequency response predicted from mathematical rotation using the two component recordings from a single azimuth.
(c) actual frequency response of physically rotated receiver.
Adapted from Ghose and Takahashi (1991).

4.3.3 Sensitivity to anisotropy

The procedures above assume that the first motion detected at the borehole is a P-wave travelling in the vertical plane containing the receiver and the source. For anisotropy within the overburden or target zone, this assumption is no longer justified except in symmetry planes, e.g., in a TIV medium (see Figure 4.28).

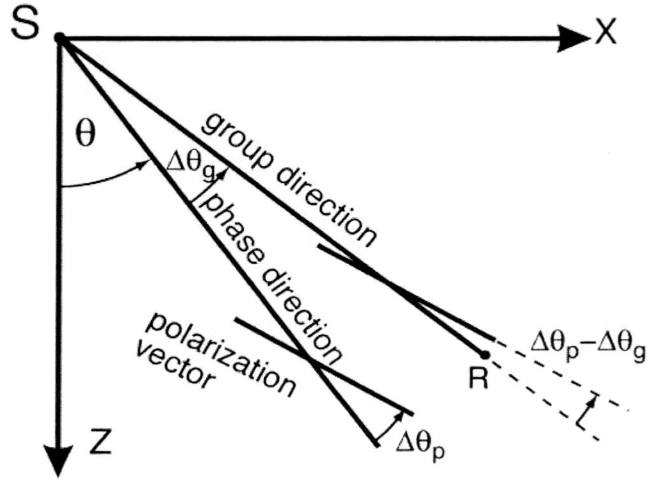

Figure 4.28. Schematic illustration of the deviation of the polarization direction and group (ray) direction from the phase (propagation) direction θ in a TIV medium. The polarization deviates from the phase direction by an angle $+\Delta\theta_p$ and the group direction deviates by an angle $+\Delta\theta_g$. As both are positive this gives a net deviation from the ray path (group) direction of $|\Delta\theta_p|-|\Delta\theta_g|$.

Chapter 3 describes the deviations that occur between the polarization, group (ray) and phase directions. Thus the recorded polarization is not directly indicative of the source-receiver direction. This direction is given instead by the (unknown) group direction. A measure of the error in using the polarization instead can be appreciated by considering a common phase direction related to both the group velocity and polarization deviations (Figure 4.28). Indeed, as both deviations have the same sign, the net deviation of group and polarization directions is relatively small. To examine the magnitude of this effect further, consider an anisotropic TIV medium, where the deviation is confined to the sagittal plane. For a specific phase direction specified by the angle of incidence θ, the qP polarization is given by angle θ_{pol}

$$\tan\theta_{pol} = \tan\theta\left(1+\frac{1}{1-\eta}\left(2\delta+4(\varepsilon-\delta)\sin^2\theta\right)\right), \tag{4.56}$$

and the group angle θ_{grp} is

$$\tan\theta_{grp} = \tan\theta\left(1+2\delta+4(\varepsilon-\delta)\sin^2\theta\right), \tag{4.57}$$

both of which have the same sign for most phase angles (Figure 4.29). Therefore, a maximum deviation of 10° degrees reduces to 2°–3°, an acceptable experimental error. A similar conclusion holds for the vertical symmetry planes for TIH and orthorhombic media. For a TIH medium, each deviation is confined to the (symmetry) plane containing both the direction of propagation and the horizontal symmetry axis. These deviations are given by the same formulae as above, with θ now referring to the angle from the symmetry axis. Thus the horizontal deviations of the group- and polarization directions are the projection of the above predictions. Thus their magnitude is similar and at most equal – to those determined for TIV symmetry.

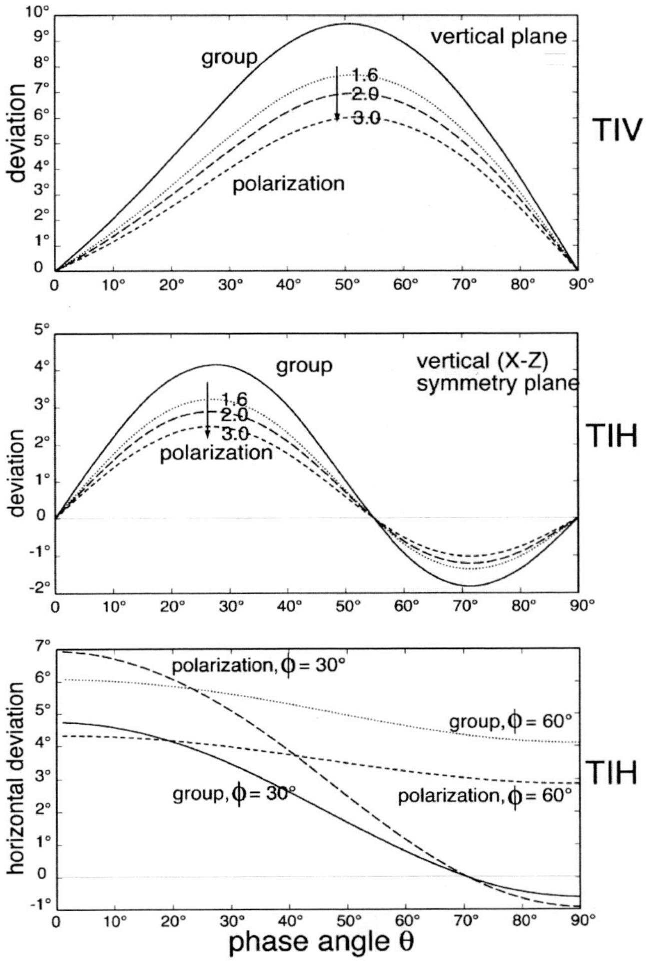

Figure 4.29. Group- and polarization deviations for TIV ($\varepsilon = 0.2$, $\delta = 0.1$) and TIH ($e_n = 0.1$, $e_t = 0.1$) media. For the TIV case – and for the vertical symmetry plane of the TIH – there is no horizontal deviation. The curves are shown for three values of V_P/V_S. For a general azimuth ϕ of propagation, each ray path possesses a deviation component out of the plane of propagation and hence in the horizontal plane. This component is shown here for two azimuths of 30° and 60°.

4.4 DISTORTION AT INTERFACES

Interfaces are unavoidable in all surveys but are bad places to do seismics. As they are the only places equipment can be placed, learning to live with them is a priority. This section relates to a natural wave phenomenon that prevents a true and faithful recording of the three-component signature of each upgoing wave.

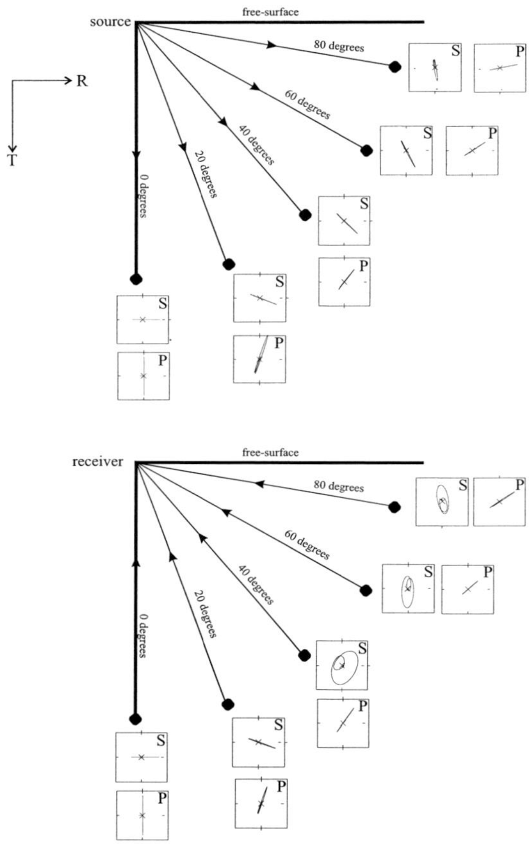

Figure 4.30. Common source and receiver gathers for a distribution of ray paths into an isotropic half-space. Note that the horizontal motion for the shear waves in the reverse VSP case is elliptical.

4.4.1 Free-surface related effects

The VSP with a surface source and downhole receiver is the most suitable configuration for anisotropy analysis. However, multi-component experiments have been shot with reverse VSP geometries (Figure 1.10) in which the receiver is placed on the free-surface and the source downhole. The fidelity of this recording is related to the nature of the boundary conditions at this interface, and is particularly important at large offsets. The term *shear-wave window* has been used to describe the cone of angles centered on the receiver marking the most oblique undistorted ray paths (Figure 4.30). Outside this window, components of the signature are differentially magnified and phase shifted. An added complication is that this regime of distortion also generates extra arrivals that interfere with the incoming signals. These effects have been considered by Nuttli (1961), Meissner (1965) and Evans (1984).

4.4.1.1 *Isotropic half-space*

Consider incident plane monochromatic P-waves from the isotropic solid onto the free surface, with propagation vectors confined to the vertical *X*-*Z* plane. The three-component signature $\mathbf{d}^{PU}(Z)$ of this upgoing P-wave with unit polarization vector at a depth *Z* is given by

$$\mathbf{d}^{PU}(Z) = \begin{pmatrix} d_X^P \\ 0 \\ d_Z^P \end{pmatrix} = \begin{pmatrix} \sin\theta_P \\ 0 \\ -\cos\theta_P \end{pmatrix} \exp\left(i\omega \left(\frac{\sin\theta_P}{V_P} X - \frac{\cos\theta_P}{V_P} Z \right) \right) \tag{4.58}$$

where the factor $\exp(i\omega t)$ has been omitted for clarity. At the free surface, this wave is reflected back as P-wave (giving $\mathbf{d}^{PD}(Z)$ but also is converted upon reflection to an SV waves (giving $\mathbf{d}^{SD}(Z)$). The motion at the free surface is the sum of these three waves. The appropriate reflection coefficients R_{PP} and R_{PS} can be obtained by solving the equation relating incident and reflected waves under the boundary condition of vanishing tangential and normal stress at the free surface. A derivation can be found in Aki and Richards (1981, p137). From these, the motion $\mathbf{d}^P(0)$ recorded exactly at the interface can be deduced

$$\mathbf{d}^P(0) = \mathbf{d}^{PU}(0) + \mathbf{d}^{PD}(0) R_{PP} + \mathbf{d}^{SD}(0) R_{PS} \ , \tag{4.59}$$

which gives, after some algebraic manipulation,

$$\mathbf{d}^P(0) = \begin{pmatrix} \chi_X d_X^P \\ 0 \\ \chi_Z d_Z^P \end{pmatrix} \exp\left(i\omega \frac{\sin\theta_P}{V_P} X \right) , \tag{4.60}$$

where

$$\chi_X = 4\sqrt{\eta} \frac{\cos\theta_P \cos\theta_S}{\Delta} \ , \tag{4.61}$$

$$\chi_Z = \frac{2\left(1 - 2\eta \sin^2\theta_P\right)}{\Delta} \ , \tag{4.62}$$

and

$$\Delta = \left(1 - 2\eta \sin^2\theta_P\right)^2 + \left(4\sqrt{\eta} \cos\theta_P \cos\theta_S\right) \sin^2\theta_S \ . \tag{4.63}$$

Here, $\eta = V_S^2 / V_P^2$, as in the entire book. Snell's law gives

$$\frac{\sin\theta_S}{V_S} = \frac{\sin\theta_P}{V_P} \ . \tag{4.64}$$

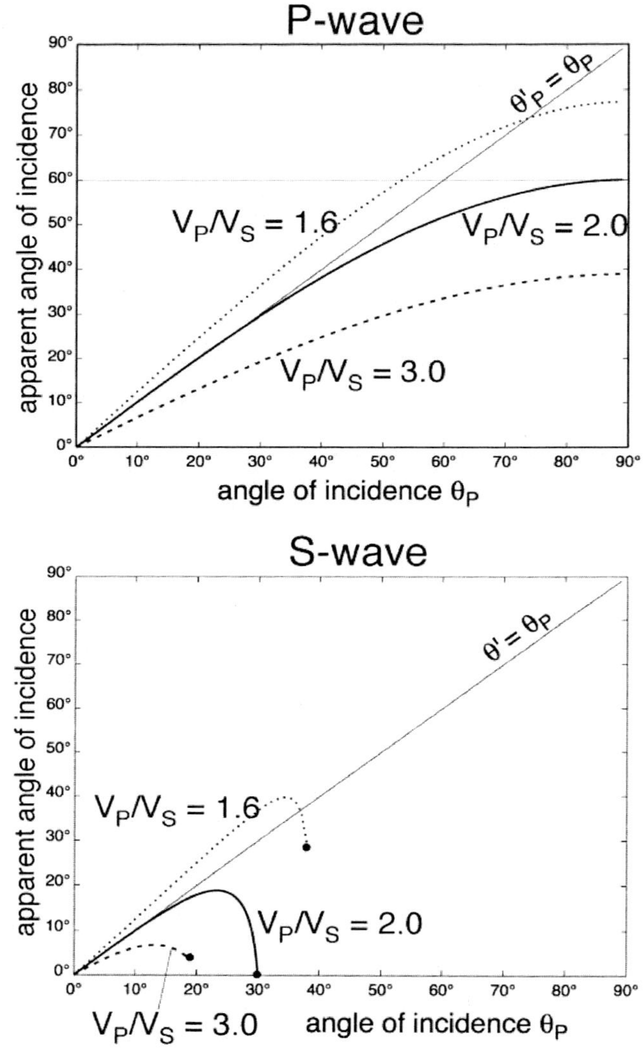

Figure 4.31. Apparent angles of incidence calculated from recordings of P- and S waves on a free surface. The results for the S waves terminate at the critical angle.

The mixing of the two P and shear waves by interaction at the free surface changes the polarization structure and overall amplitude of the recorded P-waves. Indeed, it should be noted that even at normal incidence ($\theta_P = 0$) the displacement amplitude doubles due to the summation of incident and reflected amplitudes. The amplitude approximately retains this magnitude for a wide range of incidence, decreasing slightly with increasing V_P/V_S. The recorded polarization is

$$\theta_P' = \tan^{-1}\left(\frac{\chi_X}{\chi_Z}\tan\theta_P\right) . \tag{4.65}$$

Figure 4.31 shows that this has a dramatic impact at all offsets, reaching several tens of degrees at larger offsets. Higher values of V_P/V_S result in a stronger effect and tilt the polarization more strongly towards the vertical. At offsets $\theta_P > 60°$, for which the Z-component is reinforced by P-S conversion (high value of R_{PS}), the rate of increase in divergence is enhanced (Evans 1984). There is a fixed limiting value of

$$\theta' = \tan^{-1}\left(\frac{2\sqrt{\eta(1-\eta)}}{1-2\eta}\right) \approx 60° \ , \tag{4.66}$$

with the approximation holding for $V_P/V_S = 2$. This effect must therefore be considered when trying to assess ray path directions for anisotropy analysis.

For an incident SV-wave

$$\mathbf{d}^{SU}(Z) = \begin{pmatrix} d_X^S \\ 0 \\ d_Z^S \end{pmatrix} = \begin{pmatrix} \cos\theta_S \\ 0 \\ \sin\theta_S \end{pmatrix} \exp\left(i\omega\left(\frac{\sin\theta_S}{V_S}X - \frac{\cos\theta_S}{V_S}Z\right)\right) , \tag{4.67}$$

with the reflections and conversions giving the total motion

$$\mathbf{d}^S(Z) = \mathbf{d}^{SU}(0) + \mathbf{d}^{PD}(0)R_{SP} + \mathbf{d}^{SD}(0)R_{SS} \ , \tag{4.68}$$

which, after calculation of the reflection coefficients, yields the interface motion

$$\mathbf{d}^S(0) = \begin{pmatrix} \chi_Z d_X^S \\ 0 \\ \chi_X d_Z^S \end{pmatrix} , \tag{4.69}$$

where χ_X and χ_Z are defined as above. For nearly normal incidence, there is again an overall doubling of the wave amplitudes, and for "sub-vertical" angles (between 0° and 30°) this scaling factor is approximately constant. As in the P-wave case the sensors record an apparent angle of incidence

$$\theta_S' = \tan^{-1}\left(\frac{\chi_X}{\chi_Z}\tan\theta_S\right) . \tag{4.70}$$

The variation is shown in Figure 4.31. The major difference in the S wave behavior is that with increasing offset, the P waves converted at the free-surface approach horizontal propagation relatively easily. The critical point occurs for

$$\theta_S^C = \sin^{-1}\left(\frac{V_S}{V_P}\right) . \tag{4.71}$$

Critical angles are reached at moderate offsets, as the table below shows.

Table 2. Critical offsets X_C for straight ray path and receiver at 2000 m

V_P/V_S	V_P	V_S	θ_S^C	X_C
1.60	2.5 km/s	1.56 km/s	38.68°	1601 m
1.60	3.0 km/s	1.88 km/s	38.68°	1601 m
1.60	3.5 km/s	2.18 km/s	38.68°	1601 m
2.00	2.5 km/s	1.25 km/s	30.00°	1155 m
2.00	3.0 km/s	1.50 km/s	30.00°	1155 m
2.00	3.5 km/s	1.75 km/s	30.00°	1155 m
3.00	2.5 km/s	0.83 km/s	19.47°	707 m
3.00	3.0 km/s	1.00 km/s	19.47°	707 m
3.00	3.5 km/s	1.16 km/s	19.47°	707 m

Beyond this, the cos θ_P-term in (4.54) and (4.56) becomes imaginary as

$$\eta \sin^2\theta > 1 \rightarrow \cos\theta_P = \sqrt{1-\eta\sin^2\theta_S} = i\sqrt{\eta\sin^2\theta_S - 1} \ , \tag{4.72}$$

while sin θ_P remains connected to η and sin θ_s by Snell's law. Shear waves incident at the critical angle are greatly magnified and the recorded surface motion is purely radial (due to a large R_{SP}). Beyond the critical angle, the converted P-wave is inhomogeneous (no propagation in the Z direction) and trapped at the surface – the functions $\Delta\chi_Z$ and Δ become complex. This results in a constant, frequency independent, phase shift for both components of motion

$$\mathbf{d}^S(0) = \begin{pmatrix} |\chi_Z| e^{i\phi_X} d_X^S \\ 0 \\ |\chi_X| e^{i\phi_Z} d_Z^S \end{pmatrix} . \tag{4.73}$$

The phase shifts ϕ_X and ϕ_Z produce an elliptical particle motion in the sagittal plane, where the ellipse has the same geometry at all wavelet frequencies. It can be shown that (Born and Wolf 1965) the major axis is aligned along the direction θ_E related to the incident polarization by

$$\theta_E = \frac{1}{2}\tan^{-1}\left(\tan 2\theta_S' \cos\Delta\phi\right) , \tag{4.74}$$

where $\Delta\phi = \phi_Z - \phi_X$. Interestingly, this relative phase is always equal $\pi/2$ to due to the nature of the formulation, and hence the ellipse is always oriented along the X and Z axes. The ellipticity e of this motion defined by the ratio of minor and major axis is

$$e = \tan\theta_E' \ , \tag{4.75}$$

where

$$\theta'_E = \frac{1}{2}\sin^{-1}\left(\sin 2\theta'_S \sin \Delta\phi\right) \ . \tag{4.76}$$

The ellipticity depends on the angle of incidence (Figure 4.32).

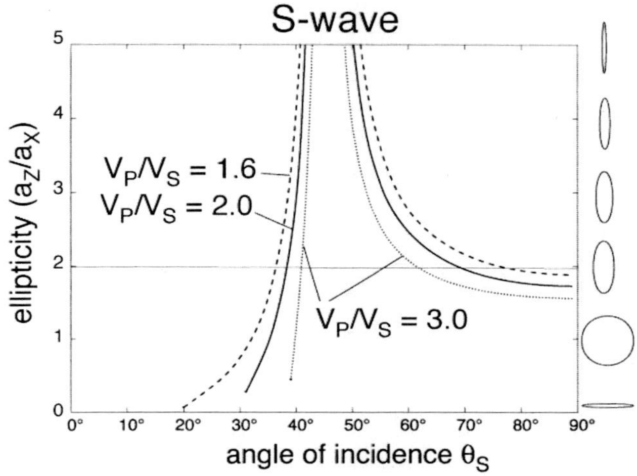

Figure 4.32. Beyond the critical angle the particle motion of the shear waves describes an ellipse in the sagittal plane. The ellipticity varies with angle of incidence, but tends to a constant value of 2 for oblique ray paths.

In conclusion, for pre-critical angles the polarization direction is preserved, while for post-critical angles the shear waves are rendered elliptical. There are, in addition, a number of further observations regarding this phenomenon:

(1) For incident SH waves there is no conversion upon reflection, and thus the reflected wave is of the same *type*, *amplitude* and *phase*. The recorded motion is twice the incident amplitude.

(2) For a general shear wave polarization (depending on the source), the motion may be decomposed into the SV and SH components and each treated separately, with the results again combined.

(3) Figure 4.32 highlights a specific range of post-critical angles at which the vertical component is almost exactly in phase and twice the incident (normal incidence) radial component. In this region the absolute phase of χ_Z is close to zero. This means that measurement of the vertical and transverse components at this range of angles is analogous to recording in the radial and transverse plane (Tsvankin and Chesnokov 1990).

(4) For curved wave fronts, each contributing plane wave is affected, but by different degrees. The result is a smoothing of the response, but the overall trend remains the same, with the sharpness of the response depending on frequency. In addition, however, shear waves incident beyond the critical angle excite a local headwave (SP) (related to the plane P wave trapped at the surface) in addition to surface waves. The latter occur re-

gardless of whether the ray paths are inside or outside the shear-wave window. The former is predominantly radially polarized SP (Gilbert and Knopoff 1961) and propagates along the free surface with a velocity V_P to arrive shortly before the direct shear-waves. The energy of the headwave is radiated into the medium as an SV wave, so that the amplitude decays rapidly as $1/r^2$ in the far field. Despite this, in the vicinity of the critical angle, the SP radial amplitude is so much larger than that of the SV that it can appear as a precursor to S. In circumstances where the radial component of SV is itself diminished, this could be misleading.

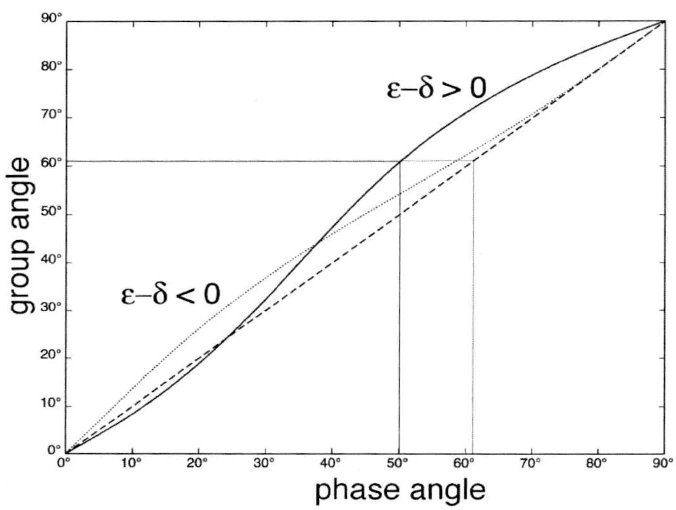

Figure 4.33. The deviation of the group direction from the phase direction means that source-receiver ray paths under-estimate the actual phase angle for media of both positive and negative anellipticity. This means that the critical angle phenomena occur at smaller offsets.

4.4.1.2 Anisotropic half-space

For an anisotropic medium the main results of the above analysis still roughly hold. Thus there is a critical angle marking the onset of the shear wave phase distortion and polarization and phase distortions. For TIV anisotropy every vertical plane is a plane of symmetry, and the qP and qSV propagation decouples for the SH wave polarized normal to the plane and may be treated as in the isotropic SH case (Chapter 3). For the incident qSV wave, the motion at the interface is governed by the vertical velocities α_v and β_v as well as the Thomsen parameters ε and δ (Chapter 2). The radial and vertical motions at the interface may be determined in a way similar to the isotropic case, by firstly calculating the reflection and conversion coefficients, and then summing the result. TIV anisotropy introduces two categories of effect:

(a) A small shift in the critical phase angle but general maintenance in the basic functional form of equations (4.54 – 4.56) and (4.62). Thus, for an incident qSV wave the new critical phase angle for weak anisotropy is

$$\theta_S' \approx \tan^{-1}\left(\frac{\beta_v}{\alpha_v}\left(1+2(1-\eta)(\varepsilon-\delta)-\varepsilon\right)\right) . \tag{4.77}$$

(b) To determine the appropriate free-surface effect, the phase angle corresponding to the group angle calculated along the source–receiver direction must be calculated. Thus, the deviation of the group (ray) angle from the phase angle is important as this contracts (and occasionally stretches) the behavior of the functions along the phase angle axis. The deviation of group angle from phase angle depends upon $\varepsilon - \delta$ and α_v/β_v (see equation 4.23). For example, a positive value of $\varepsilon - \delta$ (most shales) produces amplitude and phase behavior attributed to smaller phase angles (Figure 4.33), leading to the overall effects occurring at an earlier angle than in isotropic conditions.

4.4.2 The effect of internal interfaces

A subsurface interface may also contribute to phase distortion. To see this, consider a unit amplitude shear-wave with horizontal polarization ϕ incident upon a welded interface between two isotropic media at an angle θ_S,

$$\mathbf{d}^S = a_S\begin{pmatrix}\cos\phi\sin\theta_S\\ \sin\phi\sin\theta_S\\ \cos\theta_S\end{pmatrix} . \tag{4.78}$$

This general polarization can be decomposed into an SV component

$$\mathbf{d}^{SV} = a_S\begin{pmatrix}\cos\phi\sin\theta_S\\ 0\\ \cos\theta_S\end{pmatrix} , \tag{4.79}$$

and an SH component

$$\mathbf{d}^{SH} = a_S\begin{pmatrix}0\\ \sin\phi\sin\theta_S\\ 0\end{pmatrix} . \tag{4.80}$$

Each of these components is transmitted through the interface, with coefficients T_{SV} and T_{SH}, respectively. Expressions for the corresponding SV–SV and SH–SH transmission coefficients can be found in textbooks such as Aki and Richards (1981), Brekhovskikh (1960), and Cerveny and Ravindra (1971). Beyond normal incidence, generally $T_{SV} \neq T_{SH}$, so that

$$\mathbf{d}^S = T_{SV}d^{SV} + T_{SH}d^{SH} . \tag{4.81}$$

This recombination leads to a new horizontal polarization direction

$$\phi' = \tan^{-1}\left(\frac{|T_{SH}|}{|T_{SV}|}\tan\phi\right) , \tag{4.82}$$

and a modification in the wave amplitude. The transmission coefficient T_{SV} depends upon the angle of incidence, in addition to the elastic properties on either side of the interface.

With increasing angles of incidence the corresponding critical points are reached and exceeded. These correspond to horizontally propagating reflected and transmitted ray paths. For a "low-to-high" contrast between two media A and B, the critical angles can be grouped in descending order

$$\sin^{-1}\left(\frac{V_S^A}{V_P^A}\right) > \sin^{-1}\left(\frac{V_S^A}{V_P^B}\right) > \sin^{-1}\left(\frac{V_S^A}{V_S^B}\right) . \tag{4.83}$$

For a "high-to-low" contrast the order of the three critical angles is reversed. If any of the critical angles is exceeded, an imaginary component in the transmission coefficients and a phase shift result that produce elliptical motion. There is no conversion for the SH waves.

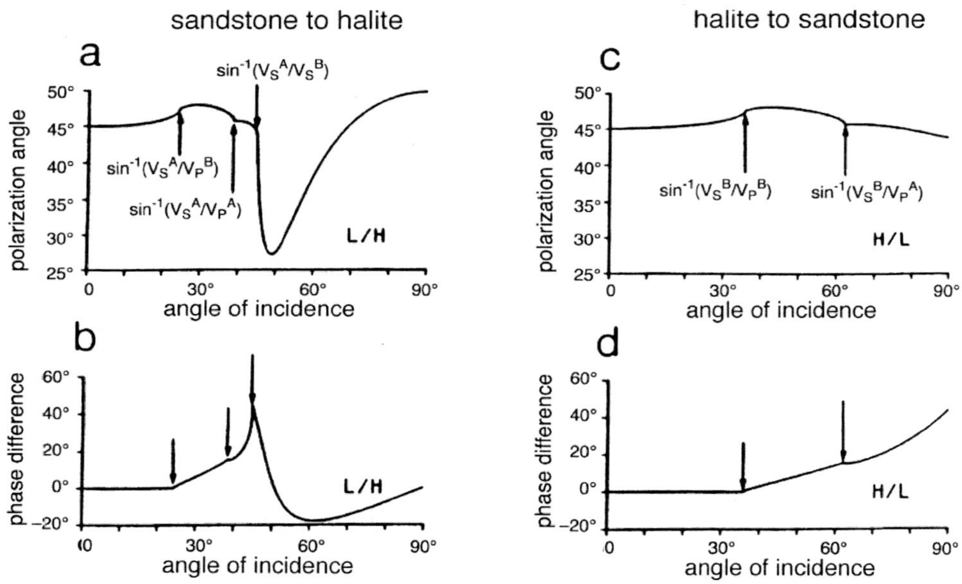

Figure 4.34. Polarization angles and differential phase rotation for horizontal shear wave components, adapted from Liu and Crampin (1990). The incident polarization is 45°. The first critical angle for the low/high contrast occurs at 24°, and the second is at 38°. Beyond the third critical angle all waves are refracted. The maximum error induced in the shear waves can reach 20° in phase and a few degrees in polarization at oblique ray paths with an angle of incidence of 30°.

The phase and amplitude distortion for SV was calculated for a sandstone-halite interface by Douma and Helbig (1987), and a halite-sandstone interface by Liu and Crampin (1990). Results for both are shown in Figure 4.34. These figures show that in both cases beyond the first critical angle the deviation of the horizontal polarization is generally less than 2°–3°, and the deviation of the phase is about 20°. For a low-to-high contrast, a second critical angle marks the point where both reflected and transmitted P-waves are totally refracted and only shear waves are transmitted. This is the region of strongest elliptical motion. When all of the critical points are exceeded, the waves become totally refracted, and all are inhomogeneous. For a wave incident upon an interbedded sequence, successive small phase shifts could accumulate to significant amounts.

4.4.3 Borehole related effects

Intuitively, the interaction with the borehole is stronger when the incident wavelength λ is comparable to – or less than – the borehole radius b, and usually rather weak for low-frequency VSPs. Such phenomena are regarded as appropriate to the kHz frequencies of cross-well experiments. They are important for uncased holes and hydrophone recordings. A practical example of borehole effects can be seen in Figure 4.35. Schoenberg (1986) gave an explicit expression for the velocity field recorded in the solid formation next to the borehole as a power series in frequency ω. The results are given in cylindrical coordinates to first order in frequency and include all axially symmetric terms. Lee (1987) gives an alternative formulation using spherical coordinates. A rough order-of-magnitude estimate of borehole coupling is obtained by considering only solutions for a vertical borehole and vertical incidence (relevant to a near-offset VSP). For an incident P-wave, the radial, tangential, and axial components of the velocity fields recorded by the wall-mounted geophone – respectively, d_r, d_ϕ, and d_z – are

$$\mathbf{d}_P = \begin{pmatrix} d_r \\ d_\phi \\ d_z \end{pmatrix} = \begin{pmatrix} i\Omega_L \chi_b \\ 0 \\ 1 \end{pmatrix}, \tag{4.84}$$

where

$$\chi_b = c_T^2 \frac{1-2\eta}{2\eta}\left(1-\frac{\alpha_f^2}{V_P^2}\right)\left(1-\frac{c_T^2}{V_P^2}\right)^{-1} \tag{4.85}$$

and $\Omega_L = \omega b / V_P$.

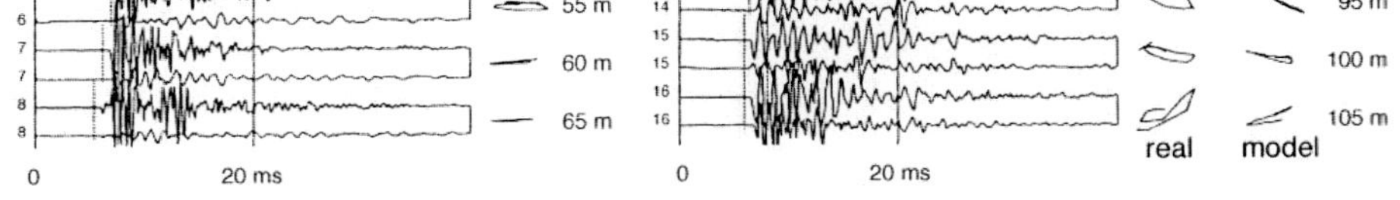

Figure 4.35. Two receiver gathers from a cross-hole experiment at the Imperial College borehole test site facility (Worthington et al. 1992) show the effects of the borehole at shallow depths. The sensor package is a 3C accelerometer allowing radial and axial motions of the P-waves to be displayed. The model result is the prediction without the borehole. The dominant frequency in the experiment is 1 kHz..

The properties of the formation are ρ, V_P and V_S, the compressional velocity for the fluid is α_f. c_T stands for the tube-wave velocity; for an uncased hole and isotropic formation it is given by

$$c_T = \alpha_f \sqrt{\frac{\rho V_S^2}{\rho V_S^2 + \rho_f \alpha_f^2}} \; . \tag{4.86}$$

For an incident SV wave at an azimuth ϕ from the X-axis, the displacement is

$$\mathbf{d}_{SV} = \begin{pmatrix} \cos\phi \\ -\sin\phi \\ i\Omega\cos\phi \end{pmatrix}, \tag{4.86}$$

with $\Omega = \omega b/V_S$. Finally, for the SH waves the displacement is

$$\mathbf{d}_{SH} = \begin{pmatrix} \sin\phi \\ \cos\phi \\ i\Omega\sin\phi \end{pmatrix}. \tag{4.87}$$

The complex quantities introduce ellipticity into the recorded motion, in the same way as in the previous sections. Incidence at other angles yields terms of similar magnitude, but with further variations with ϕ and the angle of incidence θ. Unlike the P-wave polarization, the shear wave polarization in the horizontal plane is – to first order in frequency – not affected,. For V_S = 1000 m/s, a borehole radius of 20 cm, and frequencies of 40 Hz, Ω becomes 0.01. Thus it is too small to have any effect on the recorded polarization or on estimates of the angle of incidence. Lee (1987) states that the effect is not important for $\lambda > 25\ b$. In fact, frequencies of 1 kHz–2 kHz are required to produce 10° errors in the polarization or angle of incidence. A comprehensive study of this effect by Peng et al. (1993) further showed that the strongest borehole coupling occurs for shear waves incident at normal or grazing incidence, and P-waves at 45° incidence. The effect is also stronger when the wave field is incident on the sensor side of the borehole. Motion recorded in the shadow side of the borehole decreases by a factor of 5 (Lee 1987). An increase by a factor of 2 is predicted for an empty borehole (in direct analogy with the free-surface effect – the scattered energy at the wall of the fluid-filled borehole is less than that at the wall of an empty hole). An important conclusion of this study is that for shear waves in soft formations, non-negligible effects are still expected at frequencies of 300 Hz.

CHAPTER 5 – ANISOTROPY ANALYSIS FROM NEAR-OFFSET VSP I – SYMMETRY AND UNIFORMITY

Careful estimation and interpretation of seismic anisotropy require that the procedures outlined in Chapter 4 are implemented. After correction of the acquisition elements, the resultant data matrix $\mathbf{D}^{(9)}(t)$ at each depth should now be close to the true multi-component medium response $\hat{\mathbf{M}}^{(9)}(t)$

$$\hat{\mathbf{M}}^{(9)}(t) = \begin{pmatrix} \hat{m}_{XX}(t) & \hat{m}_{XY}(t) & \hat{m}_{XZ}(t) \\ \hat{m}_{YX}(t) & \hat{m}_{YY}(t) & \hat{m}_{YZ}(t) \\ \hat{m}_{ZX}(t) & \hat{m}_{ZY}(t) & \hat{m}_{ZZ}(t) \end{pmatrix}. \tag{5.1}$$

For near-offset VSPs there is a natural separation into the P-waves (the vertical component $\hat{m}_{ZZ}(t)$) and the shear waves (the horizontal components $\hat{m}_{XX}(t)$, $\hat{m}_{XY}(t)$, $\hat{m}_{YX}(t)$, and $\hat{m}_{YY}(t)$). It appears that for most practical cases the components $\hat{m}_{XZ}(t)$, $\hat{m}_{YZ}(t)$, $\hat{m}_{ZX}(t)$, and $\hat{m}_{ZY}(t)$ are negligible (see for example, Figures 1.8 and 4.9). This is an indication that at most locations the sub-surface is isotropic, or can be approximated by TIH-, TIV- or orthorhombic symmetry (Chapter 3). Any vertical birefringence excludes isotropy and TIV. Therefore, anisotropy estimations using shear waves commonly focus on the four-component – or horizontal – subset

$$\hat{\mathbf{M}}^{(4)}(t) = \begin{pmatrix} \hat{m}_{XX}(t) & \hat{m}_{XY}(t) \\ \hat{m}_{YX}(t) & \hat{m}_{YY}(t) \end{pmatrix}. \tag{5.2}$$

This simplification should be applied with caution and requires prior inspection of the nine-component matrix. Thus, while analysis of the four-component subset provides a fairly robust and stable estimate of the major anisotropy effects, there have been data sets with indications of other symmetries or phenomena (see, for example, the deeper recordings in Figure 4.9).

For the purpose of this chapter, vertical fractures are assumed to be the direct cause of shear wave anisotropy. Thus, the result of analysis is the orientation of the horizontal symmetry axis (and hence fracture strike), and the velocities that yield c_{44} and c_{66} (and hence the fracture index e_t). For most practical purposes, the fraction of vertical birefringence is identical to e_t. No estimate of e_n is possible.

It is a strong point of near-offset VSP surveys that e_t can be optimally estimated, but a weakness that the estimates are not in themselves sufficient to distinguish it from mechanisms *not* related to fractures. The techniques discussed here must therefore be used together with supporting information from cores and appropriate fracture-sensitive logs (an example of this is shown in Section 5.2). One final assumption made in this chapter is that the orientation of the fractures varies very little with depth, and the medium is close to the idealization presented in Section 4.1. With this condition of uniformity, equation (4.13) is applicable.

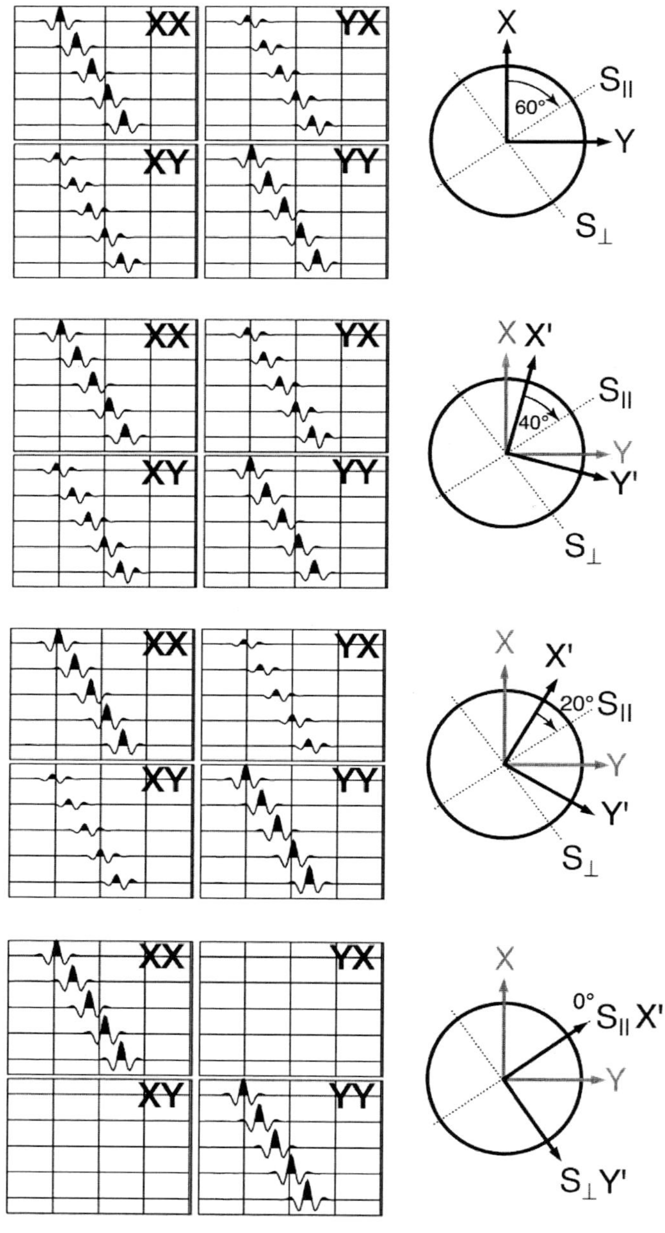

Figure 5.1 Illustration of the effect of simultaneous source-receiver rotation on the four-component data matrix. Both source and receiver axes are aligned, initially along the *X-Y* coordinate frame. Successive rotations maintain their alignment, but move their new *X'-Y'* axes closer to the principal anisotropy axes – at which point the energy on the cross-diagonals finally reduces to zero.

5.1 ESTIMATION OF SHEAR-WAVE ANISOTROPY

The assumptions of the foregoing section, together with the vector fidelity discussed in Chapter 4, allow the separation of fast ($S_\perp$) and slow ($S_\parallel$) shear waves by with the help of (4.13). This is achieved by applying the similarity transformation

$$\mathbf{M}'(t) = \mathbf{C}_Z^T(\phi)\hat{\mathbf{M}}(t)\mathbf{C}_Z(\phi) \ . \tag{5.3}$$

with the objective of diagonalization, giving an estimate of the wave field propagator convolved with the source wavelet ($\Lambda(t)*s(t)$). This matrix multiplication implies a clockwise rotation about the Z-axis by the angle ϕ. The operation is equivalent to a simultaneous rotation of the source and receiver components in which the axes remain aligned. The data are used to predict what a data set would look like had it been acquired with a different orientation of source and receiver with respect to the anisotropy. The effect of multi-component rotations on synthetic data is shown Figure 5.1 and on laboratory data in Figure 5.2. When the axes are aligned with the $S_\perp$ or $S_\parallel$ polarization, each shear wave is excited and recorded separately by the correspondingly aligned source-receiver pair, thus producing the requisite diagonal matrix. A simple, but effective, wave separation has thus been achieved. This rotation procedure is commonly known as the *Alford* rotation (Alford 1986).

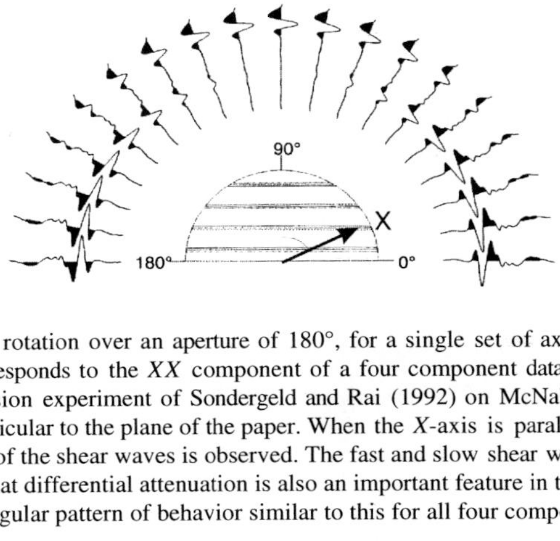

Figure 5.2. Source-receiver rotation over an aperture of 180°, for a single set of axes aligned along a rotating X-axis. This corresponds to the XX component of a four component data matrix. The data are from the pulse transmission experiment of Sondergeld and Rai (1992) on McNabb shale. Wave propagation is always perpendicular to the plane of the paper. When the X-axis is parallel or perpendicular to the bedding, only one of the shear waves is observed. The fast and slow shear waves have different amplitudes, an indication that differential attenuation is also an important feature in this shale. The Alford technique relies upon a regular pattern of behavior similar to this for all four components in the data matrix.

In application to field data (Figure 5.3), the quality of the result depends the noise level and the degree of conformance to the underlying assumptions. One indicator that often offers an immediate assessment of the final product is symmetry of the data matrix; this symmetry is necessary (but not sufficient) for successful diagonalization of $\hat{\mathbf{M}}(t)$. Alford rotation is usually applied to a set of windowed traces at each depth level j, and an estimate of the desired rotation angle f is then obtained based upon a suitable optimization criterion which defines the closeness of fit to the diagonal form within the prescribed time-window. When this optimum is found, $\hat{\phi}$ gives a direct estimate of the $S_\perp$ and $S_\parallel$ polarization directions relative to the global coordinate axes X-Y and hence the geographical axes. Inevitably, there is a multi-variant choice of optimization schemes, with the best choice largely dependent on the situation.

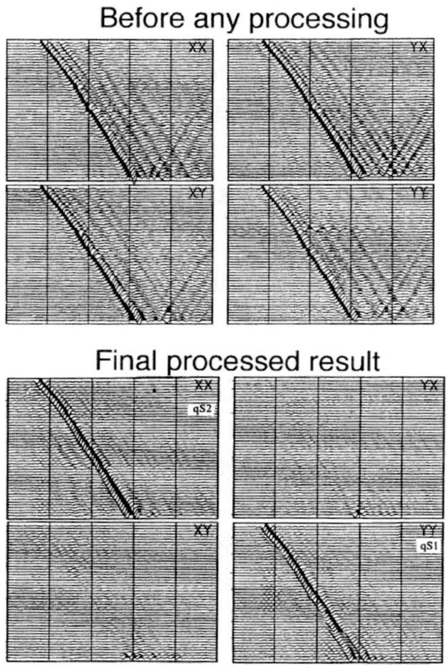

Figure 5.3. Four component subset of the data matrix described in Figure 4.6, before and after processing (tool re-orientation, near-surface correction, *f-k* filtering) to obtain the downgoing wave field, and finally an Alford rotation. Note the final form has almost no cross-diagonal energy, in agreement with theoretical predictions (the exception is some energy near the target). From MacBeth et al. 1997.

5.1.1 Numerical scan

The most direct approach is to scan a range of discrete angles ϕ ranging from 0° to 90°, and to judge the result by evaluating specific criteria. The search need not extend beyond this angular range owing to the inherent symmetry of the operation in (5.3), since for any acute angle ϕ

$$m'_{XX}(\phi+270°)=m'_{YY}(\phi+180°)=m'_{XX}(\phi+90°)=m'_{YY}(\phi)\ , \tag{5.4}$$

and

$$-m'_{YX}(\phi+270°)=m'_{XY}(\phi+180°)=-m'_{YX}(\phi+90°)=m'_{XY}(\phi)\ . \tag{5.5}$$

The following criteria have been used to evaluate the closeness to a diagonal trace matrix, for arrivals in a window of N samples width. They are plotted in Figure 5.4 for comparison.

(a) *Minimum off-diagonal energy*

$$\min\left(E_{OD}(\phi)\right)=\min\left(\sum_{i=1}^{N} m'^2_{XY}(\phi,t_i)+m'^2_{YX}(\phi,t_i)\right)\ . \tag{5.6}$$

(b) *Maximum similarity* between the predicted $S_\perp$ and $S_\parallel$ waveforms on the main diagonals, assessed by the largest maximum for the cross-correlation function

$$A_{XY}^{\max} = \max_{\phi}\left(\max_{\tau}\left(\sum_{i=1}^{N} m'_{XX}(\phi,t_i+\tau)m'_{YY}(\phi,t_i)\right)\right) . \tag{5.7}$$

Other variants on this theme have included the largest time-delay (determined by cross-correlation) between the shear waves, or the maximum symmetry of the measured correlation function (Peron 1989)).

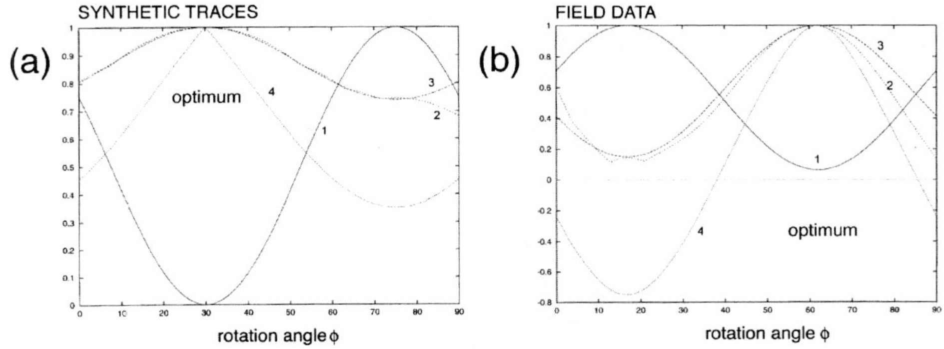

Figure 5.4. Azimuthal variation of four criteria that indicate the optimal rotation angle in the numerical scanning form of the Alford rotation. They are applied to:
(a) an ideal set of synthetic traces;
(b) field data from the Lost Hills field, California.
The criteria are:
1 - minimal cross-diagonal energy;
2 - largest maximum cross-correlation;
3 - maximum major-diagonal energy;
4 - simultaneous maximization using the $L_{1.63}$ norm.. This appears to provide the sharpest response.

(c) *Maximum of the total major-diagonal energy*

$$\max\left(E_{MD}(\phi)\right) = \max\left(\sum_{i=1}^{N} m'^2_{XX}(\phi,t_i) + m'^2_{YY}(\phi,t_i)\right) . \tag{5.8}$$

(d) *Simultaneous maximization of the major diagonal and minimization of the off-diagonal components* by minimizing the combination E_{MD}/E_{OD} or maximizing $E_{MD} - E_{OD}$.

(e) A criterion similar to (d) can be satisfied by *eigenanalysis of the covariance matrix*

$$\mathbf{A} = \sum_{i=1}^{N} \hat{\mathbf{M}}^T(t_i)\hat{\mathbf{M}}(t_i) . \tag{5.9}$$

The principal eigenvector of **A** gives the desired polarization direction, and the eigenvalues the relative energies of the fast and slow shear waves. Standard numerical routines are satisfactory for this, but provide unique answers only if the shear waves have distinctly different amplitudes.

(f) An alternative to criteria based on energy are those based on *vector space methods*, and in particular the L_p norm (Menke 1984). This may be defined for each matrix component $X = J$, $Y = K$ as

$$\left\| m_{JK}(\phi) \right\|_p = \left(\sum_{i=1}^{N} \left| m_{JK}(\phi, t_i) \right|^p \right)^{1/p} . \tag{5.10}$$

The norm determines the balance between contributions from small and large amplitudes in the signal. The sum of such norms constructed for the off-diagonal or major diagonal components can be used to determine criteria similar to (a) to (c) above. In particular, $E_{OD} - E_{MD}$ in (c) can be implemented by maximizing the difference

$$P_{SM}(\phi) = \left\| m'_{XX}(\phi, t_i) \right\|_p + \left\| m'_{YY}(\phi, t_i) \right\|_p - \left\| m'_{XY}(\phi, t_i) \right\|_p + \left\| m'_{YX}(\phi, t_i) \right\|_p . \tag{5.11}$$

This technique has been advocated by Zhang (1995), who found that an $L_{1.63}$ norm gave optimum results for surface seismic data.

5.1.2 Algebraic solution

To help speed up the process of optimization, it is possible to obtain a solution to the minimum total energy criterion in (a) above by direct calculation of the matrix product (5.3) (Zeng and MacBeth 1993b). With some manipulation, this compound matrix can be simplified to

$$\mathbf{M}'(t) = \begin{pmatrix} a(t) + \bar{a}(t)\cos 2\phi + b(t)\sin 2\phi & -\bar{a}(t)\sin 2\phi + b(t)\cos 2\phi + \bar{b}(t) \\ -\bar{a}(t)\sin 2\phi + b(t)\cos 2\phi - \bar{b}(t) & a(t) - \bar{a}(t)\cos 2\phi - b(t)\sin 2\phi \end{pmatrix} , \tag{5.12}$$

where

$$a(t) = \hat{m}_{XX}(t) + \hat{m}_{YY}(t) , \tag{5.13}$$

$$\bar{a}(t) = \hat{m}_{XX}(t) - \hat{m}_{YY}(t) , \tag{5.14}$$

$$b(t) = \hat{m}_{XY}(t) + \hat{m}_{YX}(t) , \tag{5.15}$$

and

$$\bar{b}(t) = \hat{m}_{XY}(t) - \hat{m}_{YX}(t) . \tag{5.16}$$

Note that $\mathbf{M}'(t)$ is not symmetric unless $\bar{b} = 0$ for all t, which in turn requires $\hat{\mathbf{M}}(t)$ to be symmetric. The least-squares solution criterion maximizes

$$E_{OD}(\phi) = \sum_{i=1}^{N} \left(-\bar{a}_i \sin 2\phi + b_i \cos 2\phi + \bar{b}_i \right)^2 + \sum_{i=1}^{N} \left(-\bar{a}_i \sin 2\phi + b_i \cos 2\phi - \bar{b}_i \right)^2 . \tag{5.17}$$

A stationary point in $E_{OD}(\phi)$ occurs when

$$\sum_{i=1}^{N} \left(\left(\bar{a}_i^2 - b_i^2 \right) \sin 4\phi + 2\bar{a}_i b_i \cos 4\phi \right) = 0 , \tag{5.18}$$

where $a_i = a(t_i)$, with similar definitions for the other terms. If solved directly, the equation leads to four equations and three unknowns, which can be solved to give an estimate of the rotation angle

$$\hat{\phi}_k = \tfrac{1}{4}\tan^{-1}\left(\frac{2\sum_{i=1}^{N}\overline{a}_i b_i}{\sum_{i=1}^{N}\left(\overline{a}_i^2 - b_i^2\right)}\right) + k\frac{\pi}{4} \ , \tag{5.19}$$

where $k = 0, 1, 2, 3$. There are four solutions in the range 0° to 180°, distributed at regular intervals of 45° (Figure 5.5). The choice of one of the four solutions as the minimum point depends upon the sign of the second derivative

$$E''\left(\hat{\phi}_j\right) = 16\frac{\sum_{i=1}^{N}\left(\left(\overline{a}_i^2 - b_i^2\right)^2 + = \left(2\overline{a}_i b_i\right)^2\right)}{\sum_{i=1}^{N}\left(\overline{a}_i^2 - b_i^2\right)}\cos 4\phi_k \ . \tag{5.20}$$

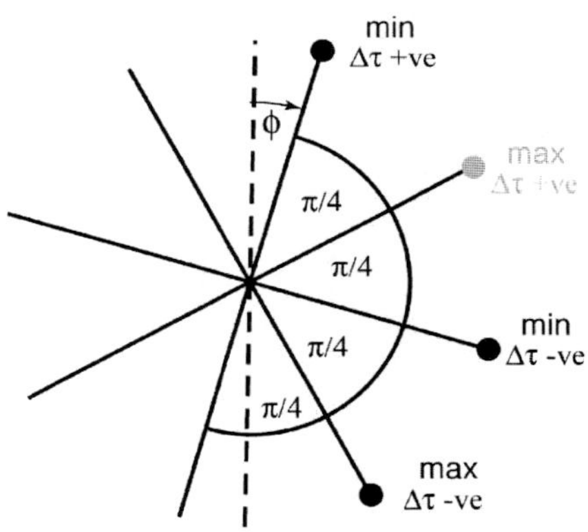

Figure 5.5. There are four possible solutions of equation (5.17), spaced 45° apart. They represent two minima and two maxima of the off-diagonal energy, these being distinguished by the sign of the second derivative. Out of the two candidates, only one leads to a positive lag in the cross-correlation function.

Solutions $\hat{\phi}_k$ must be selected such that $\cos(4\phi_k)$ and $\sum_{i=1}^{N}\left(\overline{a}_i^2 - b_i^2\right)$ have different signs. It may be shown that only one solution with a positive correlation lag can satisfy this criterion, and must be chosen for final rotation. It is interesting to note that, in principle, it is still possible to obtain a solution to (5.17) by setting the two receiver components from one source equal to zero (effectively giving a two-component data set). This yields the solution

$$\hat{\phi}_k = \tfrac{1}{4}\tan^{-1}\left(\frac{2\sum_{i=1}^{N} m_{XY}(t_i) m_{XX}(t_i)}{\sum_{i=1}^{N}\left(m_{XY}^2(t_i) - m_{XX}^2(t_i)\right)}\right) + k\frac{\pi}{4} \ . \tag{5.21}$$

5.1.3 Linear transform

The characteristics of the sums and differences of components in (5.13–16) can be more directly exploited in the linear transform procedure of Li and Crampin (1993), that provides an alternative strategy to solving the eigenproblem of (5.3) and hence another solution. In this particular form of the transform, the model prediction of (4.13) leads to the expected form

$$\mathbf{M}(t) = \begin{pmatrix} \lambda_1(t)\cos^2\phi + \lambda_2(t)\sin^2\phi & (\lambda_2(t) - \lambda_1(t)\sin\phi\cos\phi) \\ (\lambda_2(t) - \lambda_1(t)\sin\phi\cos\phi) & \lambda_1(t)\sin^2\phi + \lambda_2(t)\cos^2\phi \end{pmatrix}, \tag{5.22}$$

where the waveforms for the $S_\perp$ and $S_\parallel$ waves are defined as λ_1 and λ_2 as in (4.12).

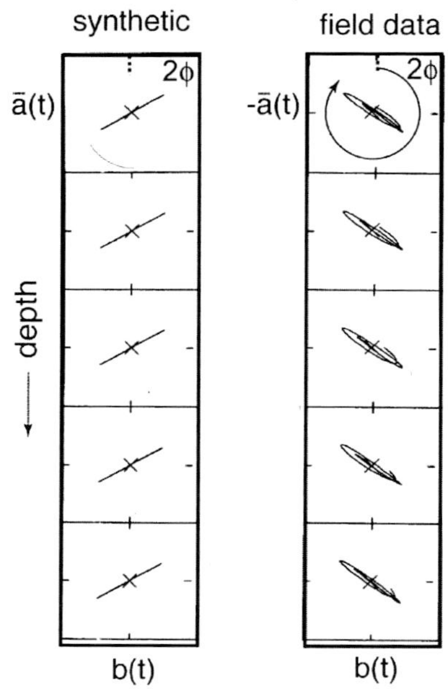

Figure 5.6. Cross-plots of linear transform variables -a(t) and b(t) for both an ideal synthetic traces set and also the field data set used in Figure 5.3. The resultant motions should be linear for a uniform medium (MacBeth et al. 1997).

With the trigonometric identities $\sin 2\phi = 2\sin\phi\,\cos\phi$ and $\cos 2\phi = \cos^2\phi - \sin^2\phi$ one can see that the sum of the cross-diagonal terms is $b(t) = (\lambda_2(t) - \lambda_1(t))$ and the difference of the terms on the main diagonal is $\bar{a}(t) = -(\lambda_2(t) - \lambda_1(t))\cos 2\phi$. If one brings the actual matrix $\mathbf{M}'(t)$ into this form, one can judge the correspondence to this prediction by plotting $-\bar{a}(t)$ against $b(t)$ (Figure 5.6). This should give a straight line; ϕ can be determined from the slope of this line, from which in turn $(\lambda_2(t) - \lambda_1(t))$ may be determined. In practice, the sums and differences may be applied to each time sample within a specified window, and the resulting transformed 'linear' motions examined visually. A least-squares estimate of ϕ is obtained, for example, by the linear regression

$$\hat{\phi} = \tfrac{1}{2}\tan^{-1}\left(\frac{N\sum_{i=1}^{N} \bar{a}_i b_i - \sum_{i=1}^{N} \bar{a}_i \sum_{i=1}^{N} b_i}{N\sum_{i=1}^{N} \bar{a}_i^2 - \left(\sum_{i=1}^{N} \bar{a}_i\right)^2} \right) \tag{5.23}$$

or by forming the covariance matrix of the two-vector $(-\bar{a}(t), b(t))^T$, which then gives (by Jacobi transformation)

$$\hat{\phi} = \tfrac{1}{2}\phi_{LT} = \tfrac{1}{4}\tan^{-1}\left(\frac{2\sum_{i=1}^{N}\bar{a}_i b_i}{\sum_{i=1}^{N}\left(\bar{a}_i^2 - b_i^2\right)}\right) . \tag{5.24}$$

Both estimates of ϕ may be compared with (5.19).

5.1.4 Confidence in the procedure

The success of these procedures can be judged qualitatively by visual examination of the rotation product $\mathbf{M}'(t)$, and more quantitatively by obtaining solutions to (5.3) for the four component traces sample by- ample, yielding the rotation angle as a function of time $\phi_{tr}(t_i)$. This gives the statistic parameter (Lefeuvre et al. 1992, Zeng 1994)

$$\sigma_1 = \frac{1}{2N}\sum_{i=1}^{N}\left|\phi_{tr}(t_i) - \hat{\phi}\right| , \tag{5.25}$$

for which a coherent signal gives the same orientation angle, and thus should be zero at a particular depth. Efficiency of the diagonalization procedure can also be measured using

$$\sigma_2 = \frac{1}{2N}\left(\frac{\sum_{i=1}^{N}\left|m'_{YX}\right|^2}{\sum_{i=1}^{N}\left|m'_{XX}\right|^2} + \frac{\sum_{i=1}^{N}\left|m'_{XY}\right|^2}{\sum_{i=1}^{N}\left|m'_{YY}\right|^2}\right) , \tag{5.26}$$

for which closeness to zero implies a good result. It should be noted that it in general a sizable delay between the fast and slow shear waves produces a more reliable estimate than a smaller delay. This results in a better agreement between σ_1 and σ_2.

5.1.5 Time-delay estimates

After rotation it is expected that the $S_\perp$ and $S_\parallel$ waves appear separately on the $X'X'$ and $Y'Y'$ components of the data matrix, so that $m'_{XX}(t) = \hat{\lambda}_1(t)$ and $m'_{TT}(t) = \hat{\lambda}_2(t)$. It is also expected that these waveforms resemble each other apart from a time-delay, so that $\hat{\lambda}_2(t) = \hat{\lambda}_1(t + \Delta\tau)$. The velocities of these waveforms together with the $\Delta\tau$ determine e_t as a function of depth. Cross-correlation is widely used to estimate this shear wave time-delay. This is achieved by locating the lag time ($\tau = \tau_{\max}$) associated with the maximum of the cross-correlation function

$$A_{XY}(\tau) = \sum_{i=1}^{N} m'_{XX}(t_i) m'_{YY}(t_i + \tau) . \tag{5.27}$$

It has often been observed that this lag determination remains remarkably stable, even when wave field separation is clearly incomplete as a consequence of an inaccurate rotation angle. This is related to the nature of the cross-correlation function in (5.27). In fact, incomplete separation of $S_\perp$ and $S_\parallel$ means that the recorded displacements are a linear superposition of the two waves

$$m'_{XX}(t) = \left(w_1\lambda_1(t) + w_2\lambda_2(t)\right)s(t) \tag{5.28}$$

and

$$m'_{YY}(t) = \left(w_3\lambda_1(t) + w_4\lambda_2(t)\right)s(t) \ , \tag{5.29}$$

where w_1, w_2, w_3 and w_4 are scalar weights, with w_1 and w_4 close to unity, and w_2 and w_3 small. The weights are functions of the true rotation and the incorrect estimate.

The cross-correlation procedure gives

$$A'_{XY}(\tau) = w_1w_2A_{11}(\tau) + w_2w_4A_{22}(\tau) + w_1w_4A_{12}(\tau) + w_2w_3A_{21}(\tau) \ , \tag{5.30}$$

where each $A_{ij}(\tau)$ is defined as (5.17) but for the individual modes. The consequence of incomplete separation is thus to superimpose two autocorrelations on top of the desired function. The cross-correlation $A_{12}(\tau)$between $\lambda_1(t)$ and $\lambda_2(t)$. $A_{12}(\tau)$ has a small amplitude in the positive lag domain as it possesses a maximum at $\tau = -\Delta\tau$. However, the interfering autocorrelations have side-lobes that may be strong enough to distort the main cross-correlation maximum. The degree of interference depends on the source function and the weighting factors. If we use the criterion $(w_1w_4)^2 > (w_1w_3 + w_2w_4)$ for "no interference", assuming that the peak $\tau = \Delta\tau$ lies close to the zero-lag time, we find that a rotation error of 35° is tolerable. This value is not too difficult to attain. Consequently, it is likely that in most cases the cross-correlation gives a satisfactory lag estimate (see Figure 5.7).

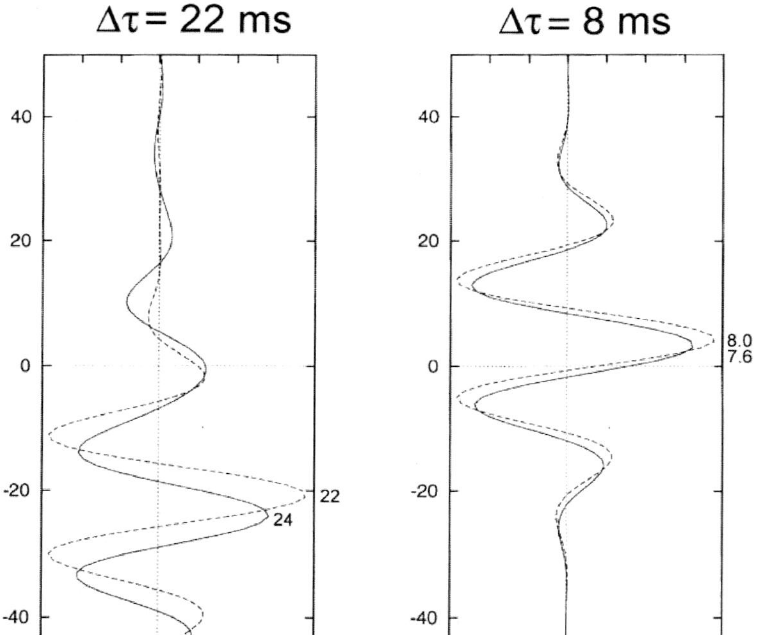

Figure 5.7. Cross-correlation functions between two split shear waves. The dashed line is the result when using exactly the same wavelets, whilst the result for the solid line inputs a mixture of the two wavelets representing a multi-component rotation error of 15°.

Another possibility for time-delay estimation is linear regression on the phase spectrum of $\hat{m}'_{XX}(\omega) / \hat{m}'_{YY}(\omega)$. This is proportional to $\omega\Delta\tau$ and thus gives a direct indication of the time-delay from its slope. However, care must be taken with this approach, as it does suffer from an undetermined cycle skip if lags are larger than the period of the wavelet. It is most useful for small values of the lag $\Delta\tau$.

5.2 DATA EXAMPLES OF ANISOTROPY ESTIMATION

5.2.1 General results

Over the past 15 years the above techniques (and those to be discussed in Chapter 6) have been applied to data from many different geographical locations and depths ranges, yielding a sizable collection of polarization and lag estimates. The degree to which these field data conform to the assumptions inherent in the Alford rotation described above is typified by the four four-component examples in Figures 5.8 to 5.15. As can be seen, application of the pre-conditioning steps in Chapter 4 followed by the coordinate transformation of (5.3) produces close agreement to the expected diagonal trace matrix. This inspires confidence in the technique, even when some of the basic assumptions are violated by the depth-dependent drift in rotation angle of some 40° and further minor fluctuations.

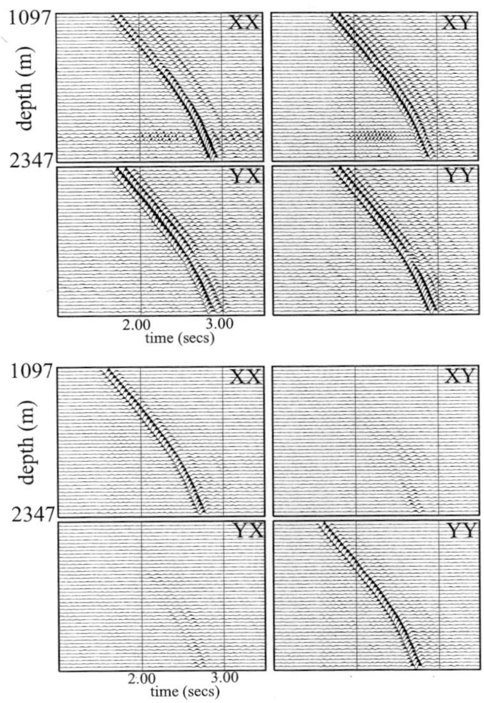

Figure 5.8. Four-component matrices from a VSP shot in the Lost Hills field, showing data immediately after tool spin correction (top) and then after near-surface correction and Alford rotation (bottom). An OMNIPULSE source is offset at 155 m, and recorded in the vertical well using single level 10 Hz geophones. The *XY* and *YX* components have been minimized, but not totally zeroed by the processing. Data courtesy of British Geological Survey and BP.

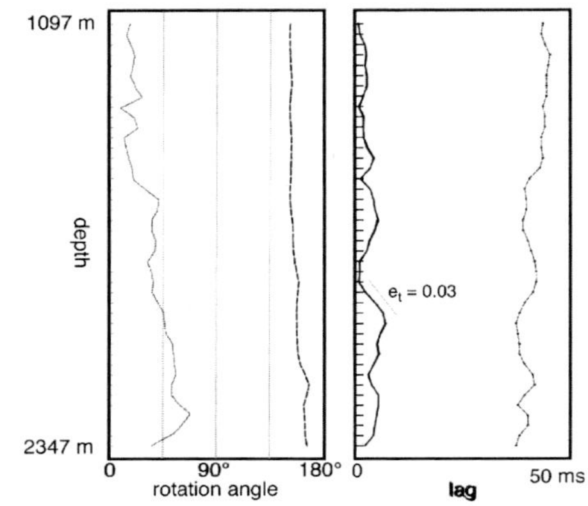

Figure 5.9. Rotation angles and shear wave lags determined from the Alford rotation procedure for the Lost Hills data in Figure 5.8. The lithology throughout the recorded section consists mainly of dipping shales, with some shallow water deposits. The overburden has a horizontally bedded mixture of siltstone, sandstone and conglomerate. Dashed lines are before near-surface correction, and demonstrate that most of the observed 42 ms lag is due to overburden layering with a different polarization from the subsurface. The actual subsurface lags are rather small, and place the reliability of the polarization estimates in question. The origin of the oscillatory behavior has been discussed by Choi and Gangi (1992) and also by MacBeth et al. (1993). If the slopes can be interpreted, they suggest sheets or layers with 3 % ($e_t = 0.03$) birefringence, and a sequence of polarization reversals.

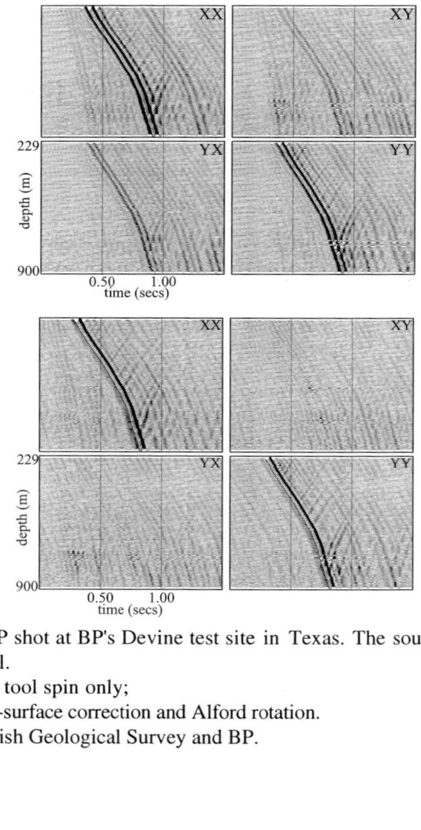

Figure 5.10. Four component VSP shot at BP's Devine test site in Texas. The source is an ARIS, offset at 181 m from the vertical well.

top: data with tool spin only;

bottom: after near-surface correction and Alford rotation.

Data courtesy of British Geological Survey and BP.

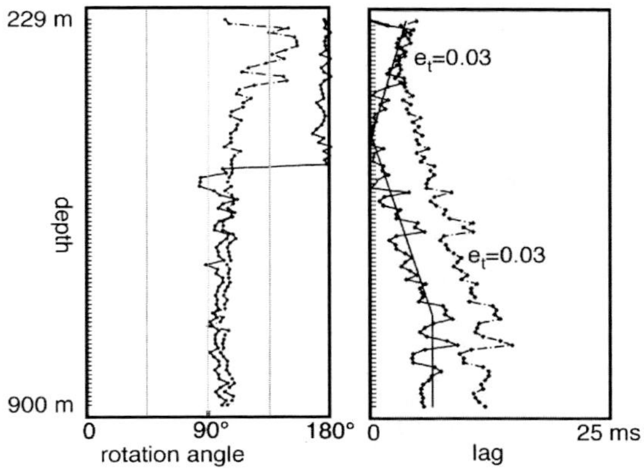

Figure 5.11. Results of Alford rotation applied before (dashed lines) and after (solid line) near-surface correction to the Devine data. There is a lag of 5 ms originating from the first few hundred meters. When corrected, the depth dependence changes and reveals a polarization jump close to a lag minimum. The slopes for both portions of the curve gives a birefringence of 3 %.

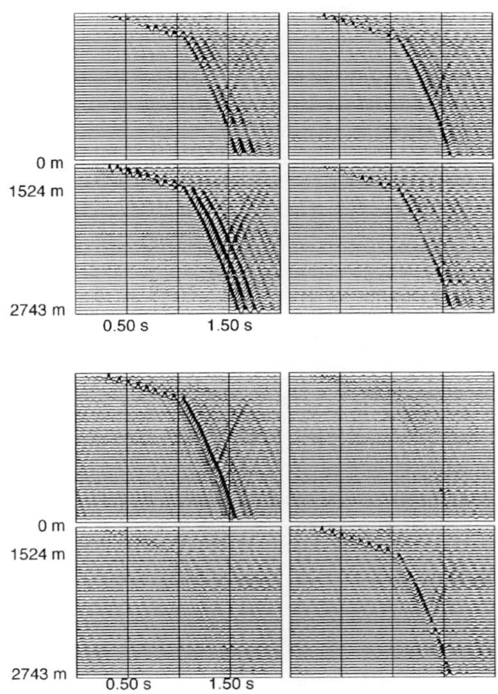

Figure 5.12. Four component VSP shot at Dimmit county, Texas, using Vibroseis technique. The in-line source is offset at 162 m, and the cross-linesource at 149 m. A single-level downhole tool is locked at intervals of 30 m, with the last ten levels having an interval of 152 m. The data are again shown before (top) and after (bottom) near-surface correction, However in this case the correction must be performed using the traces at 1524 m. The explanation for this is shown in Figure 5.13. Data courtesy of British Geological Survey and BP.

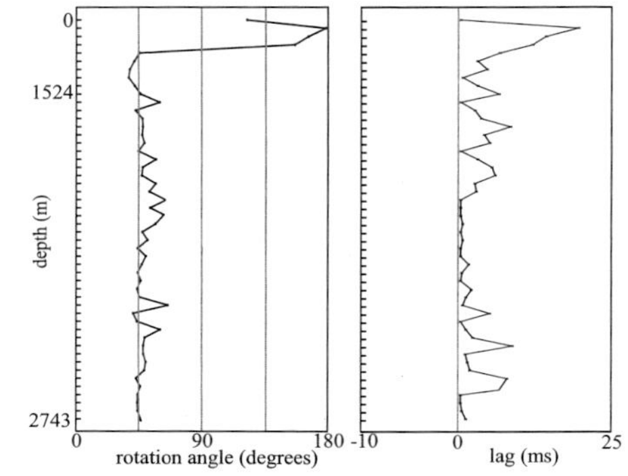

Figure 5.13. In this Dimmit dataset, there is a large lag in the shallowest layers which decreases with depth. This is due to the location of each source, and can be simulated by modeling (Yardley 1993). The subsurface birefringence value is reached after 1524 m, and rarely rises above 0.5 %. There is no change inside the target Austin Chalk layer between 1900 m and 2150 m.

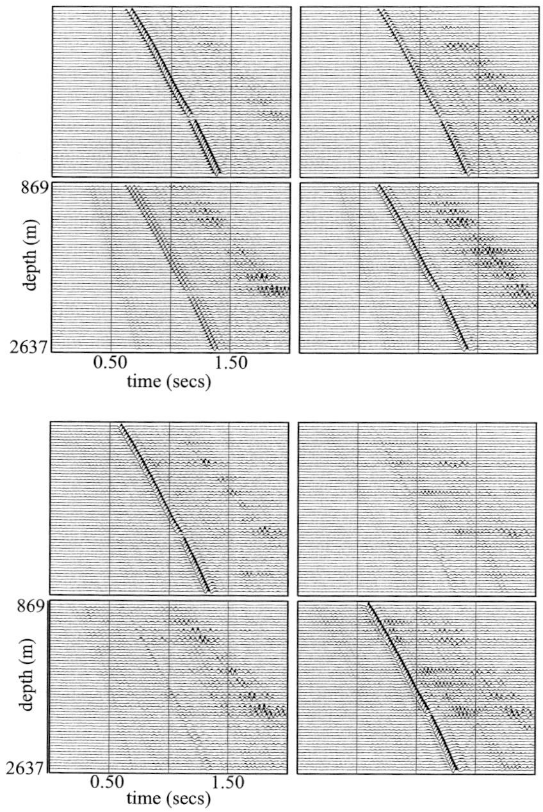

Figure 5.14. Four component VSP from the Bluebell-Altamont field in northeastern Utah. The data are acquired with the Vibroseis technique at an offset of 177 m. Further details of this work, together with the surface seismics, are given in Lynn et al. (1995). Data courtesy of Lynn Incorporated.

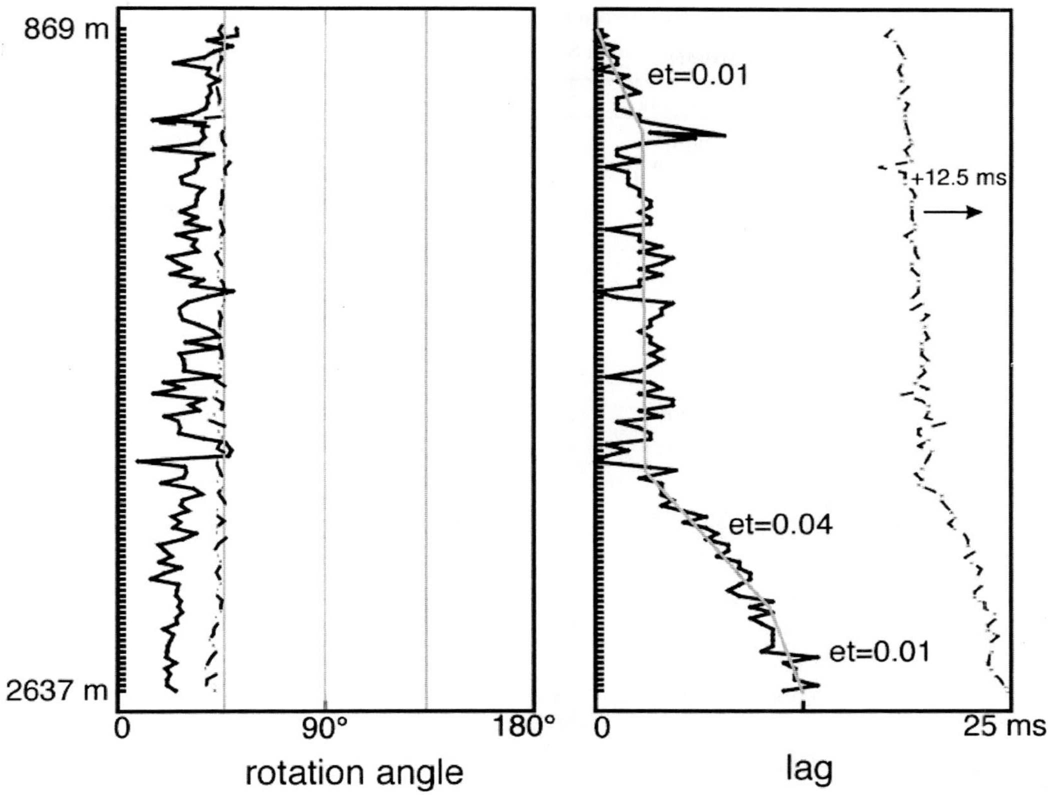

Figure 5.15. Alford rotation results for the processed data in Figure 5.14. The polarization is fairly uniform with depth. The overburden is predicted to have a vertical birefringence of 10 % or less, while the first part of the recorded interval down to 2011 m is almost isotropic. Below this interval, there is a sudden linear increase in lag, corresponding to a birefringence of 4 %. This increase coincides with the top of the Green River formation which has gas producing sandstones and carbonates. This reservoir zone is known to be fractured. The dashed lines correspond to the results before near-surface correction, and show a substantial near-surface anisotropy (note that 12.5 ms has been subtracted from the lag curve – the cause of this substantial lag is phase shift in the data from each source).

Although there is a broad agreement between the fast shear wave polarization and the maximum horizontal stress in such surveys, a more detailed picture suggests the polarization changes with depth in often quite thin zones. Chapter 6 discusses these findings in more detail, and gives further improvements beyond the current method. The estimates of shear wave lag imply intervals of both very little vertical birefringence (less than a few percent, or isotropy) and also larger birefringence values. This latter point is enlarged upon in the table of field-wide observations given below.

Table 5.1 Vertical Birefringence % Estimates from Multi-Component VSP using the techniques outlined in Section 5.1 (and also those detailed in Section 6.1).

	depth range	birefringence	rock type	location
1	≤ 259 m	high	limestones + s-s	SouthCasper Creek,Wyoming
	418–488 m	11.0–14.0	limestones + s-s	
	699–829 m	4.0–11.0*	limestones + s-s	
2	≤ 868 m	≤ 3.0		Bluebell-Altamont field, Utah
	868–2011 m	1.0		
	2011-2636 m	5.0–12.0*	s-s carbonates	
3	≤ 1000 m	≤ 3.0	s-s + conglomerates	Lost Hills, California
	1000–2000 m	≤ 1.0	shales	
4	≤ 1250 m	3.0–5.5	shales	Burleson County, Texas
5	457–732 m	8.0	dolomites + clastics	
	732–1006 m	5.0*	dolomites	
6	≤ 870 m	1.0–3.0	clays, marls, s-s	Juravkoe, Vorobievske field,
	≥ 2040	9.0*	Maikop clays	North Caucasus
	≤ 780 m	0.5–1.5		
	≤ 870 m	0.5–1.5		
7	1646 m	3.0–6.0*	gas-producing s-s	Wind River basin, Wyoming
8	??	7.0–8.0*	Austin Chalk	Marcelina Creek, Texas
9	≤ 1100 m	2.0–3.0	limestone + shale	Paris basin
	1100–1500 m	0.5	limestone + shale	
	1500–1600 m	12.0	limestone	
	> 1600 m	0.5	limestone + shale	
10	≤ 2377 m	0.3–0.5–		Pembroke field, Texas
11	≤ 228 m	3.0	–	BP's Devine test site, Texas
	228–610 m	≤ 1.0		
12	≤ 152 m	≥ 12.0		Conoco borehole test facility
	152–377 m	≥ 1.5	s-s + shale	

	377–677 m	5.0	s-s sequence	
	677-753 m	≈15.0*	s-s sequence	
	753–900 m	≈15.0*	shale	
13	300–900 m	≈1.0	shales	Natih field, Oman
	900–1050 m	isotropic	shales	
	1050–1100 m	≈20.0*	limestone	
14	≤ 750 m	≈2.0		Dimmit County, Texas
15	650–1100 m	3.0–6.0	s-s + shales	Paris basin
	1850 m	≥ 15.0*	limestone	
16	≤ 473 m	isotropic	shales + carbonates	Romashkino, Tatar Republic
	473–643 m	6.0*		
	≤ 510 m	isotropic		
	510–643 m	8.0*		
	440–553 m	isotropic		
	553–694 m	8.0*		
	694–830 m	2.0		
17	396–945 m	0.7	s-s + clay shales	Elk Hills, California
	2530–3033 m	0.0	siliceous sh, mssv s-s	
	1737–2530 m	1.4		
	1006–1494 m	3.6	mstly clay sh, some s-s	
	91–396 m	4.0	poorly sorted dirty s-s	
18	488–1981 m	0.0	evap., clastic+carbonate	Winkler County, California
19	91-914 m	0.0		McElroy, Texas
20	823–1128 m	0.0	siliceous shales	Lost Hills
	1006–1290 m	0.7	siliceous shales	
	732–975 m	0.8	siliceous shales	
	549–823 m	1.3	clastics+diatomite	
	1128–1341 m	1.6	siliceous shales	
	183–247 m	1.7	s-s	
	335–701 m	1.7	clastics+diatomite	
	320–457 m	2.4	clastics	
	61–640	m3.0	clastics+siliceous sh	

	30–152 m	5.0	s-s+conglomerate	
	46–320 m	5.2	s-s+conglomerate	
	183–247 m	1.7	s-s	
21	1200–3800 m	1.0	clastics+carbonates	W.-Central Alberta
22	1433–1469 m	0.5	clastics+carbonates	Kurten, Texas
	457–1433 m	2.6	clastics+carbonates	
23	1219–1768 m	1.0	clastics	Laredo, Texas
	419–1216 m	2.0	clastics	
	2789–3133 m	2.1	clastics	
24	1189–1615 m	3.0	s-s, claystone, siltstone	Railroad Gap, California
	1798–1981 m	4.0	siliceous shales	
	884–1128 m	4.1	s-s+clay sh	
	91–396 m	9.0	dirty s-s+conglomerate	
25	347–399 m	7.0	siliceous shale	Cymric field, California
	262–311 m	7.0	diatomite	
	189–226 m	9.0	s-s+conglomerates	
	189–238 m	9.0	s-s+conglomerates	
	250–287 m	12.0	diatomite	
	107–229 m	14.0	s-s+conglomerates	
	128–189 m	14.0	s-s+conglomerates	
	299–347 m	16.0	siliceous shale	
	40–79 m	18.0	s-s+conglomerates	
	323–384 m	20.0	siliceous shale	
26	3100–3360 m	4.0	sandstone	S.-Central Wyoming

In many fields only moderate-to-high anisotropy values have been observed, since weak anisotropy (or isotropy) can only be resolved over large intervals. The resolution limits depend on each case individually, and are often subjective and contingent upon the ability to distinguish the character of noise from signal (contrast the levels of noise in Figure 5.9, 5.11, 5.13 and 5.15). For recovery of the fine-scale fluctuations see of Chapter 6.

* Values marked with an asterisk have been correlated with a known reservoir zone, fractures or productivity.

1 - Kramer and Davis (1992);
2 - Lynn et al. (1997);
3 - Yardley (1993);
4 - Becker et al. (1989);
5 - Yardley (1993), Raikes (1991);
6 - Queen and Rizer (1990);
7 - Hake et al. 1998;
8 - Yardley (1993);
9 - Lefeuvre et al. 1992;
10 - Cliet et al. 1991; Zeng and MacBeth (1996);
11 - Yardley (1993);
12 - Baptie (1996);
13 - Slater (1997);
14 - Lynn et al. (1996);

15 - Becker et al. (1990);
16 - Bush (1990);
17 - 25 Winterstein, Meadows and De (1996);
26 - Kendall and Kendall (1996).

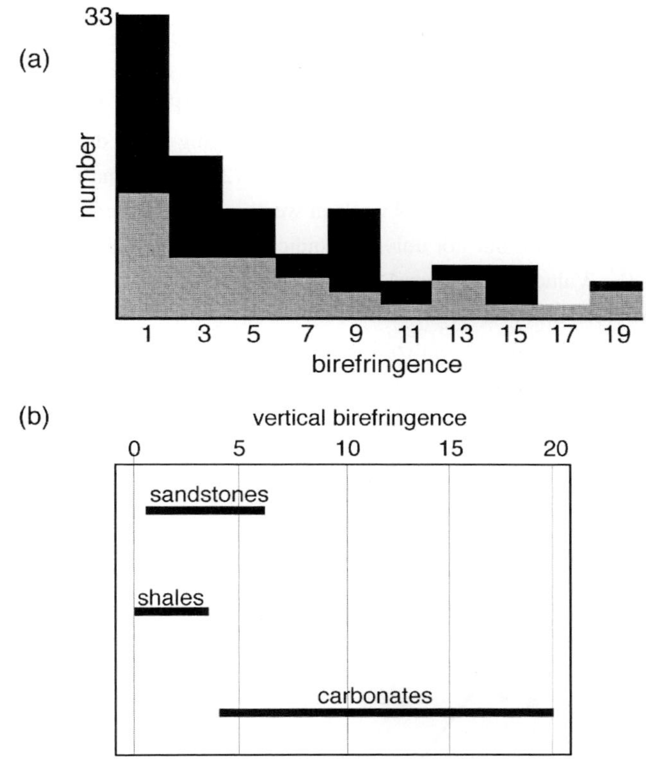

Figure 5.16. (a) Small values of birefringence are popular, but are not confined to any particular depth. The profile of the upper 500 m interval (shaded histogram) is similar to that of the total depth range.
(b) As a rough generalization, ignoring the near surface and exceptional cases, the stronger but more brittle lithologies support the greater range of birefringence values.

The data in Table 5.1 extend from isotropy to 14 %, with the range being equally well supported over all fields, depths and lithologies. Indeed, at this level there is no clear partitioning into near-surface and subsurface anisotropy, or consistency between one field and the next. The upper limit in birefringence is similar to that given in Chapter 3 for igneous and metamorphic rocks and laboratory studies, and has attracted some attention (Crampin 1994). Birefringence lows are found at all depths and appear to be common (60 % of the measurements are lower than 1 %, Figure 5.16a), whereas the highs are confined to specific lithologies. Indeed, the low values are not confined to VSP: Winterstein et al. (1996) indicate that low values of anisotropy are not confined only to VSP. Zero or near zero anisotropy has been measured in surface data (Willis et al. 1986) and also in dipole sonic imaging logs (Hatchell and Cowles 1992). In general, the shales display a smaller range of weaker birefringence, with the largest values often in the carbonates (Figure 5.16b). This is consistent with the stronger rocks also being brittle and more easily fractured (Nelson 1985). The exception to this rule are the unconsolidated sediments in the top few hundred meters (such as the loose conglomerates) for which birefringence highs are to be found in almost every field and across lithologies. Winterstein et al. (1996) suggests a possible correlation between vertical birefringence and unequal horizontal principal stresses.

Most moderate-to-high values of birefringence appear inversely proportional to the layer thickness, with many of the thinner zones being reservoir targets (Figure 5.17). Interestingly, there is evidence for much thinner zones of strong anisotropy emerging from dipole sonic imaging studies (see Chapter 9). More recent work by Winterstein et al. (2001) summarizes the analysis of multi-component VSP for 54 depth intervals in 19 wells in western North America. They conclude that vertical birefringence is a common but not universal indicator of azimuthal anisotropy, with magnitude varying from 0–21 %. Values larger than 4 % occur only in formations above 1200 m and within 40 km of the San Andreas fault. At larger scales, the birefringence tends to be blocky, with its magnitude and the associated shear wave polarization azimuth being consistent over large depth intervals, then changing abruptly by often large amounts. As a conclusion they state that most reservoirs are birefringent, but in their study there is no instance of increased birefringence confined solely to the reservoir.

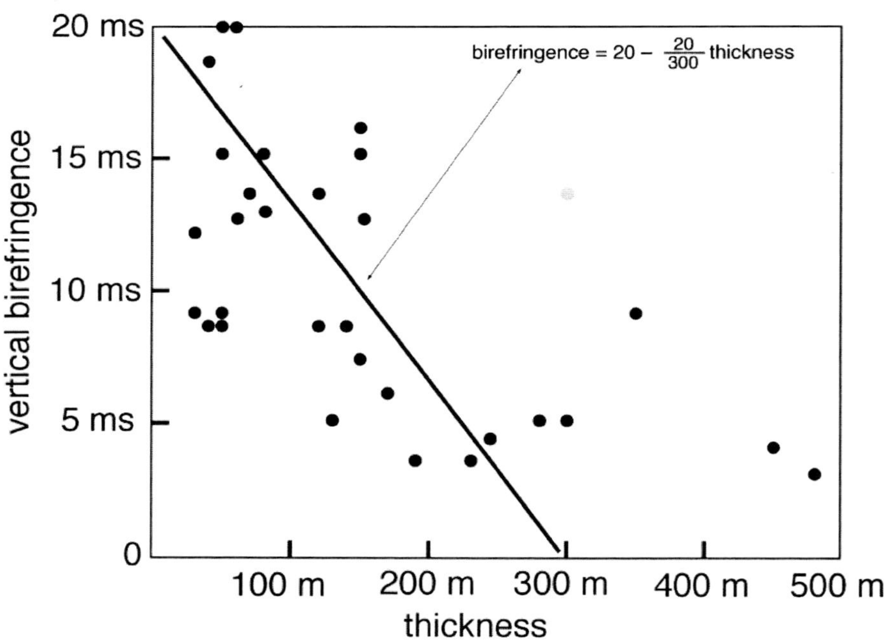

Figure 5.17. Highly birefringent layers tend to be thinner, regardless of the lithology. Here, all layers above 3 % birefringence are plotted to reveal an approximately linear trend. The gray shaded outlier is at a shallow depth.

At present the interpretation of birefringence values remains a major challenge so far unmet. For exploration purposes, it is satisfying that weighty evidence supports correlation between a productive reservoir zone and seismic anisotropy. Calibration of this response to confirm a fracture origin requires careful control using hard information from cores and fracture-related logs.

5.2.2 A specific lithology correlation

The lithology column is an essential guide to interpreting fine-scale birefringence fluctuations. This is illustrated by an example from the Conoco borehole test facility, Kay County, Oklahoma attributed to Horne (1995). The processing for one of the two VSPs has already been shown in Figures 4.7, 4.9 and 5.3, while for the second the final data matrix is shown in Figure 5.18. Processing steps identical to those outlined in Chapter 4 are applied to each VSP, followed by the algebraic rotation technique and cross-correlation analysis of Section 5.1.

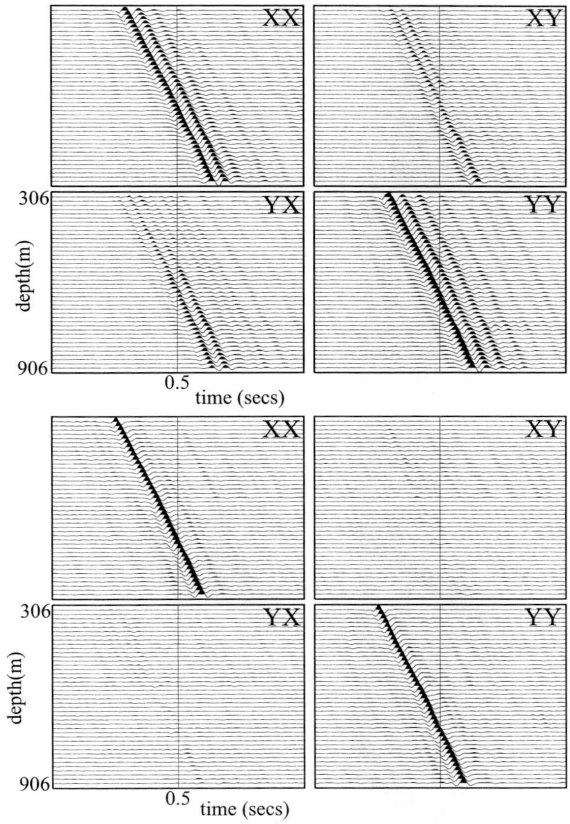

Figure 5.18. Four-component data from the Peel well of the Conoco borehole test facility, acquired with a vibrator positioned at 65 m from the wellhead along a azimuth diametrically opposite to that of Figure 5.3.
top: before processing;
bottom: after near-surface correction and Alford rotation.
Data courtesy of British Geological Survey.

The results of Alford rotation for each VSP are plotted in Figure 5.19. The polarization azimuth determined by the rotation angle is approximately constant, but the gradient of the lag estimates changes with depth in an excellent correlation with the lithology log. The lag profiles from both VSPs evolve in four stages, each corresponding to a different overall birefringence regime. The first VSP reveals a weak birefringence of approximately 1 % in the shallowest section where there is a mixture of shales, limestones, and sandstones. A second anisotropic zone lies between 422 m (the top of the sandstone sequence) and 617 m, corresponding to 5 % birefringence. The top and bottom of this zone are marked by two thin sandstone layers of approximately 10 % birefringence. A third zone is nearly isotropic, with a slight lag reversal and 0.2 % percent birefringence. The fourth zone is between 677 m and 767 m. This sandstone is characterized by a strong birefringence of 15 %. Below 767 m, the gradient decreases steeply in the shales, corresponding to a birefringence of 15 %.

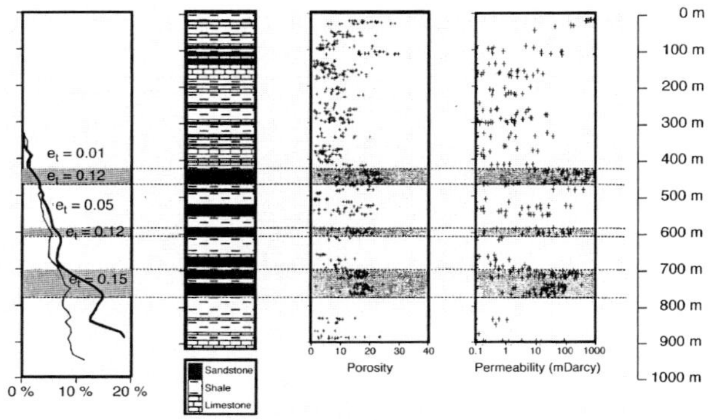

Figure 5.19 Comparison between lag profiles derived from two multi-component VSPs at the Conoco borehole test facility, with lithology log and core measurements. The stratigraphy consists of a mixture of shales, limestones, and sandstones. The shaded regions correspond to highly permeable and porous zones of sandstone for which the birefringence is pronounced. Diagram adapted after Horne (1995).

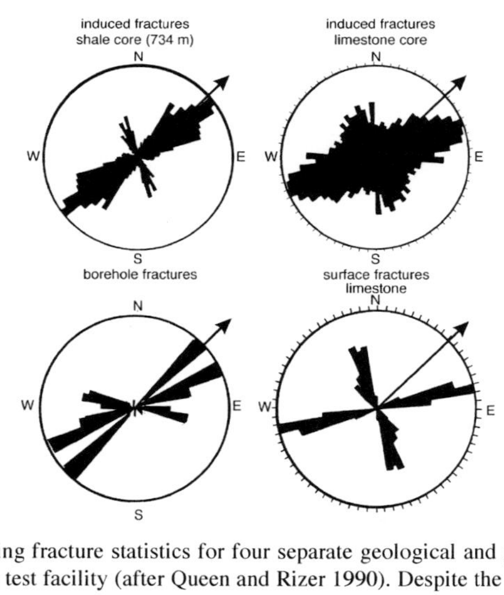

Figure 5.20. Rose plots showing fracture statistics for four separate geological and geophysical measurement at the Conoco borehole test facility (after Queen and Rizer 1990). Despite the structural simplicity of the area, the stress-fracture system is rather complicated, with several distinct directions. The maximum compressive stress was between N55°E and N80°E, determined through borehole breakouts and impression packers of induced hydraulic fractures. The direction determined from the VSP seismic is indicated by the arrow in each plot.

All three highly birefringent sandstones show a strong correlation of birefringence with the core-derived horizontal permeability and porosity (Figure 5.19). More specifically, the origin of the anisotropy in the 700 m sandstone are (presumably) near-vertical open fractures; these fractures. These fractures are (presumably) the cause of a fluid loss while drilling (Queen, personal communication). This link between anisotropy and fractures has been further investigated by comparing the $S_\perp$ polarization estimate (which remained at a near constant N55°E), with hard information from other geological and geophysical measurements. These included borehole televiewer images, borehole breakouts, outcrop analyses, thin section analysis, point load tests on core sections, and visual inspection of the core (Queen and Rizer 1990; Figure 5.20). The seismic data correlate with one of the three predominant fracture strikes present in the core data, suggesting a possible fracture origin for the measurements. The $S_\perp$ polarization is also close to the maximum horizontal compressive stress determined from borehole breakouts.

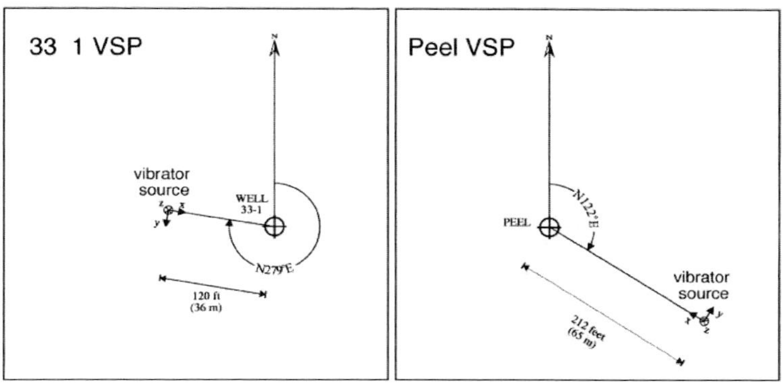

Figure 5.21. Acquisition geometry for the two VSPs acquired at the Conoco borehole test facility, showing the location of the near-offset vibrator. The sources are roughly diametrically opposed. Extracted from Horne et al. (1997).

The lag profile for the second VSP shows a similar – though less pronounced – trend. Horne (1995) interpreted this feature as an inclined crack system. The smaller lags are assumed to arise because the offset azimuth of this second VSP is directed away from the inclination direction and towards the line singularity. The other VSP has greater lags, as ray paths are closer to the plane of the fracturing. For a fracture set inclined by θ_F and striking at an azimuth ϕ_F to the offset direction, the birefringence reduction is predicted from Chapter 3 to be

$$e_t \rightarrow e_t - \zeta e_n \sin^2 \phi_F \sin^2(\theta - \theta_F) \, , \tag{5.31}$$

where θ is the (small) incidence angle determined by the offset. This reduction is plotted as a function of inclination in Figure 5.22. Horne (1995) suggested that such diametrically opposed azimuth surveys could be used to estimate the inclination angle θ_F from this relationship.

By using a known relationship between e_t and e_n (say, for example, thin fluid-filled cracks for which $3e_t = e_t$ for $V_P/V_S = 2$), the ratio of the birefringence values for different surveys but at the same depth is an indicator of the inclination. To interpret the lag behavior accurately, Horne (1995) developed a genetic algorithm inversion scheme based on forward modeling by ray tracing. The procedure predicts a fracture system striking N50°E and dipping 18° from the vertical to the south-east. This result supports the conclusions reached by Horne and MacBeth (1994) and also Liu et al. (1991) for other studies at the same site. It was noted in this study that a further improvement to the fit could be found by including sedimentary layering. However, the balance between over-parameterization and goodness-of-fit is inconclusive. The study provides further evidence that the fine-scale polarization and lag variations output in the Alford rotation technique require a careful and sophisticated technique for complete interpretation.

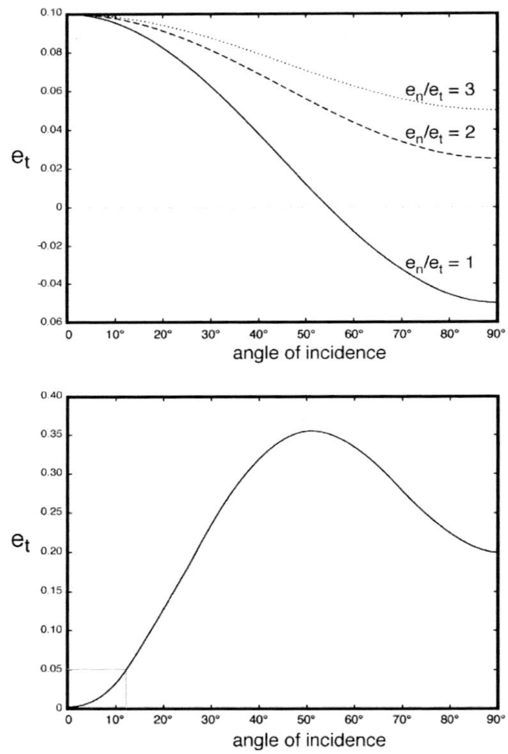

Figure 5.22. Top: reduction in vertical birefringence caused by inclined fracturing with a varying en but fixed $e_t = 0.20$. Bottom: predicted vertical birefringence (e_t) for dipping Dog Creek shale with Thomsen parameters $\varepsilon = 0.225$, $\delta = -0.02$, and $\gamma = 0.1$. It shows that a dip of 12° is required to give an apparent 5 % birefringence.

One final possibility often raised during interpretation is that the lag variations are caused by dipping beds. In this specific example this is highly unlikely, as the geology of the area is relatively simple, despite being bordered by several large-scale tectonic features. In fact, it may be quite accurately represented as a plane-layered structure with a dip less than 1° and with little lateral variation. Considering the expected response from dipping beds helps to make this explanation more unlikely. For a bed dipping at θ_B the vertical birefringence is predicted as

$$e_t \approx 2\gamma \sin^2 \theta_B + \frac{\varepsilon - 2\delta}{2\eta} \sin^2 2\theta_B \ , \tag{5.32}$$

where $\eta = V_S^2 / V_P^2$, and ε, δ and γ are Thomsen's parameters (Chapter 2). The variation with incidence angle is plotted in Figure 5.22. The increase is not dramatic, due to the dependence on $\sin^2\theta$. However, with a known high anisotropy in the shales, observed birefringence of 5 % or less can be attained with bedding dips of at most 12°. This point is underlined by the synthetic seismograms of Figure 5.23, where it is shown that the result of Alford rotation is, in this case, identical to the effect of vertical fractures.

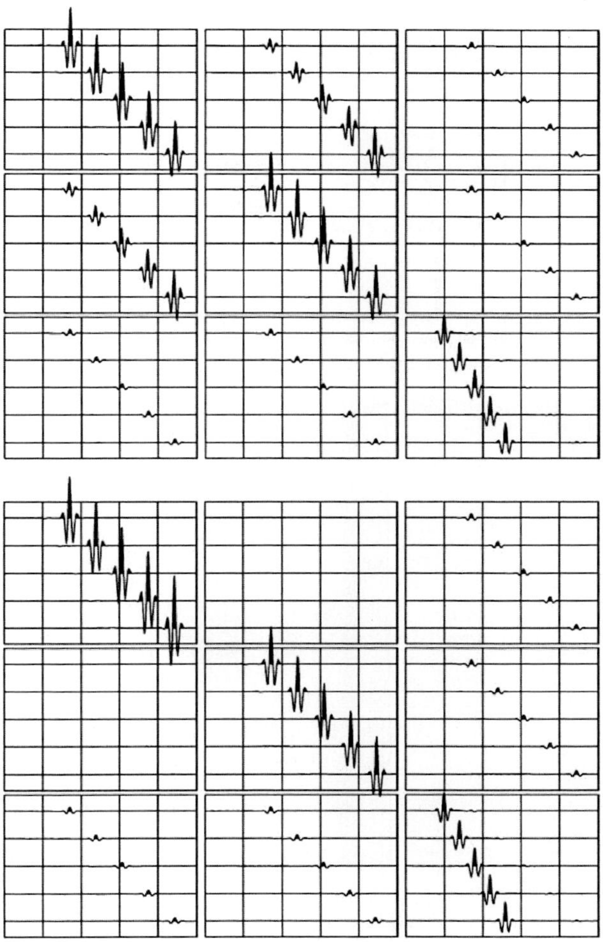

Figure 5.23. The effect of a dipping shale bed on nine-component VSP. The dip is only 15°, but produces a sizable off-diagonal component. As the offset is not perpendicular to the strike, there are also additional off-diagonal components. These could be used as indicators of the bedding angle. Before (top) and after (bottom) Alford rotation.

5.3 NORMAL INCIDENCE REFLECTIVITY

5.3.1 Multi-component analysis

An additional output from processing is the upgoing multi-component wave field $\mathbf{R}_{LL}^{U}(t)$ such as that shown in Figure 4.10. This yields reflectivity, and can be used to support the analyses outlined in Sections 5.1 and 5.2 in cases when the lag is small and resolution of the reservoir anisotropy is a particularly difficult. Correlation lag analysis provides results directly proportional to the average anisotropy, whereas the difference in shear wave reflectivities relates to the local anisotropy differences, and thus is more relevant to reservoir analysis.

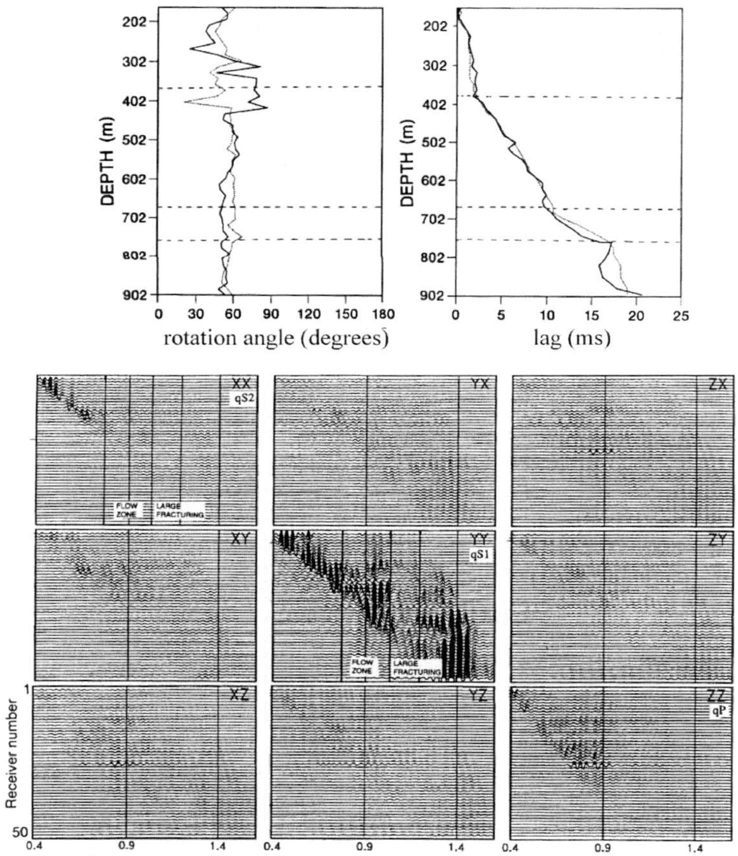

Figure 5.24. Final results from an Alford rotation applied to the upgoing wave field of Figure 4.10. The rotation angle and lags (shown on top) agree closely with those calculated for the downgoing wave field (solid lines). In addition, the off-diagonal matrix components have little energy, thus verifying the validity of the approach. There is an amplitude reduction in the $S_{\parallel}$ (qS_2) wave beyond 677 ms. This corresponds to the highly birefringent, permeable thick sandstone described in Figure 5.19. The dimming of the $S_{\perp}$ (qS_1) amplitudes beyond 1000 ms is indicative of larger-scale fracturing (MacBeth et al. 1997).

The idealized model of (4.13) can also be applied to this upgoing wave field, with identical underlying assumptions, but with the extra constraint of identical orientation in the overburden and the substratum. Thus, energy decomposes into the fast and slow shear wave directions at the source, propagates to the reflector and then back to the receiver, where it is recorded in the receiver frame. The decomposition into the principal shear wave axes at the source level means that only pure mode coefficients for the downgoing $S_\perp$, $S_\parallel$ and P waves need be treated at the reflector. The resultant data matrix is an extension of (4.13):

$$\mathbf{D}^{(9)}(t) = \mathbf{C}_Z(\phi)\Lambda(t)\mathbf{R}\Lambda(t)\left(\mathbf{C}_Z(\phi)\mathbf{S}^{(9)}(t)\right) , \tag{5.33}$$

where

$$\mathbf{R} = \begin{pmatrix} R_\parallel & 0 & 0 \\ 0 & R_\perp & 0 \\ 0 & 0 & R_P \end{pmatrix} . \tag{5.34}$$

The formal similarity of (5.33) and (4.13) means that the technique of multi-component rotation in (5.3) can be applied directly to the "upfield" without modification (see Figure 5.24 for an example). Indeed, the original form of the Alford rotation (Alford 1986) was introduced to treat surface seismic data by this procedure. If the overburden is isotropic and the target TIH, then

$$\mathbf{D}^{(9)}(t) = \Lambda^{iso}(t)\mathbf{C}_Z(\phi)\mathbf{R}\mathbf{C}_Z(\phi)\Lambda^{iso}(t)\left(\mathbf{S}^{(9)}(t)\right) , \tag{5.35}$$

where in $\Lambda^{iso}(t)$ $\lambda_\perp(t) = \lambda_\parallel(t) = \lambda_S(t)$. The energy can now be visualized to propagate to the target level with polarization fixed by the specific source motion, before decomposing into the principal shear wave axes at the reflector level, reflecting, and then reconstituting for the return leg. In both (5.34) and (5.35) the physical act of rotating the source-receiver frame to align with the anisotropy yields the reflectivity matrix **R** directly. After near-surface correction, the four component subset predicted for an anisotropic overburden is

$$\mathbf{D}^{(4)}(t) = \begin{pmatrix} \cos^2\phi\lambda_\parallel^2 R_\parallel + \sin^2\phi\lambda_\perp^2 R_\perp & \lambda_\perp\lambda_\parallel(R_\perp - R_\parallel)\sin\phi\cos\phi \\ \lambda_\perp\lambda_\parallel(R_\perp - R_\parallel)\sin\phi\cos\phi & \sin^2\phi\lambda_\parallel^2 R_\parallel + \cos^2\phi\lambda_\perp^2 R_\perp \end{pmatrix} , \tag{5.36}$$

and for isotropic overburden it is

$$\mathbf{D}^{(4)}(t) = \lambda_S^2(t)\begin{pmatrix} \cos^2\phi R_\parallel + \sin^2\phi R_\perp & (R_\perp - R_\parallel)\sin\phi\cos\phi \\ (R_\perp - R_\parallel)\sin\phi\cos\phi & \sin^2\phi R_\parallel + \cos^2\phi R_\perp \end{pmatrix} . \tag{5.37}$$

In both cases, the upgoing wave field can be treated using the Alford rotation in exactly the same way as the downgoing wave field. After rotation, the respective reflection coefficients are obtained on the main diagonals, and these shear wave amplitudes are sensitive to the fracturing. The expected sensitivity to a fractured target with an isotropic overburden (velocity V_S and density ρ) is shown in Figure 5.25 for a low-to-high impedance contrast representing a carbonate reservoir. The $S_\parallel$-$S_\parallel$ reflection coefficient is given by

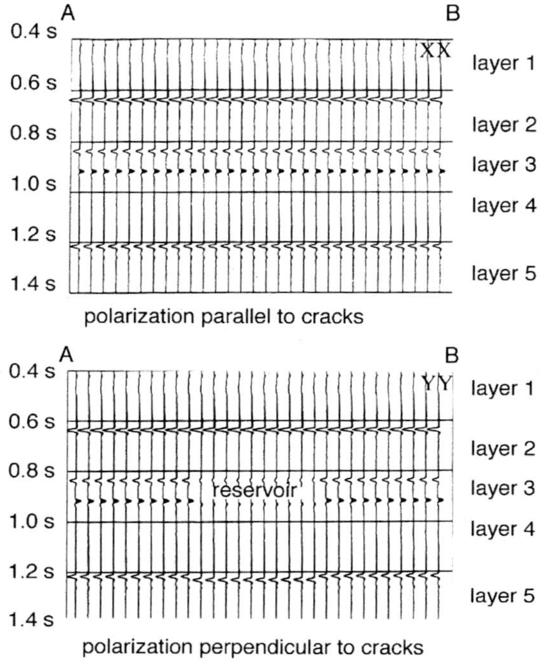

Figure 5.25. Illustration of the expected sensitivity of the amplitudes of reflected shear waves to reservoir fracturing (15 % birefringence). Only the principal components after Alford rotation are relevant, and these show the expected dimming of the $S_{\parallel}$ (qS_2) amplitudes in the central reservoir zone, while the $S_{\perp}$ (qS_1) reflectivity shows no effect and responds to the matrix properties (Yardley 1993).

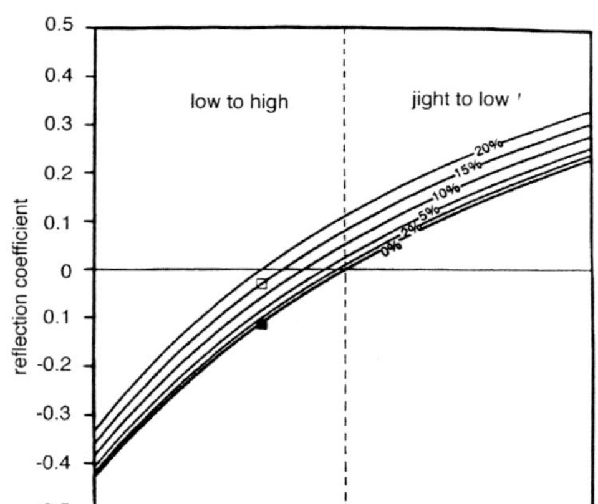

Figure 5.26. Normal incidence qS_2 reflection coefficients for a range of birefringence and reservoir contrasts (after Yardley 1993). The contrast variation is created by fixing the reservoir (layer 2) velocity at 2.5 km/s varying the overburden (layer 1) velocity. The $S_{\perp}$ (qS_1) reflectivity corresponds to the 0 % line. The lines indicate that the most significant change takes place for the weaker contrasts, and predict both dimming and brightening.

$$R_{\parallel} = \frac{\rho V_S - \rho^F V_{\parallel}^F}{\rho V_S + \rho^F V_{\parallel}^F}, \tag{5.38}$$

whereas for the $S_{\perp}$-$S_{\perp}$ coefficient it is

$$R_{\perp} = \frac{\rho V_S - \rho^F V_{\perp}^F}{\rho V_S + \rho^F V_{\perp}^F}, \tag{5.39}$$

where ρ^F, $V_{\parallel}^F$, and $V_{\perp}^F$ are density, slow and fast wave velocity in the target. For a fractured medium, $V_{\perp}^F > V_{\parallel}^F$, and hence the low-to-high contrast gives $R_{\parallel} > R_{\perp}$. The dimming of events on the slow section relative to those on the fast section has been observed by Mueller (1991), who successfully enhanced production by drilling clusters with high (several percent) anisotropy in the Austin Chalk. The technique was first applied by Yardley (1993) to VSP shot in the Austin Chalk. In the latter case, the amplitude ratio of the top target reflection and downgoing waves at each depth level was plotted (see for example Figure 5.26).

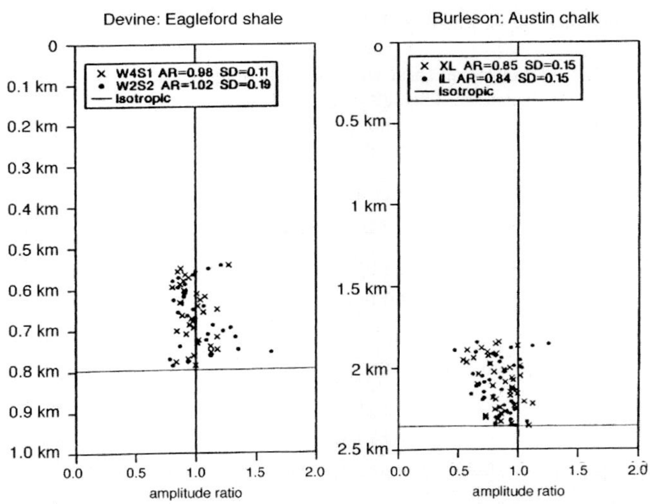

Figure 5.27. The ratio of the upgoing to downgoing shear wave amplitudes can be used to detect anisotropy, based on the principles in Figure 5.25 and 5.26. Here, two examples are shown. As expected from the transmission results, the left hand one shows an isotropic response (Yardley 1993).

The overall response is determined by the interplay between target contrast and anisotropy, and therefore the approach requires careful interpretation (Figure 5.27). Indeed, for a high to low contrast, the same relative velocities can also create a brightening of the slow amplitudes. This effect has been utilized for a coal bed methane reservoir by Shuck et al. (1996) (maximum anisotropy of 8 %). Use of this amplitude-ratio technique is limited by the quality of seismic amplitudes. For this reason, many surface seismic studies have continued to used travel-time delays: Martin and Davis (1987) (1 %, 4 %); Lynn and Thomsen (1990) (1.4 %); Lewis et al. (1991) (5 %); and Potters et al. (1999) (15 %). However, the greater amplitude control in VSP makes the technique an essential contribution in anisotropy analysis.

5.3.2 Use of a single source

An alternative way of interpreting the isotropic/anisotropic case was offered by Spencer and Chi (1991). For a single horizontal source motion along the X-axis in an isotropic overburden, the two-component, horizontal recordings along the X and Y axes are

$$\mathbf{d}(t) = \begin{pmatrix} R_{\parallel}\cos^2\phi + R_{\perp}\sin^2\phi \\ (R_{\perp} - R_{\parallel})\sin\phi\cos\phi \end{pmatrix} . \tag{5.40}$$

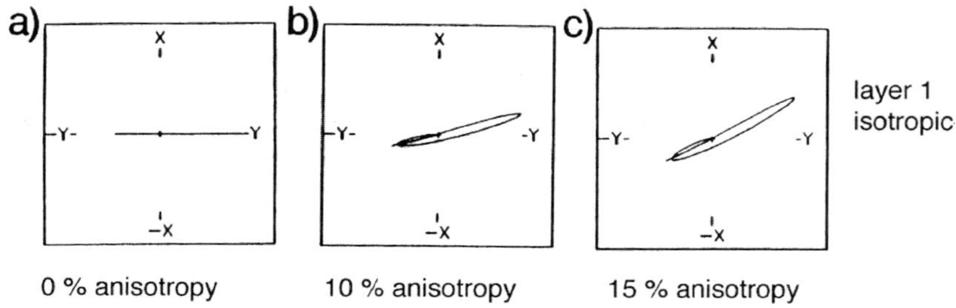

Figure 5.28. The polarization of an incident shear wave, linearly polarized along the Y-axis, rotates when the reservoir birefringence increases. This effect is predicted by Spencer and Chi (1991) for an isotropic overburden. When the reservoir zone is thin, the rotation is accompanied by an ellipticity formed from the interfering arrivals. Diagram extracted from Yardley (1993).

This means that the transverse component recorded along the Y-axis will not be zero even for this "normal incidence, isotropic" case, due to the differing reflection coefficients. The pure linearly polarized S wave returns from the target as a linearly polarized mixture of the two component waves. Physically, this is necessary as the displacements in the lower medium are fixed by the anisotropy, not the acquisition geometry. Thus, the polarizations in the upper medium rotates in order to make the displacements and stresses continuous across the interface.

The polarization direction given by the eigenanalysis of the upgoing particle motion $\hat{\phi}$ (see 4.45) is not the same as for the incident shear waves (determined by the source motion). The reflected wave has a linear motion (sum of two linear waves with no time-delay), rotated through an angle ψ with respect to the incident wave (Figure 5.28). Relative to the X-axis it is

$$\tan\psi = \frac{(1-r)\tan\phi}{1+r\tan\phi} , \tag{5.41}$$

where $r = R_{\perp}/R_{\parallel}$. This angle is shown for different anisotropies in Figure 5.29a, where it is observed to be a strong effect for a weak contrast. A robust estimate of the fracture strike ϕ_F can be made by rotating the source polarizations at the same location, giving

$$\psi = \tan^{-1}\frac{(1-r)\tan(\phi_F - \delta_i)}{1+r\tan(\phi_F - \delta_i)} , \tag{5.42}$$

where δ_i (with i = 1, ..., N) are different source motions relative to some fixed X-axis. Through comparison of the predictions with the results from eigenanalysis, with known δ_i, the unknown strike

ϕ_F and the ratio r can be determined by a non-linear optimization procedure using the method outlined in Spencer and Chi (1991). Another example of this type of technique, but using only the transverse component, will be shown in Chapter 7.

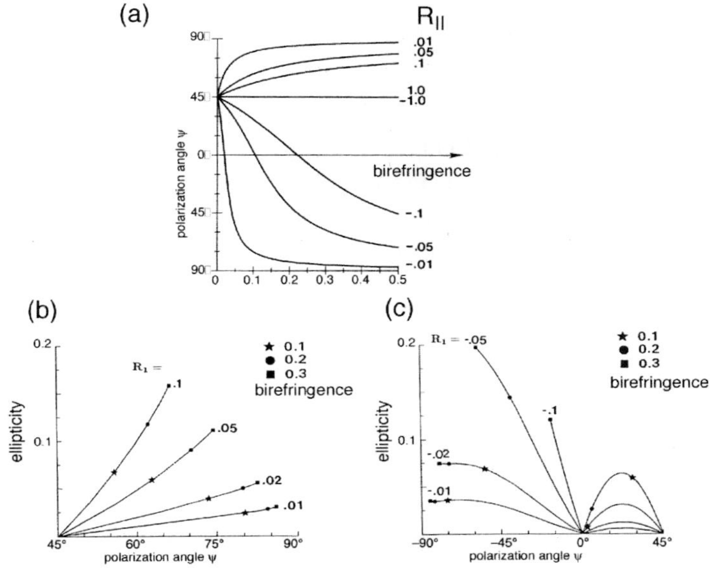

Figure 5.29. (a) The sensitivity of angle ψ, defined in equation (5.41), to reservoir reflectivity and also birefringence. For isotropy, the angle is 45° as both reflection coefficients are identical.
(b, c) ellipticity against axis orientation for a range of birefringence, and different reflection coefficients. All diagrams are extracted from Spencer and Chi (1991).

5.3.3 Multi-component thin-bed response

It is difficult to detect thin fractured layers, as the shear-wave lags induced by such layers are small. A novel approach to this was considered by Spencer and Chi (1991), who proposed using a plot of the multi-component components XX against YY, after Alford rotation. For a very thick reservoir, this ratio applied to the top event yields the pseudo-angle

$$\tan\Omega = \frac{R_\perp}{R_{||}} . \tag{5.43}$$

For a thinly fractured bed (i.e., facture frequency above the tuning frequency) encased in isotropic surroundings, with equal and opposite $S_\perp$ and $S_{||}$ coefficients, interference from the top and bottom of the layer gives an elliptical motion in the plot of XX versus YY (see Figure 5.28). The major axis of this ellipse is oriented at angle Ω with

$$\tan\Omega \approx \frac{R_\perp}{R_{||}} \frac{\tau_\perp}{\tau_{||}} , \tag{5.44}$$

where $\tau_\perp$ and $\tau_\parallel$ are the two-way travel-times of the slow and fast shear waves through the bed. The angle Ω is independent of bed thickness, but very sensitive to the anisotropy in the target. In fact, changes of 120° or more are possible with up to 20 % birefringence. The ellipticity E of this motion is given by

$$E = \tfrac{1}{4}\omega(\tau_\perp - \tau_\parallel)\sin 2\Omega \ . \tag{5.45}$$

The ellipticity decreases with layer thickness and is sensitive to birefringence and reservoir reflectivity in a complicated fashion. Spencer and Chi (1991) suggest plotting Ω against E to recover the anisotropy (Figures 29b,c). This requires a calibration estimate for thickness of the reservoir interval and $R_\parallel$. From this, the anisotropy can be evaluated and subsequently mapped over the reservoir area.

CHAPTER 6 – ANISOTROPY ANALYSIS FROM FAR-OFFSET VSP II – ASYMMETRY AND NON-UNIFORMITY

The techniques described in Chapter 5 provide an essential step to anisotropy interpretation. They are robust and apparently quite effective. However, some cross-diagonal energy still remains unexplained. In practice, it is tempting to treat this energy as a visual measure of the quality of the Alford diagonalization. This would ignore important details for which a more careful analysis is necessary. This requires a better understanding of the acquisition error and the departures from the assumptions inherent in equation (4.13).

6.1 ACQUISITION INACCURACY

Errors arising from acquisition inaccuracy can be assessed by generalizing the original Alford rotation (an eigenanalysis) to a *singular value decomposition* (SVD). Important categories of error arising from both the source and receiver can be treated with this extension. The procedure yields a quality control indicator that helps to isolate acquisition elements that affect the data matrix. It should be noted that most problems originating from the source can be eliminated by the near-surface correction described in Section 4.2 (an example is shown in Section 4.2.4).

6.1.1 Source or receiver mis-orientation

The simplest cause of asymmetry arises from imperfect alignment of the source and receiver axes with the acquisition *X-Y* frame (Figure 6.1). It was observed in Section 4.3 that errors of up to 15° could remain, even after tool spin correction or attempts at source alignment in the field. To understand the overall impact on fracture analysis, first consider an anticlockwise misorientation error in the receiver axes of δ_R. Assuming vector fidelity, the predicted relationship between the actual earth response $\mathbf{D}^{(9)}(t)$ and the recorded data matrix, $\mathbf{D}^{(9)\prime}(t)$ is

$$\mathbf{D}^{(9)\prime}(t) = \mathbf{C}_Z(\delta_R)\mathbf{D}^{(9)}(t). \tag{6.1}$$

An Alford rotation is now applied to the idealization of (4.13), with this additional rotation included. The off-diagonal energy remaining after using the algebraic solution (5.19) or the linear transform approach (5.24) is minimized (but not removed completely). The result departs from the ideal case: the revised fracture strike estimate is now

$$\hat{\phi} = \phi + \frac{\delta_R}{2}, \tag{6.2}$$

and the best diagonalization

$$\hat{\Lambda}(t) = \mathbf{C}_Z\left(\frac{\delta_R}{2}\right)\Lambda(t)\mathbf{C}_Z\left(\frac{\delta_R}{2}\right), \tag{6.3}$$

in terms of the actual diagonal trace matrix (Figure 6.2). In view of the arguments outlined in Section 5.1.5, this type of asymmetry does not alter the correlation lag substantially. Thus, it is expected that errors in tool spin correction are directly added to the final polarization estimates, but their influence halved. With an additional anticlockwise error δ_S in the source axes, the techniques in Section 5.1 produce a total fracture strike bias of

$$\hat{\phi} = \phi + \left(\frac{\delta_R}{2} - \frac{\delta_S}{2} \right) . \tag{6.4}$$

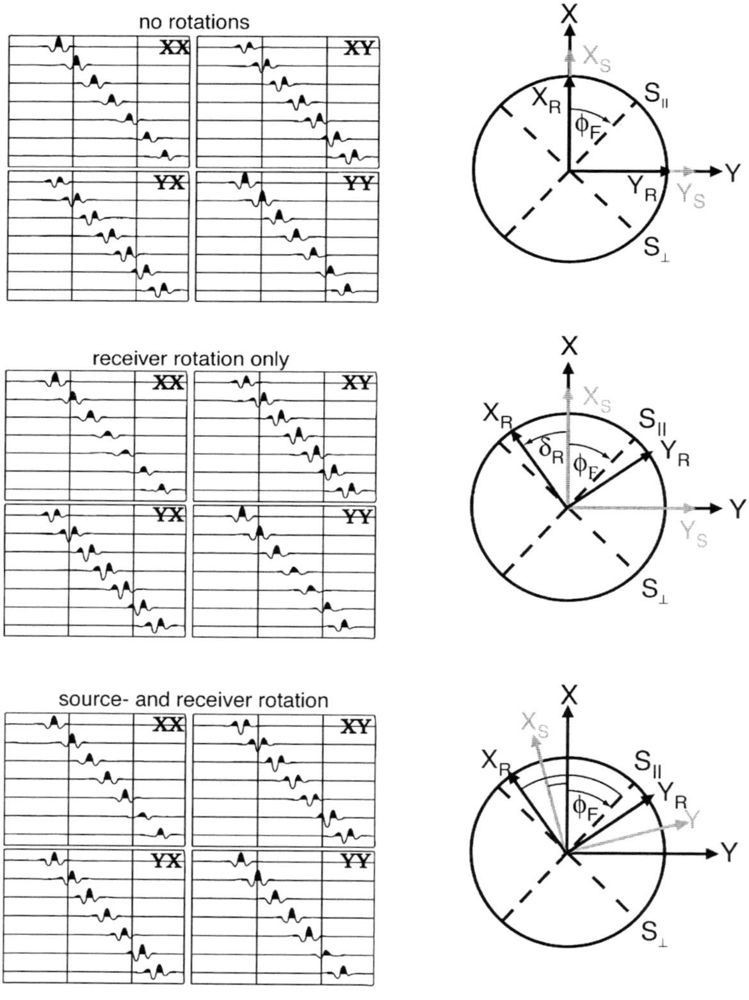

Figure 6.1. Illustration of the effect of source (X_S-Y_S axes) and receiver (X_R-Y_R axes) orientation errors on the data matrix. The X-Y axes represent the global acquisition frame. Note the apparent changes in polarity and phase on the off diagonal trace components and the relative stability on the main diagonals. This asymmetry is detected using the SVD algorithm.

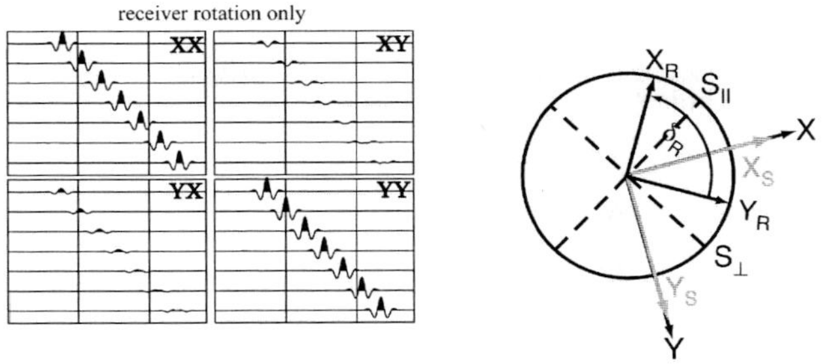

Figure 6.2. An Alford rotation does not totally minimize the data matrix when a receiver is incorrectly re-oriented. In this case the best diagonalization produces anti-symmetric off-diagonals.

Thus, the source and receiver errors may enhance or cancel each other depending upon their sense of rotation. These errors are likely to vary with depth if receiver contributions are significant, while the source contribution remains constant over a group of levels. The matrix combination of (4.13) and the rotation above suggest a generalization of the Alford rotation in (5.3) to accommodate the asymmetry arising from source and receiver errors (MacBeth et al. 1994):

$$\mathbf{M}'(t) = \mathbf{C}_Z^T(\phi_R)\mathbf{D}^{(9)\prime}(t)\mathbf{C}_Z(\phi_S) \;, \tag{6.5}$$

where ϕ_R and ϕ_S each provide a measure of the fracture strike plus the respective receiver and source errors (Figure 6.3). The equation has the form of a singular value decomposition (Lawson and Hanson 1974). As before, the best estimates of ϕ_R and ϕ_S that diagonalize the recorded data matrix can be determined by numerical search (over the two angles) using selected criteria, or an adaptation of the algebraic solution or linear transform technique (Section 5.1).

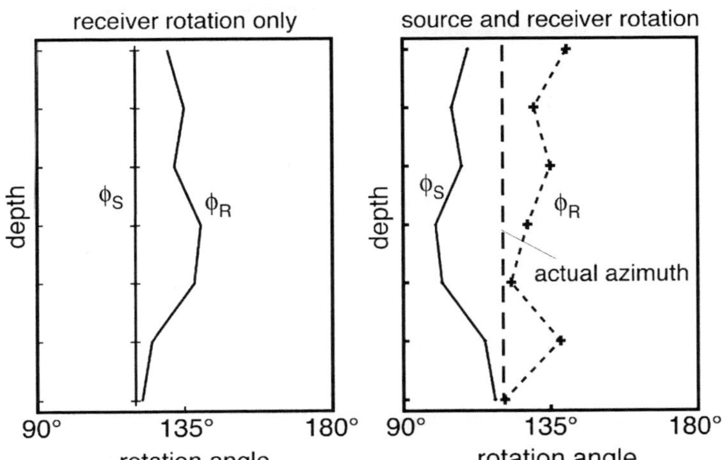

Figure 6.3. The results of application of the SVD algorithm to the data matrices in Figure 6.1. The method correctly picks up the actual variation in the rotation angles, and completely diagonalizes the matrices.

(a) *Algebraic solution* – Setting $\phi = \phi_R + \phi_S$ and $\overline{\phi} = \phi_R - \phi_S$, the right hand side of (6.5) can be written in terms of the transformed components of (5.4) as

$$\mathbf{M}'(t)=\frac{1}{2}\begin{pmatrix} a(t)\cos\overline{\phi}+\overline{a}(t)\cos\phi-b(t)\sin\phi+\overline{b}(t)\sin\overline{\phi} & -a(t)\sin\overline{\phi}+\overline{a}(t)\sin\phi+b(t)\cos\phi+\overline{b}(t)\cos\overline{\phi} \\ a(t)\sin\overline{\phi}+\overline{a}(t)\sin\phi+b(t)\cos\phi-\overline{b}(t)\cos\overline{\phi} & a(t)\cos\overline{\phi}-\overline{a}(t)\cos\phi+b(t)\sin\phi+\overline{b}(t)\sin\overline{\phi} \end{pmatrix} \quad (6.6)$$

The off-diagonal energy is

$$E_{OD}(\phi_R,\phi_S)=\sum_{i=1}^{N}\left(-a_i\sin\overline{\phi}+\overline{a}_i\sin\phi+b_i\cos\phi+\overline{b}_i\cos\overline{\phi}\right)^2+$$

$$\sum_{i=1}^{N}\left(a_i\sin\overline{\phi}+\overline{a}_i\sin\phi+b_i\cos\phi-\overline{b}_i\cos\overline{\phi}\right)^2 . \quad (6.7)$$

The minimum energy criterion of (5.16) yields

$$\sum_i\left(\overline{a}_i\sin\phi+b_i\cos\phi\right)\left(\overline{a}_i\cos\phi-b_i\sin\phi\right)=0 \quad (6.8)$$

and

$$\sum_i\left(-a_i\sin\overline{\phi}+\overline{b}_i\cos\overline{\phi}\right)\left(a_i\cos\overline{\phi}+\overline{b}_i\sin\overline{\phi}\right)=0. \quad (6.9)$$

Solutions are

$$\hat{\phi}_R=\frac{1}{4}\tan^{-1}\left(\frac{2\sum_i\overline{a}_ib_i}{\sum_i\left(\overline{a}_i^2-b_i^2\right)}\right)-\frac{1}{4}\tan^{-1}\left(\frac{2\sum_i a_i\overline{b}_i}{\sum_i\left(a_i^2-\overline{b}_i^2\right)}\right)+k_1\frac{\pi}{4} \quad (6.10)$$

and

$$\hat{\phi}_S=\frac{1}{4}\tan^{-1}\left(\frac{2\sum_i\overline{a}_ib_i}{\sum_i\left(\overline{a}_i^2-b_i^2\right)}\right)+\frac{1}{4}\tan^{-1}\left(\frac{2\sum_i a_i\overline{b}_i}{\sum_i\left(a_i^2-\overline{b}_i^2\right)}\right)+k_2\frac{\pi}{4}, \quad (6.11)$$

where k_1 and k_2 are integers. The angles ϕ_R and ϕ_S each have four possible solutions, giving sixteen possible extrema. As before, the actual solution must be selected by examining the sign of the double derivative and the correlation lag time. Comparison of (5.19) with the solutions (6.10) and (6.11) shows that the fracture strike determined by the algebraic version of the Alford rotation is

$$\hat{\phi}=\frac{\hat{\phi}_R+\hat{\phi}_S}{2}, \quad (6.12)$$

from which the relation (6.4) can also be deduced. It is important to note that the SVD technique permits separate estimation of ϕ_R and ϕ_S, and hence a possibility to monitor the apportionment of errors between the source and receiver components. While it cannot be used to determine the unknown fracture strike – for which further a priori information or assumptions are necessary – the closer φ_R and ϕ_S, the better their average will estimate the fracture strike.

(b) Linear *transform approach* – Assuming that the form (4.13) is valid, inspection of the transform components yields

$$a(t) = (\lambda_2(t) + \lambda_1(t))\cos(\phi_R - \phi_S) \ , \tag{6.13a}$$

$$\bar{a}(t) = -(\lambda_2(t) - \lambda_1(t))\cos(\phi_R + \phi_S) \ , \tag{6.13b}$$

$$b(t) = (\lambda_2(t) - \lambda_1(t))\sin(\phi_R + \phi_S) \ , \tag{6.13c}$$

and

$$\bar{b}(t) = -(\lambda_2(t) + \lambda_1(t))\sin(\phi_R - \phi_S) \ . \tag{6.13d}$$

These equations predict a direct linear relationship between $a(t)$ and $\bar{b}(t)$ and between $\bar{a}(t)$ and $b(t)$, from which it is possible to determine $\hat{\phi}_R - \hat{\phi}_S$ and $\hat{\phi}_R + \hat{\phi}_S$. These lead in turn to separate determinations of ϕ_R, ϕ_S, $\lambda_2(t) - \lambda_1(t)$ and $\lambda_2(t) + \lambda_1(t)$.
(c) An alternative is to form the covariance matrices $\sum_{i=1}^{N} \mathbf{D}^{(9)}(t_i)\mathbf{D}^{(9)}(t_i)^T$ and $\sum_{i=1}^{N} \mathbf{D}^{(9)}(t_i)' \mathbf{M}^{(9)}(t_i)$. Application of eigenanalysis to the single resultant matrix from each summation provides an alternative estimate of both ϕ_R and ϕ_S.

6.1.2 Polarity errors

Accurate knowledge of polarities is not guaranteed, even when standards set for consistent multi-component recording and processing (Pruett 1989, Landrum et al. 1994) are carefully followed. A field test of surface three-component geophones by Lawton and Bertram (1993) underlined the practical problems associated with polarity reversals, incorrect coding and couplings. In particular, polarity reversals are fairly common in multi-component acquisition and have been notoriously difficult to eliminate, especially with phase shifts or narrow-bandwidth data. Such reversals can cause asymmetry, and the SVD method helps to detect their presence. Four categories of polarity change can be considered

(a) a single source component is reversed, producing a left-hand source frame – this gives $\phi_R = \phi$, $\phi_S = -\phi$, and an Alford rotation result of 0° or 90°.
(b) a single geophone component is reversed, producing a left-hand receiver frame – this gives $\phi_R = -\phi$, $\phi_S = \phi$, and an Alford rotation result of 0° or 90°.
(c) both a source and a geophone component are reversed – this gives $\phi_R = -\phi$, $\phi_S = -\phi$, with an Alford rotation result of $-\phi$.
(d) both components are reversed for either the sources or the geophones, equivalent to a rotation of 180° – this gives $\phi_R = \phi$, $\phi_S = \phi$, and an Alford rotation result of ϕ.

The most obvious situations that can be detected are (a) and (b), for which the presence of a polarity change will be immediately obvious, but not its origin. For (c) and (d) there is no opportunity to detect a change and the fracture strike could remain inaccurate. The results in (d) also suggest that if there is a constant phase shift between the in-line and cross-line sources, the estimate of the fracture strike remains accurate.

6.1.3 Source or receiver imbalance/misalignment

The in-line and cross-line sources often have different strengths, either due to a difference in coupling or due to the stacking procedure. Assuming identical source wavelets, the recorded four-component data matrix is now

$$\mathbf{D}^{(4)\prime}(t) = \mathbf{D}^{(4)}(t)\begin{pmatrix} 1 & 0 \\ 0 & \kappa_S \end{pmatrix} \tag{6.14}$$

or

$$\mathbf{D}^{(4)\prime}(t) = \mathbf{D}^{(4)}(t)\begin{pmatrix} m_{XX} & \kappa_S m_{XY} \\ m_{YX} & \kappa_S m_{YY} \end{pmatrix}, \tag{6.15}$$

where κ_S is a scaling factor with values between 0 and 1. The value of κ_S can be estimated using the recorded energies from each source:

$$\hat{\kappa}_S^2 = \frac{\sum_{i=1}^{N}\left(d_{XX}^2(t) + d_{YX}^2(t)\right)}{\sum_{i=1}^{N}\left(d_{XY}^2(t) + d_{YY}^2(t)\right)}. \tag{6.16}$$

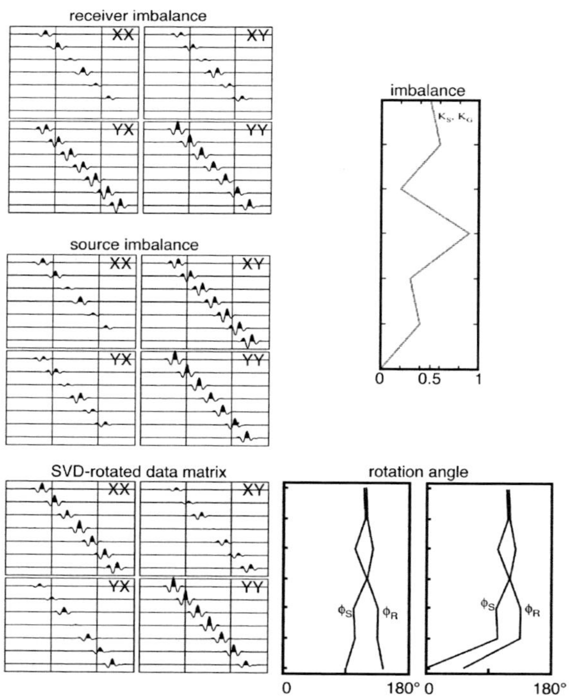

Figure 6.4. The effect of imbalance in either the source or the receiver components is to create an asymmetry in the data matrix. These synthetic seismograms have been computed using the imbalance profile shown top right, and the SVD results are shown in the lower diagrams, where the best diagonalization for the source case is also shown. Note that although divergence of the two angles is symptomatic of this particular situation, unlike the imperfect re-orientation, a residual off-diagonal energy remains.

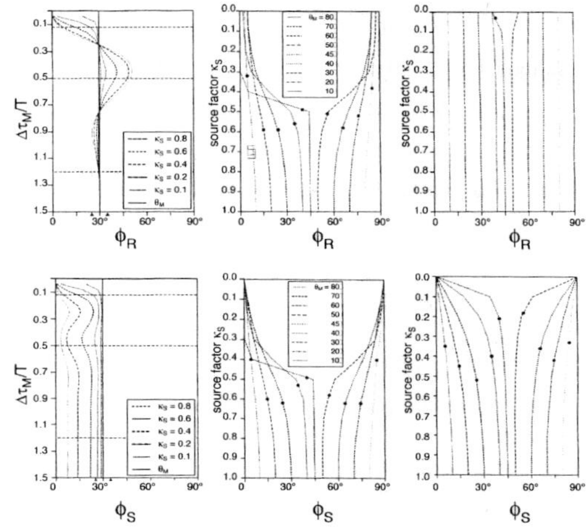

Figure 6.5. Variation of the asymmetry indicators from SVD, for different source imbalance (κ_S), birefringence (lag time $\Delta\tau_M$ normalized by the peak period T), and principal directions (θ_M). From Zeng and MacBeth (1993a).

It has been shown that such source balancing can substantially reduce fracture strike fluctuations (Choi and Gangi 1991). Any remaining imbalance leads to an asymmetry that can be analysed using (6.10) and (6.11), but the final relationship to the angles ϕ_R and ϕ_S is complicated. In general, the divergence of ϕ_R and ϕ_S increases with the mismatch between the sources (Figure 6.4). The magnitude of the divergence depends on the shear wave birefringence (Figure 6.5). Zeng and MacBeth (1993a) have shown that a 40 % reduction in cross-line source strength can be tolerated when a fracture strike accuracy of 5° or less is necessary. This requires a shear wave birefringence with a lag of more than 0.22 T (where T is the peak period of the data) – for example, 11 ms at 20 Hz. For smaller lags, the condition reduces to 80 %. It thus appears that accuracy is guaranteed by both source balancing prior to analysis and an inherent robustness against source imbalance.

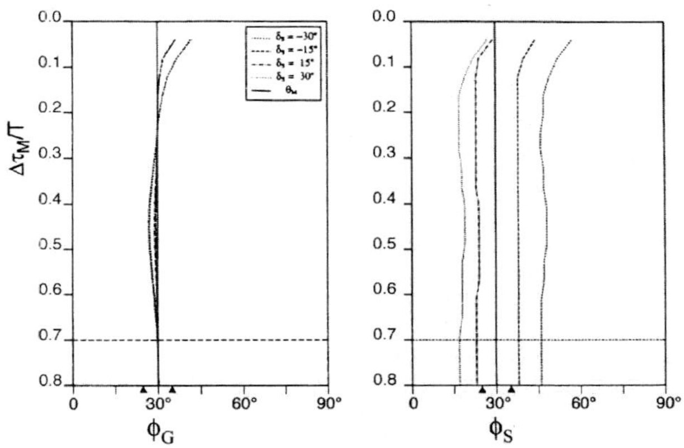

Figure 6.6. Asymmetry indicators from SVD analysis for different misalignment factors (δ_S), and birefringence values (defined as in Figure 6.5). From Zeng and MacBeth (1993a).

Zeng and MacBeth (1993a) have also demonstrated a further insensitivity to single source misalignment against the orthogonal frame (Figure 6.6). In particular, for up to 20° misalignment, the fracture strike is in error by only 5°, provided birefringence values are larger than 0.12 T. The near-surface correction of Section 4.2 renders balancing or alignment correction unnecessary, and significantly improves the accuracy of the anisotropy analysis. The overall effect of receiver imbalance is similar, although the differential coupling response in the downhole package is expected to be smaller.

6.1.4 Data examples

6.1.4.1 Four-component data from the Romashkino field

Figures 6.7 and 6.8 show the results of quality control analysis using SVD. The original Alford rotation results appeared reasonable, with the final data matrix only little residual energy on the off-diagonal traces. However, the apparent symmetry of ϕ_R and ϕ_S about a common azimuth of 60° looks suspicious (Figure 6.7a).

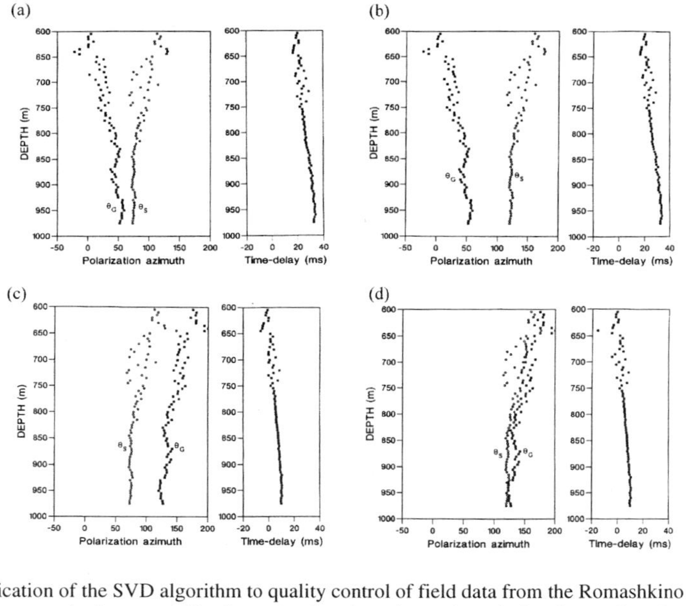

Figure 6.7. Application of the SVD algorithm to quality control of field data from the Romashkino field. This particular example shows combined source mis-orientation and a polarity change. Results are for original data (a), (b) correction for source mis-orientation, (c) correction for polarity reversal of the in-line geophone, and (d) correction for the source misalignment and the polarity reversal (MacBeth et al. 1994).

Unfortunately, taking into account the known source misalignment has the unexpected effect of increasing ϕ_R by 47°, thus making $\phi_R - \phi_S$ even larger (Figure 6.7b). The diagonalization is now considerably worse (Figures 8a and 8b), although the lag estimates remain the same. These observations are due to the opposite polarities of the X and Y receivers. A left-handed set used in the acquisition explains these polarities. This was confirmed by analysis of the P-waves from an offset source (and also by the contractor). Correction for this produces almost parallel ϕ_R and ϕ_S estimates separated on average by 47° (Figures 7c and 7d), which is the value of the source misalignment. The lag estimates increase linearly from –6 ms to 10 ms, with the cross-correlation algorithm now picking an earlier half-cycle (giving about 20 ms difference). Further inspection reveals that the off-diagonal energy could be reduced in the shallower section by another source rotation of 10° combined with some source balancing.

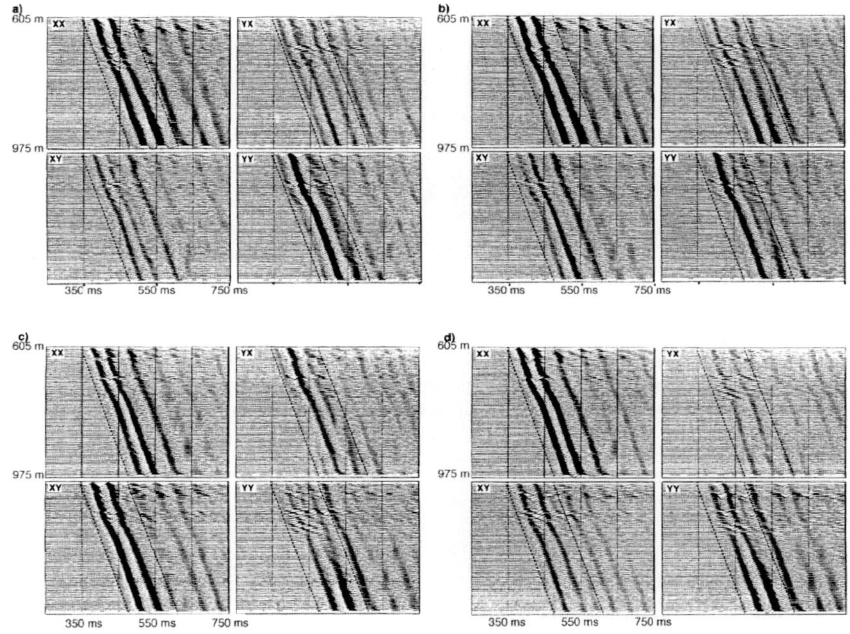

Figure 6.8. Attempts at the diagonalization of the recorded data matrices from the field data analysed in Figure 6.7, after various stages involving correction for the acquisition errors (MacBeth et al. 1994).

6.1.4.2 Lost Hills field

Figures 6.9 and 6.10 show the results of SVD analysis applied to the four-component data of Figure 5.8 from the Lost Hills field, California. The SVD analysis was originally applied in an attempt to understand the complicated results (Figure 5.9). The ϕ_R and ϕ_S estimates lie together only at isolated depths and exhibit an oscillatory behavior with mirror symmetry. If the data matrix is corrected for the known misalignment between the source and receiver, the ϕ_R and ϕ_S estimates move closer together. The mirror symmetry remains however, suggesting a polarity reversal. Unfortunately, there is no further evidence to support this polarity reversal. Instead, it is found that layer stripping of the overburden anisotropy (almost equivalent to a near-surface correction in this case) brings the SVD estimates together, but produces a slightly unusual variability in polarization which does coincide with source pad changes. The lags remain oscillatory in nature, with only a marginal improvement in the off-diagonal energy.

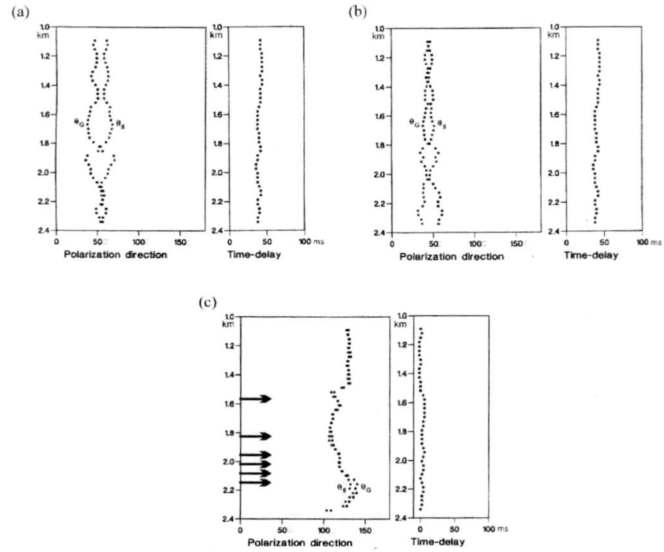

Figure 6.9. Application of the SVD algorithm to field data from the Lost Hills field. Results are for: (a) original data; (b) correction for source misorientation; and (c) correction for overburden anisotropy. Horizontal arrows mark points at which the pad position was changed to preserve coupling (MacBeth et al. 1994).

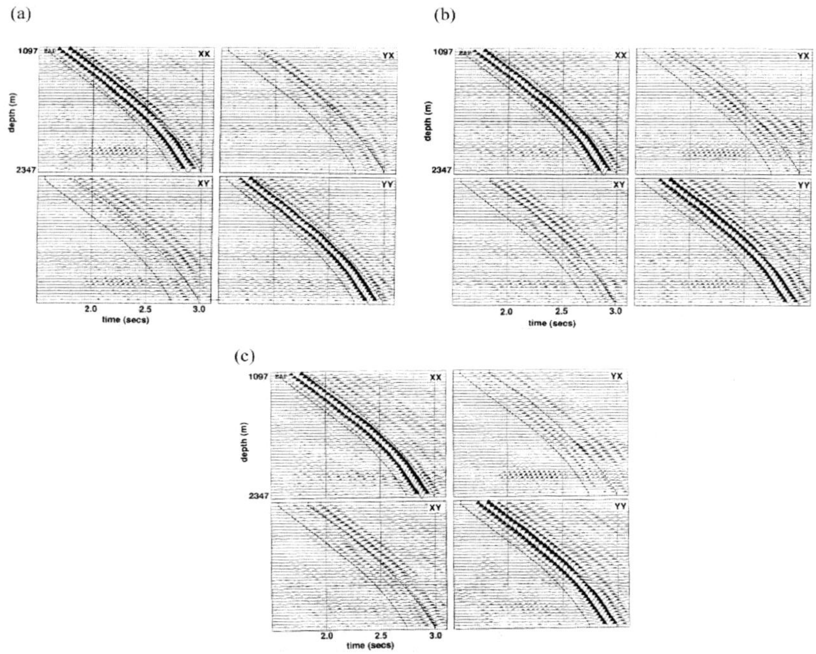

Figure 6.10. Attempts to diagonalize the data in Figure 6.9. The changes on the off-diagonal components are very subtle and could give a false impression if visual examination alone is used as a guide to the final quality (MacBeth et al. 1994).

6.2 NON-UNIFORMITY WITH DEPTH

There has been growing observational evidence to suggest that the orientation of anisotropy changes with depth. The consequence is shear wave polarization rotates, which in turn results in multiple shear-wave splitting – each shear wave separately splits into further two (Figures 6.11 and 6.12). This phenomenon leads to a departure from the idealization in (4.13), and – if not properly taken into account – to an erroneous interpretation. Figure 6.13 shows examples of the expected effect on the four-component data matrix.

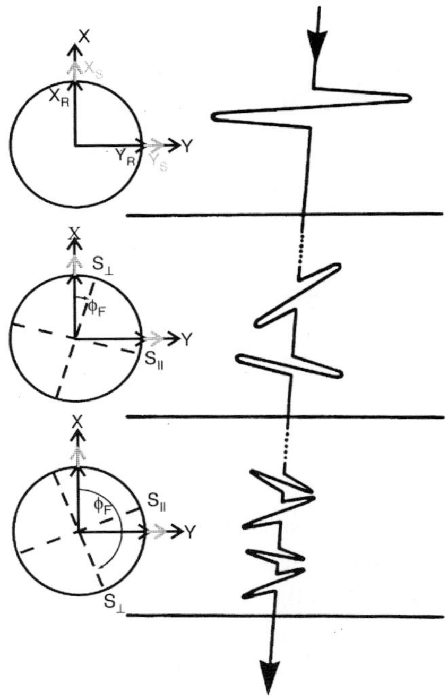

Figure 6.11. Illustration of the multiple shear wave birefringence that causes asymmetry in the data matrix. In this case there are a total of four shear wave arrivals, the first of which is polarized parallel to the fracture strike in the second medium. Adapted from Yardley (1993).

6.2.1 Discrete polarization change at depth

The simplest adaptation of (4.13) is accomplished by sub-dividing the subsurface into two uniform TIH media A and B, with fracture strikes along ϕ_A and ϕ_B. Each source motion is decomposed as before into independent shear waves propagating through medium A. Upon transmission across the interface, the energy then propagates as a different set of shear wave modes to the receivers for final projection onto the receiver axes. This entire process can be written as

$$\mathbf{D}(t) = \left(\mathbf{C}_Z(\phi_B)\Lambda_B(t)\right) * \left(\mathbf{T}_{AB}\Lambda_A(t)\right) * \mathbf{C}_Z^T(\phi_A)\mathbf{S}(t) \ . \tag{6.17}$$

which reduces to the four-component form of (4.13) when $\phi_A = \phi_B$. Transmission across the interface mixes energies of the two pairs of pure modes in each medium. The inter-conversion coefficients are shown in Figure 6.14 and calculated in the Appendix. For small changes $\Delta\phi = \phi_B - \phi_A$, they are approximately

$$\mathbf{T} \approx \begin{pmatrix} t_{11} & 0 \\ 0 & t_{22} \end{pmatrix} \mathbf{C}_Z(\Delta\phi) \ . \tag{6.18}$$

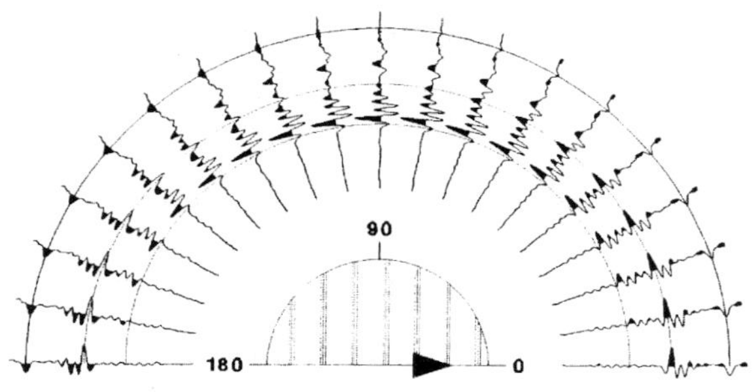

Figure 6.12. The evolution of double birefringence as a function of receiver rotation but fixed transmitter polarization, for a stack of two samples of McNabb shale. Bedding planes of the lower shale are vertical (out of the plane of the page) but rotated at 45° to the transmitter polarization (arrow). At 0°, the second and fourth shear waves are prominent, and at 90° the first and third. Four separate arrivals can be seen on the other azimuths. The reverberations are caused by interference of the second and third arrivals. After Sondergeld and Rai (1992).

This result can be visualized by considering an infinitesimal isotropic layer sandwiched between the two media, with the same background properties as medium A. Modes propagating through medium A are transmitted across the anisotropic/isotropic interface with only a small loss. In the isotropic material, the shear waves are projected onto the principal directions of medium B, and propagate across the isotropic/anisotropic interface with coefficients t_{11} and t_{22} determined mainly by the impedance contrast. The dominating influence on the transmission is the change in polarization (Figure 6.14). Asymmetry of the data matrix can be immediately observed, as the matrices of the group do not commute.

If the polarization change is exactly 90°, multiple splitting does not occur, but instead there is a linear decrease in lag (see, e.g., Figure 5.11). The model in (6.17) can predict this by firstly giving $t_{11} \approx t_{22}$ and hence

$$\mathbf{D}_{90}(t) \approx \left(\mathbf{C}_Z(\phi_B)\mathbf{\Lambda}_B(t)\right) * \left(\mathbf{C}_Z^T(\phi_B)\mathbf{C}_Z(\phi_B + \tfrac{\pi}{2})\mathbf{\Lambda}_A(t)\right) * \left(\mathbf{C}_Z^T(\phi_B + \tfrac{\pi}{2})\mathbf{S}(t)\right) , \tag{6.19}$$

or

$$\mathbf{D}_{90}(t) \approx \left(\mathbf{C}_Z(\phi_B)\mathbf{\Lambda}_B(t)\right) * \left(\mathbf{C}_Z(\tfrac{\pi}{2})\mathbf{\Lambda}_A(t)\right) * \left(\mathbf{C}_Z^T(\tfrac{\pi}{2})\mathbf{C}_Z^T(\phi_B)\mathbf{S}(t)\right) . \tag{6.20}$$

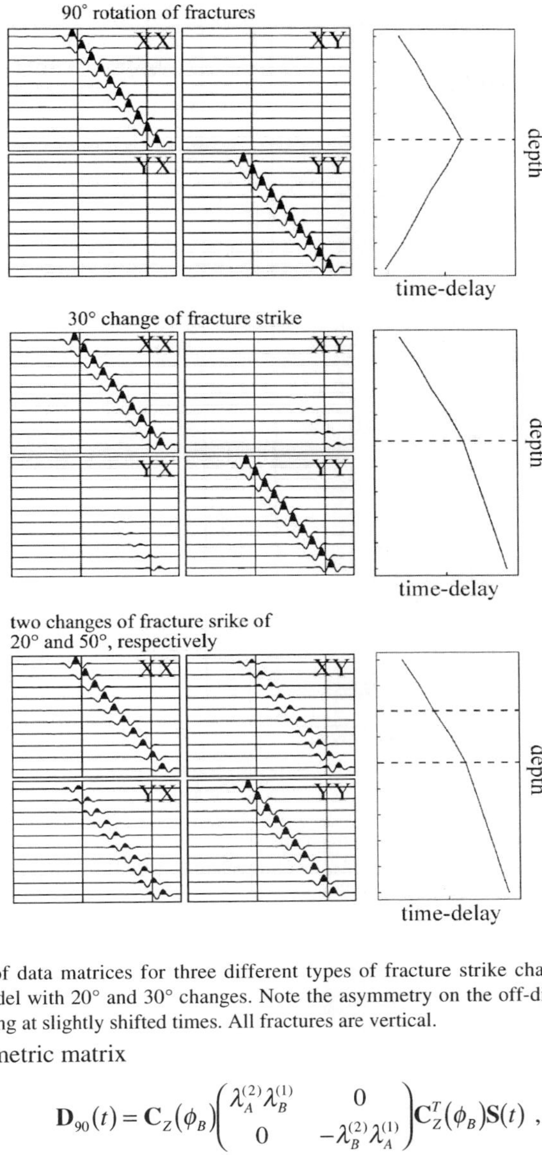

Figure 6.13. Examples of data matrices for three different types of fracture strike change, 90°, 30°, and a three-layered model with 20° and 30° changes. Note the asymmetry on the off-diagonal data, with the wavelets appearing at slightly shifted times. All fractures are vertical.

This leads to the symmetric matrix

$$\mathbf{D}_{90}(t) = \mathbf{C}_Z(\phi_B)\begin{pmatrix} \lambda_A^{(2)}\lambda_B^{(1)} & 0 \\ 0 & -\lambda_B^{(2)}\lambda_A^{(1)} \end{pmatrix}\mathbf{C}_Z^T(\phi_B)\mathbf{S}(t) \ , \tag{6.21}$$

The principal diagonal contains the product of the fast and slow operators for both media; they reduce the travel time difference between the shear waves, and hence the net lag. Medium B effectively reverses the overall lag built up in medium A. The Alford rotation does not see this change of polarization, but can detect the lag reduction with depth. The "faster" wave may eventually overtake the "slower"; as algorithms are usually set to output only positive lag, they compensate by recording a change in the polarization.

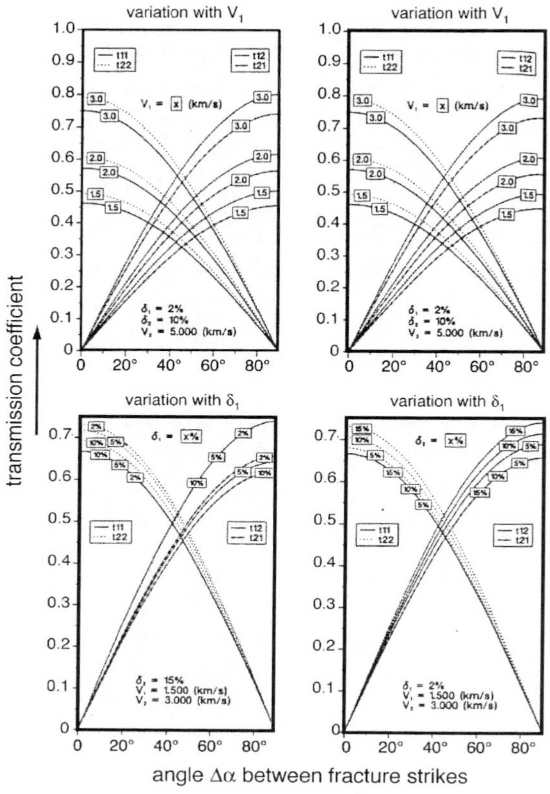

Figure 6.14. Transmission coefficients for intermodal shear wave mixing at an interface between two media with different fracture strikes. The coefficients are calculated from the accurate solution of (6.52) for a fracture change $\Delta\alpha$, birefringences (δ_1 and δ_2), and various interface contrasts. After Li (1992).

6.2.2 Estimation of non-uniformity of properties

6.2.2.1 Layer stripping

The layer-stripping procedure accounts for the polarization change by unraveling multiple shear wave splitting. The method was developed by Winterstein and Meadows (1991a, b) by extending Alford rotation. It works by repeated application of Alford rotation followed by lag reversal, and can be exemplified by application to the model of (6.17). Assuming ideal sources, the first Alford rotation gives

$$\mathbf{M}'(t) = \mathbf{C}_Z^T(\phi)\mathbf{D}(t)\mathbf{C}_Z(\phi) \ , \tag{6.22}$$

which yields

$$\mathbf{M}'(t) = \left(\mathbf{C}_Z(\phi_B - \phi)\mathbf{\Lambda}_B(t)\right) * \left(\mathbf{C}_Z^T(\phi_B)\mathbf{C}_Z(\phi_A)\mathbf{\Lambda}_A(t)\mathbf{C}_Z^T(\phi_A - \phi)\right) \ . \tag{6.23}$$

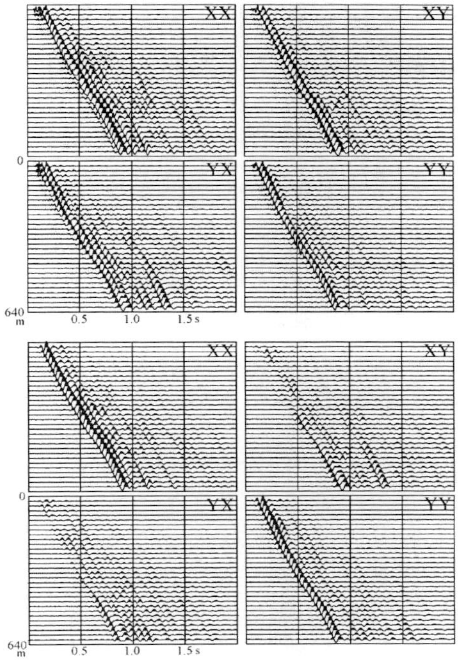

Figure 6.15. Four-component data acquired in the Cymric field, California. Above: raw stacked data; Below: band-pass filtered at 30 Hz, tool-spin corrections, and Alford rotation. Note the strong residual cross-diagonal energy and its increase with depth – this is symptomatic of polarization changes. Data courtesy of the British Geological Survey.

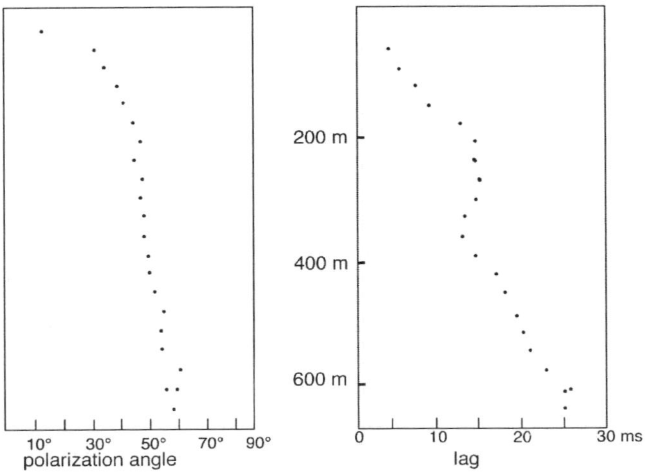

Figure 6.16. Polarization and lag determined from initial Alford rotation of the data in Figure 6.15. The systematic increase in angle is due to thin layers with change in polarization. The lag profile has been shifted to compensate for a small near-surface effect. Note the distinct reversal in the lag gradient at 210 m, a clear indication of a major polarization change. Re-drawn from Winterstein and Meadows (1991b).

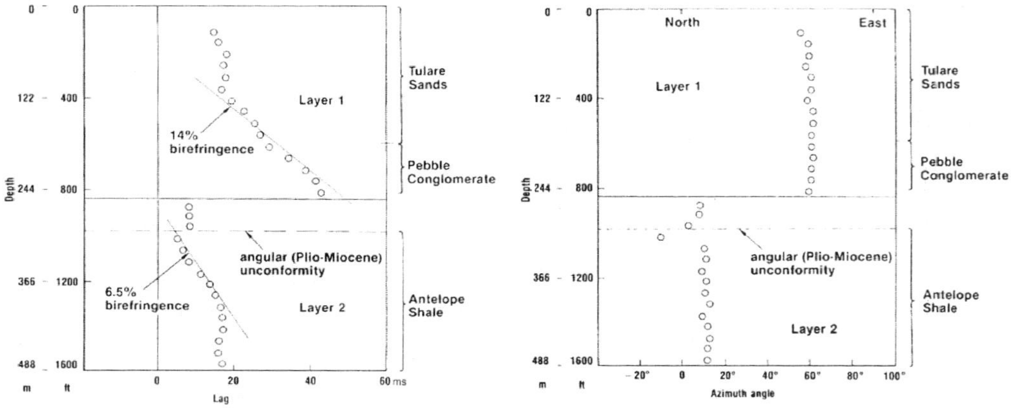

Figure 6.17. Final results of layer stripping by Winterstein and Meadows (1991b) for the Cymric VSP data. The polarization change occurs at an unconformity with a 35° dip.

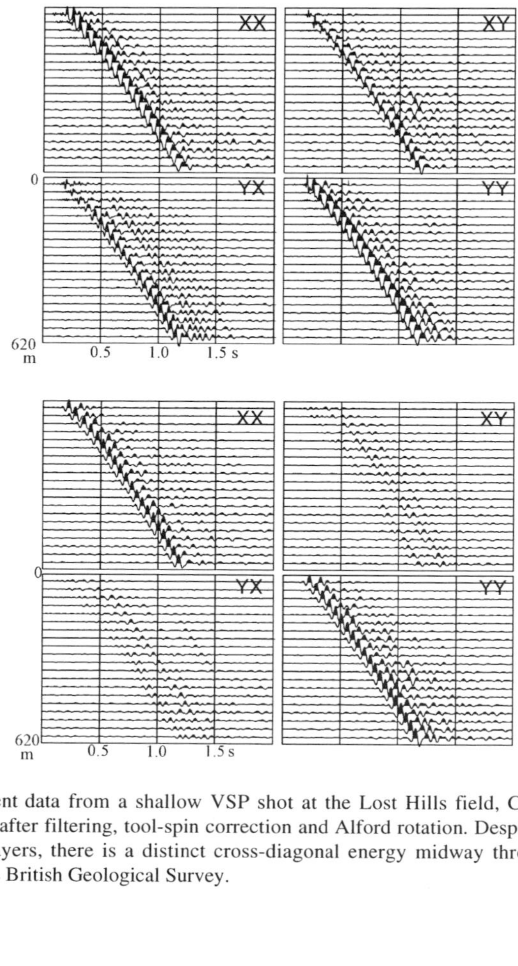

Figure 6.18. Four-component data from a shallow VSP shot at the Lost Hills field, California. Top: raw stacked data; Bottom: after filtering, tool-spin correction and Alford rotation. Despite a satisfactory rotation for the shallow layers, there is a distinct cross-diagonal energy midway through the depth range. Data courtesy of the British Geological Survey.

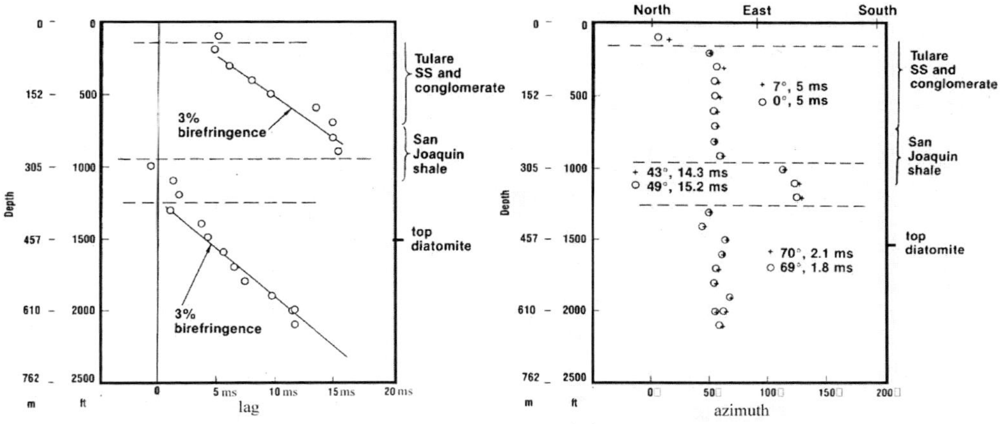

Figure 6.19. The results of a final layer stripping analysis by Winterstein and Meadows (1991a) for the data in Figure 6.18. The lags indicate an anomalous zone between 210 m and 370 m, but an otherwise linear increase in the two media. Results are shown for a near-surface with two different polarization angles and a 5 ms lag.

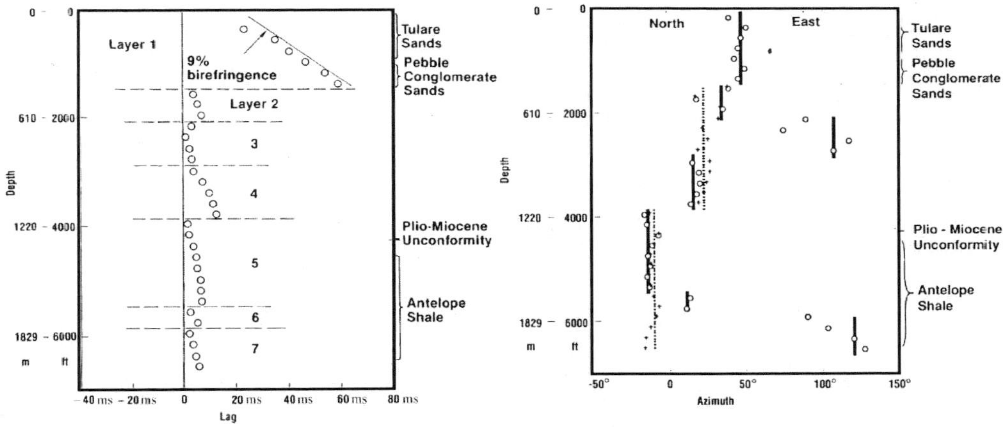

Figure 6.20. Final detailed layer stripping for multi-component data from the Railroad Gap field, California (extracted from Winterstein and Meadows 1991b). Seven layers are required to accommodate the polarization changes (circles and solid vertical lines). All lags increase linearly with depth within each layer.

A numerical search over ϕ to minimize the off-diagonal energy sets ϕ to ϕ_A and produces the form

$$\mathbf{M}'(t) = \left(\mathbf{C}_Z(\phi_B - \phi_A)\mathbf{\Lambda}_B(t)\right) * \left(\mathbf{C}_Z^T(\phi_B)\mathbf{C}_Z(\phi_A)\mathbf{\Lambda}_A(t)\right) . \tag{6.24}$$

This is then followed by a re-adjustment of the lag times (post-multiplication by the operator $\mathbf{\Lambda}_A^{-1}(t)$) to give

$$\mathbf{M}''(t) = \mathbf{C}_Z(\phi_B - \phi_A)\mathbf{\Lambda}_B(t)\mathbf{C}_Z^T(\phi_B - \phi_A) . \tag{6.25}$$

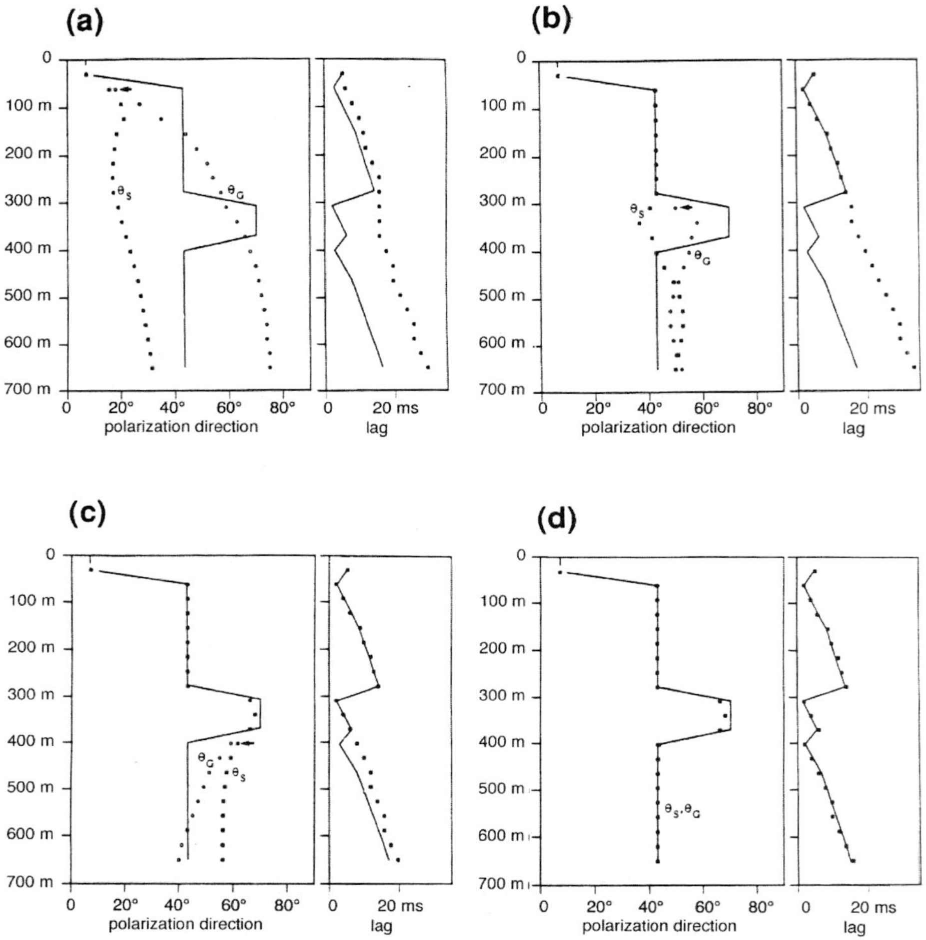

Figure 6.21. Synthetic seismogram example of layer stripping guided by SVD asymmetry indicators. The model values are shown by the solid lines, with each diagram representing a successive strip to greater depths (MacBeth et al. 1994).

In practice this operation can be performed as a shift in the time domain. **M″** is a symmetric matrix, for which a second Alford rotation provides a lag estimate for the lower medium. The technique is not limited to one polarization change, and multiple applications are possible by further iteration of this cycle. Winterstein and Meadows (1991a) define a strategy for when to layer strip, and a recipe for decision making during this interactive process. Field examples of the procedure are given in Figures 6.15 to 6.20. The approach requires the data bracket the depth at which the change occurs, which in turn means that the VSP must extend to the near-surface layers for a complete interpretation. Near-surface correction helps to evolve the method at a subsurface datum, and thus enhances accuracy. The asymmetry of the data matrix provides a robust indicator that a zone of polarization change has been encountered, and may be used to guide the layer stripping (Figures 21 and 22).

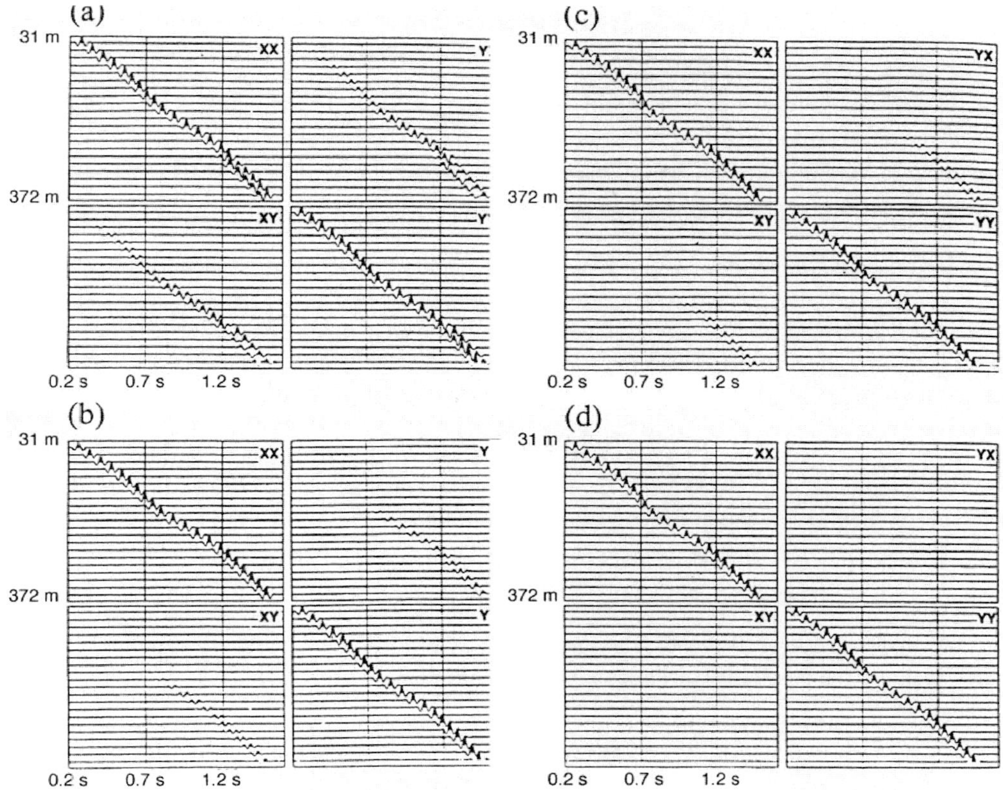

Figure 6.22. Best Alford rotation results after different stages of the layer stripping procedure in Figure 6.21 (MacBeth et al. 1994).

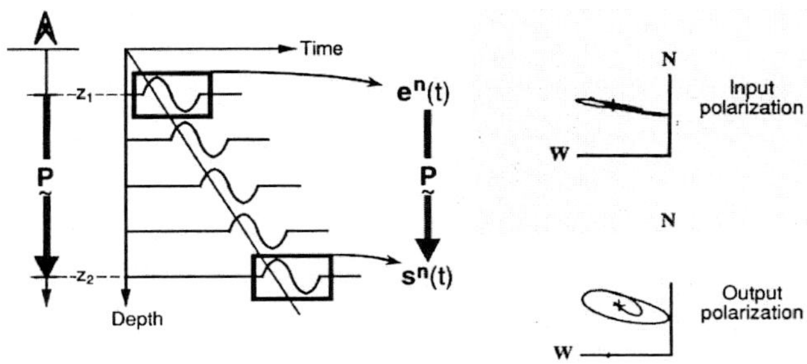

Figure 6.23. The propagator matrix **P** connects two sets of multi-component traces. It is determined by using all source motions and wave types available (Lefeuvre et al. 1992).

6.2.2.2 Wavefield extrapolation

This approach was adopted by MacBeth and Yardley (1992); MacBeth, Li and Queen (1993); and Ohanian and Beckham (1992). Here, a trial inverse operator

$$\mathbf{N}(\omega) = \mathbf{C}_Z^T(\phi_A)\mathbf{\Lambda}_A^{-1}(\omega)\mathbf{C}_Z(\phi_A) \tag{6.26}$$

directly post-multiplies the data matrix in the frequency domain. The parameters for ϕ_A and $\mathbf{\Lambda}_A(\omega)$ are taken from an Alford rotation performed on the layer to be stripped. As the near-surface region is unlikely to have uniform anisotropy, the near-surface correction of Section 4.2 is recommended as a first step before repeated application at each depth.

6.2.2.3 Interval estimates

A further possibility is to calculate a propagator or transfer matrix $\mathbf{M}_{int}(t)$ connecting the horizontal recordings $\mathbf{d}_j(t)$ at different depths (Figure 6.23) (Naville, 1986; Nicoletis et al. 1988; Lefeuvre et al. 1992), for example

$$\mathbf{d}_2(t) = \mathbf{M}_{int}(t) * \mathbf{d}_1(t) \ , \tag{6.27}$$

or groups of depth levels and/or different source realizations

$$(\mathbf{d}_{i+1}(t) \,|\, \mathbf{d}_{i+2}(t) \,|\, \ldots\ldots \,|\, \mathbf{d}_{i+n-1}(t)) = \mathbf{M}_{int}(t) * (\mathbf{d}_i(t) \,|\, \mathbf{d}_{i+1}(t) \,|\, \ldots\ldots \,|\, \mathbf{d}_{i+n-2}(t)) \ . \tag{6.28}$$

The inverse problem is identical to that described in Section 4.2, and with a multi-component source includes the possibility of direct application of the multi-component VSP recordings at levels i and i+1

$$\mathbf{D}_2(t) = \mathbf{M}_{int}(t) * \mathbf{D}_1(t) \ , \tag{6.29}$$

or for a group of levels

$$(\mathbf{D}_{i+1}(t) \,|\, \mathbf{D}_{i+2}(t) \,|\, \ldots\ldots \,|\, \mathbf{D}_{i+n-1}(t)) = \mathbf{M}_{int}(t) * (\mathbf{D}_i(t) \,|\, \mathbf{D}_{i+1}(t) \,|\, \ldots\ldots \,|\, \mathbf{D}_{i+n-2}(t)) \ . \tag{6.30}$$

The solution to these equations can be obtained by least-squares inversion. Writing (6.27) in the form

$$\begin{pmatrix} X_2(\omega) \\ Y_2(\omega) \end{pmatrix} = \begin{pmatrix} P_{XX} & P_{XY} \\ P_{YX} & P_{YY} \end{pmatrix} \begin{pmatrix} X_1(\omega) \\ Y_1(\omega) \end{pmatrix} , \tag{6.31}$$

Lefeuvre et al. (1992) suggest one specific solution for the propagator components

$$P_{XX}(\omega) = \frac{\langle X_1^* X_2 \rangle \cdot \langle Y_1^* Y_1 \rangle - \langle Y_1^* X_2 \rangle \cdot \langle X_1^* Y_1 \rangle}{\langle X_1^* X_1 \rangle \cdot \langle Y_1^* Y_1 \rangle - \langle Y_1^* X_1 \rangle \cdot \langle X_1^* Y_1 \rangle} , \tag{6.32}$$

$$P_{XY}(\omega) = \frac{\langle X_1^* X_1 \rangle \cdot \langle Y_1^* X_2 \rangle - \langle Y_1^* X_1 \rangle \cdot \langle X_1^* X_2 \rangle}{\langle X_1^* X_1 \rangle \cdot \langle Y_1^* Y_1 \rangle - \langle Y_1^* X_1 \rangle \cdot \langle X_1^* Y_1 \rangle} , \tag{6.33}$$

$$P_{YX}(\omega) = \frac{\langle X_1^* Y_2 \rangle \cdot \langle Y_1^* Y_1 \rangle - \langle Y_1^* Y_2 \rangle \cdot \langle X_1^* Y_1 \rangle}{\langle X_1^* X_1 \rangle \cdot \langle Y_1^* Y_1 \rangle - \langle Y_1^* X_1 \rangle \cdot \langle X_1^* Y_1 \rangle}, \tag{6.34}$$

$$P_{YY}(\omega) = \frac{\langle X_1^* X_1 \rangle \cdot \langle Y_1^* Y_2 \rangle - \langle Y_1^* X_1 \rangle \cdot \langle X_1^* Y_2 \rangle}{\langle X_1^* X_1 \rangle \cdot \langle Y_1^* Y_1 \rangle - \langle Y_1^* X_1 \rangle \cdot \langle X_1^* Y_1 \rangle}, \tag{6.35}$$

where the angular brackets indicate an average over different source realizations or groups of levels. This solution is a function of the auto- and cross-power spectra between the components of the input and output polarizations. Since an alternative estimate can be obtained by interchanging the depth levels, Lefeuvre et al. (1992) suggest a scheme for calculating a unique estimate by firstly normalizing the auto-correlations of the input and output matrices (thus transforming them to a unit matrix) prior to inversion of (6.31).

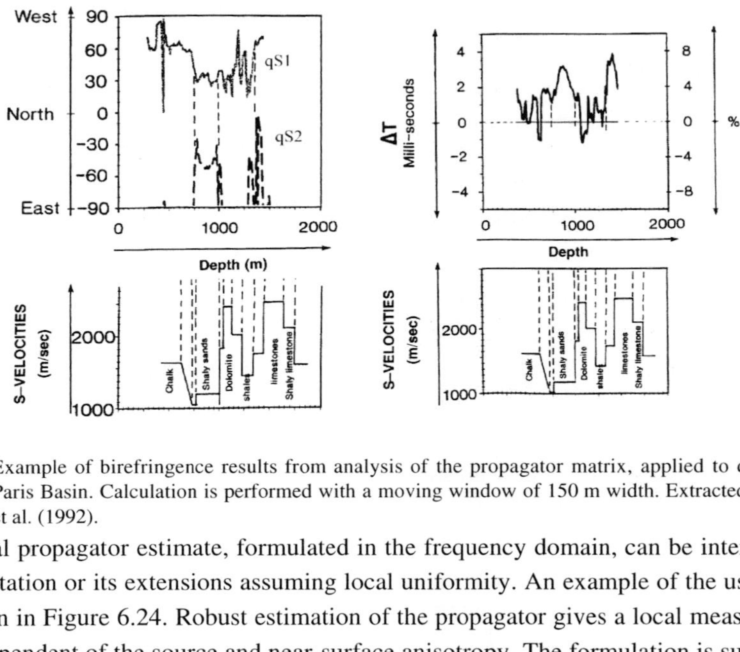

Figure 6.24. Example of birefringence results from analysis of the propagator matrix, applied to data shot in the Paris Basin. Calculation is performed with a moving window of 150 m width. Extracted from Lefeuvre et al. (1992).

The final propagator estimate, formulated in the frequency domain, can be interrogated using the Alford rotation or its extensions assuming local uniformity. An example of the use of this technique is given in Figure 6.24. Robust estimation of the propagator gives a local measure of the anisotropy, independent of the source and near-surface anisotropy. The formulation is sufficiently general to include a group of recordings from a variety of source realizations, and also independent shear-wave conversions from the same marine VSP (Lefeuvre and Queen 1992). Inherent in the procedure is a correction for tool spin, and an assumption of balanced components. Optimum choice of the depth window for the analysis depends upon the noise in the records, bandwidth of the shear waves, and the anisotropic properties. The final choice must be made by the user on a trial and error basis. This interval technique is particularly suited to VSP operations using multi-level sondes, which considerably reduce errors arising from source variability. Layer stripping and propagator matrix methods compare favorably in application to multi-component VSP data from California (Figure 6.25).

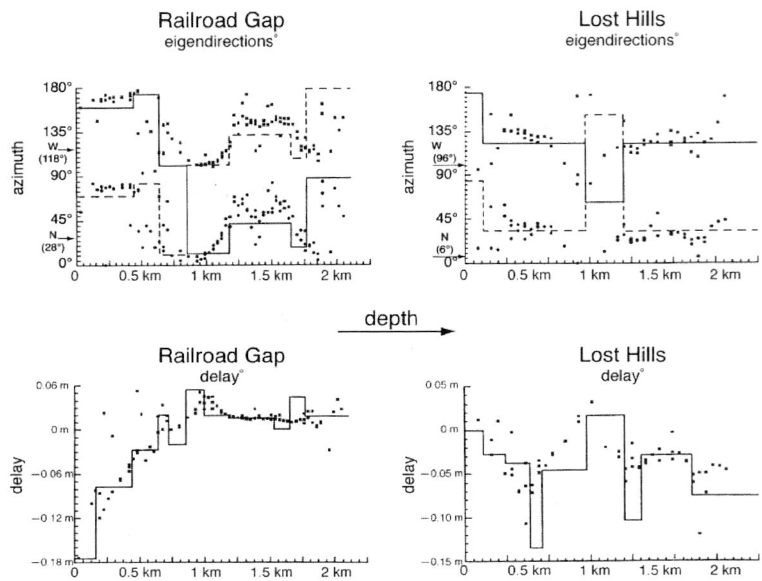

Figure 6.25. A comparison between the layer stripping (solid lines) approach of Winterstein and Meadows (1991a, b) and the propagator matrix method (dots) of Lefeuvre et al.(1992).

6.2.3 Causes of polarization change

While the general correlation between the fast shear wave polarization and the maximum horizontal stress remains broadly valid, results of fine-scale analyses using the techniques above also reveal many clear indications of polarization changes with depth. These changes often occur over small intervals: Winterstein, Meadows and De (1996) cite a polarization change of 50° over only 20 m, and are now frequently quoted in studies of near-offset VSP. Indeed, polarization changes are the rule and not the exception. Polarization may be also constant for large blocks in the subsurface, changing across a discrete interface with another block, but there is considerable inconsistency between geographical areas. Many observations for the near surface have already been documented (Section 4.2). Table 6.1 catalogues the additional sub-surface estimates from multi-component VSP.

Table 6.1. Polarization changes in near-offset VSP

reference	depth range	polarization change	Rock type	location
1	760 m	–	shale	Conoco BTSF, Oklahoma
2	690 m	–	Top Bunter shale	North Sea
	950 m	–	Base Bunter shale	North Sea
3	2040 m	–	Maikop clays	N. Caucasus
4	750 m	30°	chalk/shaly sands	Paris Basin
	1000 m	15°	shaly sands/dolomite	Paris Basin
5	≤ 1850 m	many	sand/shales/pebble conglomerate	Railroad Gap field
6	210–370 m	60°	Base shale–Top diatomite	Lost Hills field
7	250 m	50°	Base pebble conglomerate/angular unconformity	Cymric field

When the technique employed is layer stripping or interval estimation, a quantitative assessment of the change can be made. Results from the Alford rotation detect lag reversal relating to the changes, but are unable to provide an accurate magnitude.

1 - MacBeth et al. (1995); 2 - MacBeth et al. (1997);
3 - Slater (1997); 4 - Lefeuvre et al. (1992);
5, 7 - Winterstein and Meadows (1991a); 6 - Winterstein and Meadows (1991b).

An interpretation of these effects is a major challenge so far unmet. Causes are uncertain, although candidates are:

a) changes in stress conditions due to local structure and lithology (Bruno and Winterstein 1994);
b) change in pore pressure (see, e.g., Crampin, Zatsepin, Slater and Brodov 1996 – Chapter 3);
c) changes in fracture strike (Figure 6.26), with increasingly exotic possibilities;
d) additional intersecting (non-parallel) set of fractures propped open by mineralized fill (the effect can be seen in Liu, Crampin, Queen, and Rizer 1993a; MacBeth 1996; or Sayers 1998 – Chapter 3);
e) change in saturant (such as liquid to a gas) (Figure 6.26);
f) change in V_P/V_S of the matrix.

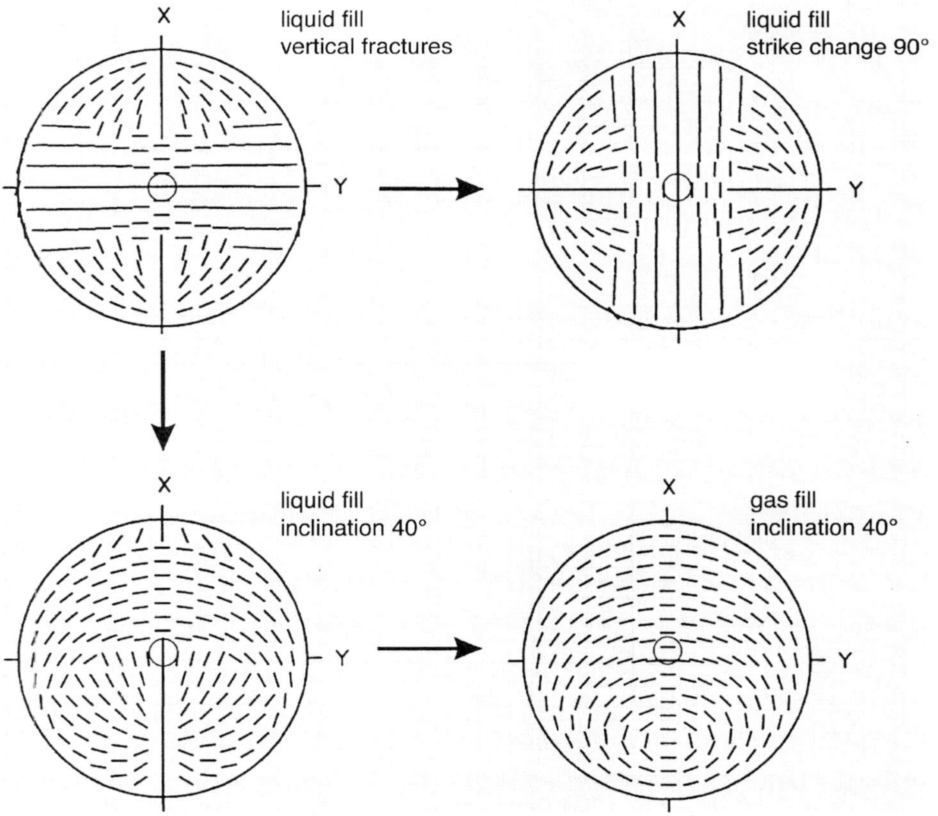

Figure 6.26. Three simple scenarios that can cause an abrupt change in polarization. Near-offset VSP samples the central portion of this plot (indicated by the circle).

6.3 INHERENT NON-ORTHOGONALITY

A departure from the idealization inherent in (4.13) occurs if there is a region of strong anisotropy arising from inclined fractures, dipping beds, or some exotic combination. The shear wave polarizations then do not in general lie in the horizontal plane (Chapter 3), and their horizontal projections are no longer orthogonal. A further complication arises if the anisotropy is strong enough to produce cusps (Chapter 3): this leads to more than two shear wave arrivals propagating along the ray path.

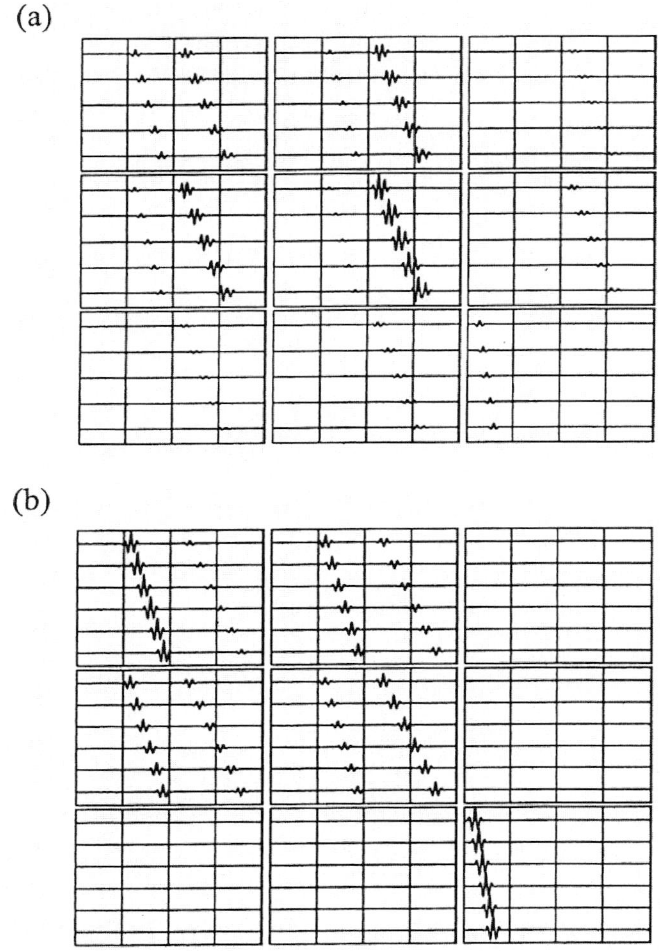

Figure 6.27. Nine-component data calculated to simulate a near-offset VSP in a strongly anisotropic, tilted orthorhombic medium. The level of non-orthogonality induced is 15°, but it has a noticeable effect: the off-diagonal energy is higher than normal (compare the examples of Chapter 5 and Chapter 4). Ncte that the entire data matrix remains symmetric.

The deviation of the qP-wave polarization is generally small (see Section 4.3.3), and normal incidence remains a good assumption for this wave type. The effect on the shear waves can be immediately observed by the presence of energy on the *XZ*, *YZ*, *ZX* and *ZY* off-diagonals of the nine-component VSP (Figure 6.27). This effect could be the explanation for the feature on the deeper traces of Figure 5.3. The deviation from orthogonality (in degrees) can be several times greater than the maximum shear wave anisotropy quoted (in percent). Thus, in the most common case of weak (up to 3 %) anisotropy and ray paths far from singularities (Chapter 3), the assumption of orthogonality is adequate. However, when the anisotropy rises above 5 %, the effects start to become significant. Such non-orthogonality reduces the value of results interpreted using the Alford rotation, and may be another cause of incomplete diagonalization. The impact of this was examined by Dellinger and Nolte (1997) and Dellinger et al. (1998); the foundation for their analysis was the principle of seismic reciprocity. One interpretation of this principle is that the roles of sources and receivers can be interchanged, with the same resultant signal (Aki and Richard 1981) regardless of anisotropy, attenuation or heterogeneity. Dellinger and Nolte (1997) suggest this implies the off-diagonal components of the data matrix must remain symmetric, even for non-orthogonal shear wave polarizations. As a consequence, they propose an adaptation of (4.13):

$$\mathbf{D}(t) = \mathbf{C}(\phi,\Delta)\mathbf{\Lambda}(t)\mathbf{C}^T(\phi,\Delta) \tag{6.36}$$

with a revised Alford rotation

$$\mathbf{M}'(t) = \mathbf{C}^{-1}(\phi,\Delta)\mathbf{M}(t)\mathbf{C}^T(\phi,\Delta)\ , \tag{6.37}$$

where

$$\mathbf{C}(\phi,\Delta) = \begin{pmatrix} \cos\phi & -\sin(\phi+\Delta) \\ \sin\phi & \cos(\phi+\Delta) \end{pmatrix} . \tag{6.38}$$

The angles ϕ and $\phi+\Delta$ are the horizontal polarization directions of the qS_1 and qS_2 waves relative to the acquisition *X*-axis (Figure 6.28). Although (6.36) is an approximation and does not adequately account for the necessary coupling between the modes, the matrix reduces to the original Alford rotation when $\Delta = 0$ and honors reciprocity. Estimates of ϕ and Δ can be found, following previous examples, by minimizing the off-diagonal energy within a time window of *N* samples. Dellinger et al. (1998) note that it is critical for this time window to be long enough to entirely enclose the split shear waves.

Although the approach was designed for surface seismics and cross-dipole data, it is also applicable to near-offset VSP after near-surface correction. In practice, it is important to apply this general form of the Alford rotation when one expects anisotropy values above 5 %. Indeed, application of (5.3) still produces a minimum in the off-diagonal energy despite non-orthogonality, but with misleading results. The smaller the time window, the more likely the diagonalization will be achieved. The procedure will output orthogonal polarizations with a qS_1 direction of $\phi+\Delta/2$.

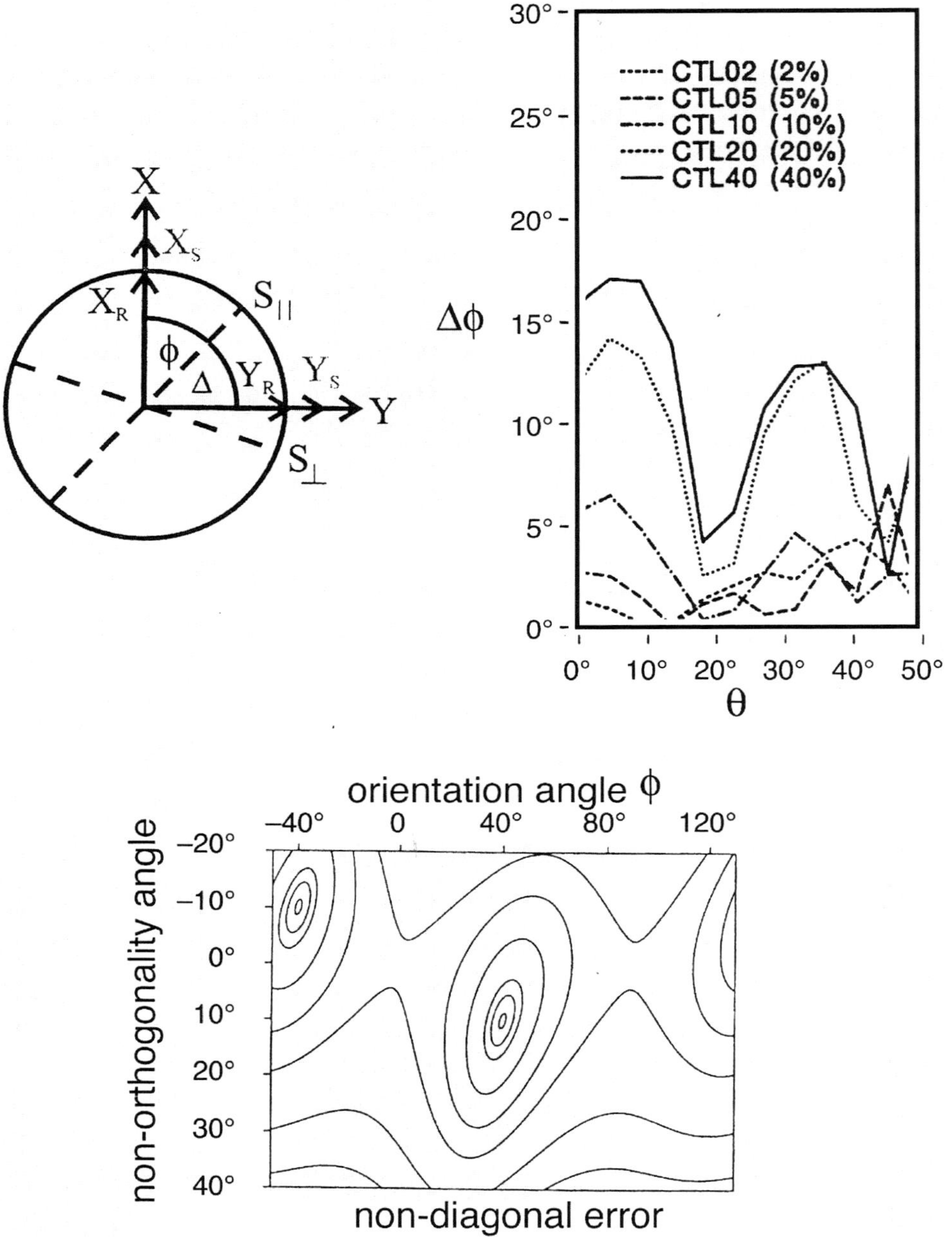

Figure 6.28. Top: strong anisotropy produces shear wave polarizations that deviate from the horizontal plane. Their horizontal projection is non-orthogonal, and the waves no longer propagate as pure modes. The degree of non-orthogonality is shown here for increasing anisotropy in a tilted orthorhombic medium. Bottom: Contour plot of the off-diagonal energy from the modified Alford rotation of Dellinger et al. (1997) designed to cope with non-orthogonality.

APPENDIX: TRANSMISSION AND REFLECTION COEFFICIENTS AT A POLARIZATION CHANGE

Consider plane shear waves of arbitrary polarization, incident vertically on a flat horizontal interface. Axes x and y lie parallel to the interface, and the z-axis points vertically downwards. The media on either side of the interface are TIH, defined by stiffnesses c_{11}, c_{33}, c_{44}, c_{66}, and c_{12} (not all of which are relevant to this particular problem). At the interface ($z = 0$) there is continuity of both the velocity vector $(v_x, v_y, v_z)^T$ and the stress vector $(\sigma_{zx}, \sigma_{zy}, \sigma_{zz})^T$. In the TIH media at vertical incidence, the shear-waves are decoupled from the P-wave motion. The shear wave polarizations lie within the horizontal plane and normal to the vertical symmetry plane. Thus the quantities v_z and σ_{xy} are unnecessary. This leads to the continuity relations

$$\begin{pmatrix} v_x \\ v_y \end{pmatrix}^A = \begin{pmatrix} v_x \\ v_y \end{pmatrix}^B \tag{6.39}$$

and

$$\begin{pmatrix} \sigma_{zx} \\ \sigma_{zy} \end{pmatrix}^A = \begin{pmatrix} \sigma_{zx} \\ \sigma_{zy} \end{pmatrix}^B . \tag{6.40}$$

The notation of Schoenberg and Protazio (1992) provides a definition for the velocity field at the interface due to transmitted, reflected and incident shear waves in medium A:

$$\begin{pmatrix} v_x \\ v_y \end{pmatrix}^A = i_1^A \begin{pmatrix} e_x^{(1)} \\ e_y^{(1)} \end{pmatrix}^A \mathrm{e}^{i\omega q_1^A z} + r_1^A \begin{pmatrix} e_x^{(1)} \\ e_y^{(1)} \end{pmatrix}^A \mathrm{e}^{-i\omega q_1^A z} + i_2^A \begin{pmatrix} e_x^{(2)} \\ e_y^{(2)} \end{pmatrix}^A \mathrm{e}^{i\omega q_2^A z} + r_2^A \begin{pmatrix} e_x^{(2)} \\ e_y^{(2)} \end{pmatrix}^A \mathrm{e}^{-i\omega q_2^A z} , \tag{6.41}$$

where the unit vectors $\left(e_x^{(1)}, e_y^{(1)}\right)^T$ and $\left(e_x^{(2)}, e_y^{(2)}\right)^T$ are the qS_1 and qS_2 polarizations, and q_i represents the vertical slowness for each shear waves. The horizontal polarizations for both up and downgoing waves are identical. A similar expression exists for medium B where reflected amplitudes are now set to zero and the incident amplitudes i_i replaced by the transmitted amplitudes t_i. The stress field is given by

$$\begin{pmatrix} \sigma_{zx} \\ \sigma_{zy} \end{pmatrix}^A = i_1^A q_1^A \begin{pmatrix} c_{66} e_x^{(1)} \\ c_{44} e_y^{(1)} \end{pmatrix}^A \mathrm{e}^{i\omega q_1^A z} - r_1^A q_1^A \begin{pmatrix} c_{66} e_x^{(1)} \\ c_{44} e_y^{(1)} \end{pmatrix}^A \mathrm{e}^{-i\omega q_1^A z} + i_2^A q_2^A \begin{pmatrix} c_{66} e_x^{(2)} \\ c_{44} e_y^{(2)} \end{pmatrix}^A \mathrm{e}^{i\omega q_2^A z} - r_2^A q_2^A \begin{pmatrix} c_{66} e_x^{(2)} \\ c_{44} e_y^{(2)} \end{pmatrix}^A \mathrm{e}^{-i\omega q_2^A z} . \tag{6.42}$$

For incident qS_2 waves with unit amplitude, $i_1^A = 0$ and $i_2^A = 1$, r_1^A becomes r_{21} and r_2^A becomes r_{22}. In the lower medium, t_1^B / i_2^A and t_2^B / i_2^A become t_{21} and t_{22}, respectively. The continuity relations are now

$$\mathbf{X}_A \left(\begin{pmatrix} 0 \\ 1 \end{pmatrix} + \begin{pmatrix} r_{21} \\ r_{22} \end{pmatrix} \right) = \mathbf{X}_B \begin{pmatrix} t_{21} \\ t_{22} \end{pmatrix} , \tag{6.43}$$

and

$$\mathbf{X}_A\left(\begin{pmatrix}0\\1\end{pmatrix}+\begin{pmatrix}r_{21}\\r_{22}\end{pmatrix}\right)=\mathbf{X}_B\begin{pmatrix}t_{21}\\t_{22}\end{pmatrix}, \tag{6.43}$$

and

$$\mathbf{Y}_A\left(\begin{pmatrix}0\\1\end{pmatrix}-\begin{pmatrix}r_{21}\\r_{22}\end{pmatrix}\right)=\mathbf{Y}_B\begin{pmatrix}t_{21}\\t_{22}\end{pmatrix}, \tag{6.44}$$

where

$$\mathbf{X}_{A,B}=\begin{pmatrix}e_x^{(1)} & e_x^{(2)}\\ e_y^{(1)} & e_y^{(2)}\end{pmatrix}^{A,B} \tag{6.45}$$

and

$$\mathbf{Y}_{A,B}=\begin{pmatrix}c_{66}^{A,B} & 0\\ 0 & c_{44}^{A,B}\end{pmatrix}\mathbf{X}_{A,B}\begin{pmatrix}q_1^{A,B} & 0\\ 0 & q_2^{A,B}\end{pmatrix}. \tag{6.46}$$

The reflection coefficients r_{11} and r_{12} and transmission coefficients t_{11} and t_{12} are generated in a similar way by setting $i_1^A=1$ and $i_2^A=0$:

$$\mathbf{X}_A\left(\begin{pmatrix}0\\1\end{pmatrix}+\begin{pmatrix}r_{11}\\r_{12}\end{pmatrix}\right)=\mathbf{X}_B\begin{pmatrix}t_{11}\\t_{12}\end{pmatrix}, \tag{6.47}$$

and

$$\mathbf{Y}_A\left(\begin{pmatrix}0\\1\end{pmatrix}-\begin{pmatrix}r_{11}\\r_{11}\end{pmatrix}\right)=\mathbf{Y}_B\begin{pmatrix}t_{11}\\t_{12}\end{pmatrix}. \tag{6.48}$$

From these, it is possible to write a matrix equation for the reflection- and transmission matrices

$$\mathbf{X}_A(\mathbf{I}+\mathbf{R})=\mathbf{X}_B\mathbf{T} \tag{6.49}$$

and

$$\mathbf{Y}_A(\mathbf{I}-\mathbf{R})=\mathbf{Y}_B\mathbf{T}, \tag{6.50}$$

and extract the individual reflection and transmission matrices as

$$\begin{pmatrix}r_{11} & r_{12}\\ r_{21} & r_{22}\end{pmatrix}=\left(\mathbf{Y}_B^{-1}\mathbf{Y}_A+\mathbf{X}_B^{-1}\mathbf{X}_A\right)^{-1}\left(\mathbf{Y}_B^{-1}\mathbf{Y}_A-\mathbf{X}_B^{-1}\mathbf{X}_A\right) \tag{6.51}$$

and

$$\begin{pmatrix}t_{11} & t_{12}\\ t_{21} & t_{22}\end{pmatrix}=2\mathbf{X}_B^{-1}\mathbf{X}_A\left(\mathbf{Y}_B^{-1}\mathbf{Y}_A+\mathbf{X}_B^{-1}\mathbf{X}_A\right)^{-1}\mathbf{Y}_B^{-1}\mathbf{Y}. \tag{6.52}$$

A simplification is possible by orienting the local coordinate system such that x and y are parallel to the fracture strike (vertical symmetry plane) of medium A, for which $e_y^{(1)} = 0$ and $e_x^{(2)} = 0$, and hence $\mathbf{X}_A$ is now a unit matrix and $\mathbf{Y}_A$ diagonal. Using this fixed coordinate frame, and considering now a relative rotation of $\Delta\phi = \phi_2 - \phi_1$ between media A and B that rotates the polarization vectors and the stresses in medium B

$$\begin{pmatrix} e_x \\ e_y \end{pmatrix}^B \rightarrow \mathbf{C}(\Delta\phi)\begin{pmatrix} e_x \\ e_y \end{pmatrix}^B , \tag{6.53}$$

and

$$\begin{pmatrix} \sigma_{zx} \\ \sigma_{zy} \end{pmatrix}^B \rightarrow \mathbf{C}(\Delta\phi)\begin{pmatrix} \sigma_{zx} \\ \sigma_{zy} \end{pmatrix}^B . \tag{6.54}$$

This leads to the transformation of the boundary conditions such that

$$\mathbf{X}_B \rightarrow \mathbf{C}(\Delta\phi)\mathbf{X}_B \tag{6.55}$$

and

$$\mathbf{Y}_B \rightarrow \mathbf{C}(\Delta\phi)\mathbf{Y}_B . \tag{6.56}$$

The latter follows from the transformation of stiffnesses

$$\begin{pmatrix} c_{66} & 0 \\ 0 & c_{44} \end{pmatrix} \rightarrow \mathbf{C}(\Delta\phi)\begin{pmatrix} c_{66} & 0 \\ 0 & c_{44} \end{pmatrix}\mathbf{C}^T(\Delta\phi) . \tag{6.57}$$

Insertion of values for stiffnesses in terms of the densities ρ_A and ρ_B, material velocities V_A and V_B, and the anisotropy coefficients ε_A and ε_B (defined by normalized velocity differences) results in

$$\mathbf{X}_A = \mathbf{I} , \tag{6.58}$$

$$\mathbf{X}_B = \mathbf{C}(\Delta\phi) , \tag{6.59}$$

$$\mathbf{Y}_A = \begin{pmatrix} \rho_A V_A & 0 \\ 0 & \rho_A V_A(1-\varepsilon_A) \end{pmatrix} , \tag{6.60}$$

and

$$\mathbf{Y}_B = \mathbf{C}(\Delta\phi)\begin{pmatrix} \rho_B V_B & 0 \\ 0 & \rho_B V_B(1-\varepsilon_B) \end{pmatrix} . \tag{6.61}$$

For an isotropic upper medium, $\varepsilon_A = 0$ and

$$\mathbf{R}_0(\phi) = \mathbf{C}(\Delta\phi) \begin{pmatrix} \dfrac{\rho_A V_A - \rho_B V_B}{\rho_A V_A + \rho_B V_B} & 0 \\ 0 & \dfrac{\rho_A V_A - \rho_B V_B(1-\varepsilon_B)}{\rho_A V_A + \rho_B V_B(1-\varepsilon_B)} \end{pmatrix} \mathbf{C}^T(\Delta\phi) \tag{6.62}$$

and

$$\mathbf{T}_0(\phi) = \begin{pmatrix} \dfrac{2\rho_A V_A}{\rho_A V_A + \rho_B V_B} & 0 \\ 0 & \dfrac{2\rho_A V_A}{\rho_A V_A + \rho_B V_B(1-\varepsilon_B)} \end{pmatrix} \mathbf{C}^T(\Delta\phi) \ . \tag{6.63}$$

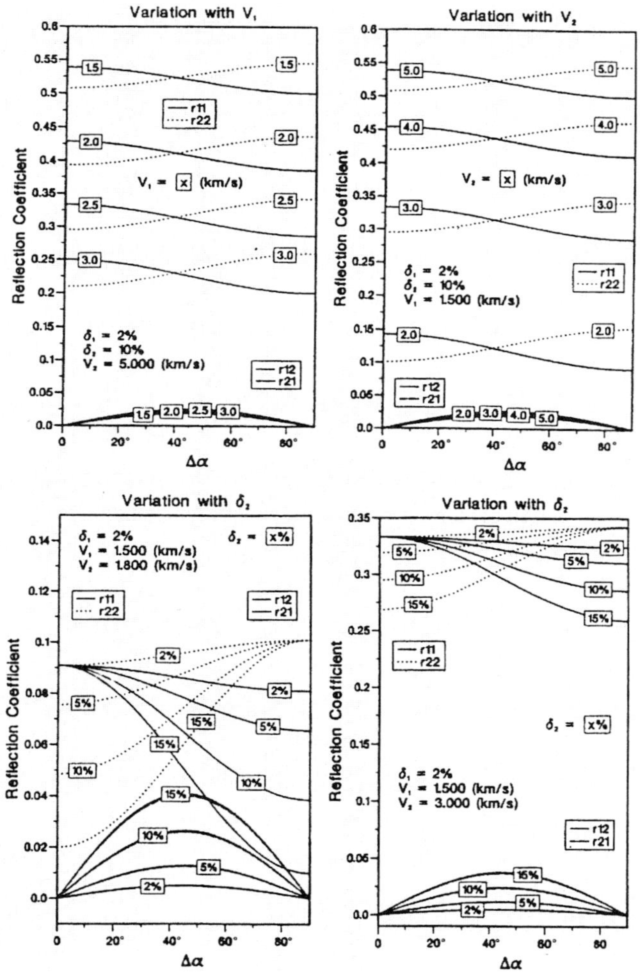

Figure 6.29. Reflection coefficients for a range of polarization changes $\Delta\alpha$ as a function of impedance contrast and birefringences δ_1 and δ_2. The larger the contrast, the smaller is the effect of anisotropy on the cross terms (Li 1992).

gesting the inter-mixing causes a reduction in the dimming/brightening phenomenon. For an upper medium with TIH symmetry but small changes ($\Delta\phi$)

$$\mathbf{R}_\phi \approx \mathbf{C}(\Delta\phi)\begin{pmatrix} \dfrac{\rho_A V_A - \rho_B V_B}{\rho_A V_A + \rho_B V_B} & 0 \\ 0 & \dfrac{\rho_A V_A(1-\varepsilon_A) - \rho_B V_B(1-\varepsilon_B)}{\rho_A V_A(1-\varepsilon_A) + \rho_B V_B(1-\varepsilon_B)} \end{pmatrix}\mathbf{C}^T(\Delta\phi) \tag{6.64}$$

and

$$\mathbf{T}_\phi \approx \begin{pmatrix} \dfrac{2\rho_A V_A}{\rho_A V_A + \rho_B V_B} & 0 \\ 0 & \dfrac{2\rho_A V_A(1-\varepsilon_A)}{\rho_A V_A(1-\varepsilon_A) + \rho_B V_B(1-\varepsilon_B)} \end{pmatrix}\mathbf{C}^T(\Delta\phi)\ , \tag{6.65}$$

and when $\Delta\phi$ is close to 90°

$$\mathbf{R}_{90} \approx \mathbf{C}(\Delta\phi)\begin{pmatrix} \dfrac{\rho_A V_A - \rho_B V_B(1-\varepsilon_B)}{\rho_A V_A + \rho_B V_B(1-\varepsilon_B)} & 0 \\ 0 & \dfrac{\rho_A V_A(1-\varepsilon_A) - \rho_B V_B}{\rho_A V_A(1-\varepsilon_A) + \rho_B V_B} \end{pmatrix}\mathbf{C}^T(\Delta\phi) \tag{6.66}$$

and

$$\mathbf{T}_{90} \approx \begin{pmatrix} \dfrac{2\rho_A V_A(1-\varepsilon_A)}{\rho_A V_A(1-\varepsilon_A) + \rho_B V_B} & 0 \\ 0 & \dfrac{2\rho_A V_A(1-\varepsilon_A)}{\rho_A V_A + \rho_B V_B(1-\varepsilon_B)} \end{pmatrix}\mathbf{C}^T(\Delta\phi)\ . \tag{6.67}$$

For arbitrary $\Delta\phi$, the equations are more complicated; it is best to compute (6.51) and (6.52) numerically.

CHAPTER 7 – MULTIPLE-OFFSET VSP – KINEMATICS

Anisotropy analysis of fractured formations using near-offset VSP is confined to estimation of the fast and slow shear wave velocities (giving $a_{44} = c_{44}/\rho$ and $a_{66} = c_{66}/\rho$), from which the birefringence and hence tangential fracture index e_t can be determined. The P-wave measurements give only the vertical velocity ($a_{33} = c_{33}/\rho$), and hence no measure of the normal fracture index e_n. Additionally, for the analysis of shales and sedimentary sequences, P- and S-wave behavior at oblique incidence is required. VSPs with large offset ranges provide this additional information about the TIV parameters ε, δ, and γ, and the fracture index e_n, by using techniques that sense variations of local phase slowness and moveout velocity. Acquisitions suited to these analyses are

(a) a *walk-away* VSP (with a line of several hundred shots along a specific azimuth, passing as close as technically possible to the wellhead),
(b) a *walkaround* or *azimuthal* VSP (with a circle of shots), and
(c) a *3D-VSP* (with a surface spread of up to a thousand shots);
(d) dual acquisition using a rig and boat (normal incidence) source.

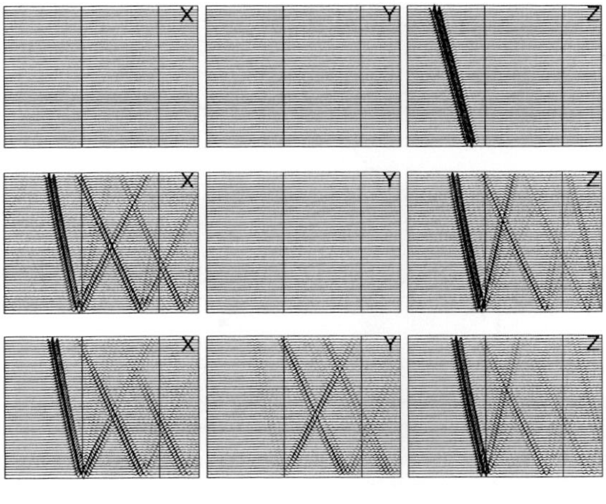

Figure 7.1. Simulation of three-component recordings from an explosive source in a fixed-offset VSP. Upper panel: zero-offset and isotropic medium; middle panel: moderate offset and isotropic medium; lower panel: moderate offset and TIH medium with arbitrary orientation of the acquisition *X-Y* axes. There are converted shear waves form a near-surface layer and also a subsurface reflector. The identity of these waves is confirmed by the energy on the *Y* (transverse) component of the lower set of traces.

Standard acquisition practice is to shoot these VSPs with multiple level receiver tools. With the use of a multi-component source at far offsets now being less common, sources are usually vibrators or dynamite charges for land surveys, and airguns for offshore surveys. These sources generate shear waves due to the radiation pattern and/or through conversion in the subsurface. Unfortunately, there is no partitioning into purely P- and S-wave subsets in the multi-component recordings, thus more sophisticated techniques for wavefield separation are now necessary (Figure 7.1).

7.1 WAVEFIELD SEPARATION USING A MULTI-COMPONENT RECEIVER

Separation of the up- and downgoing P- and S waves is necessary for anisotropy estimation. This separation is a prerequisite for analyses of multiple-offset VSP data for phase slowness, velocity and amplitude. Use of three-component data is essential, as increasing offset makes separation based on moveout alone progressively more difficult, with the P- and S-waves overlapping on both the vertical and horizontal components. Indeed, a major effect can be achieved by rotation of the horizontal recordings about the vertical to minimize the transverse component energy in the direct P-wave. This approach uses as its basis estimates of polarization direction taken using the techniques described in Section 4.3. Horizontal rotation analysis ensures that the receiver components are now aligned along radial and transverse directions, with the transverse component containing only shear waves. Further receiver rotation about the transverse component helps provide additional separation, whilst an extension of this using polarization filtering gives refinement (Section 7.1.1).

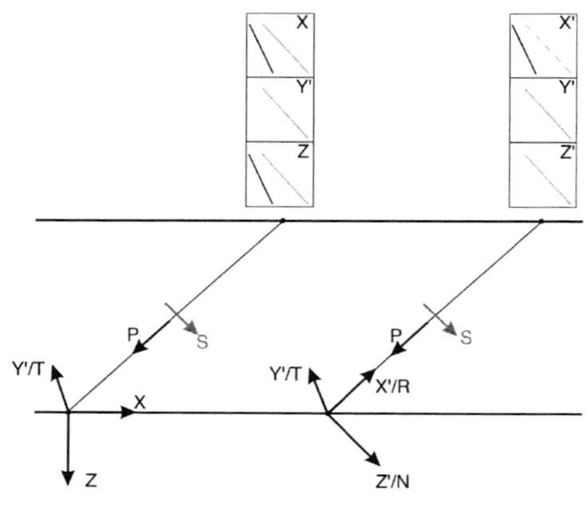

Figure 7.2. Rotation into the dynamic axes is achieved by using the initial onset of the direct P-wave. The X-axis is rotated to align with the radial (R) direction thus detecting the maximum P-wave amplitude, while the Z-axis is now normal (N) to the ray path and detects mainly the S-waves. In practice, the S-waves are unlikely to travel the same ray path as the P-wave if they are generated by conversion. The transverse (T) component provides an immediate partitioning into the shear wave energy, but not a complete separation.

As in the near-offset case, up/downgoing separation can be achieved by application of a velocity (dip) filter to each component separately. This can be implemented by the widely used f-k and τ-p techniques. These can be applied to a large and well-populated number of depth levels and moderate impedance contrasts. These domains also permit the application of filtering procedures based on a prediction of the polarizations; these require knowledge of the in situ elastic stiffnesses (Section 7.1.3). However, for a more accurate representation in the presence of abrupt impedance jumps or inhomogeneity, a local operator such as the median filter (Hardage 2000) or optimal filters (Kommedal and Tjostheim 1989) are better suited. These methods shift processing emphasis to the treatment of a small number of receiver levels and are the foundation for the local plane-wave expansion of Esmersoy (1990a), Leaney and Esmersoy (1989), and Leaney (1990). Local plane-wave expansion can include polarization information in an up/downgoing, P- and S-wave separation procedure (Section 7.1.2).

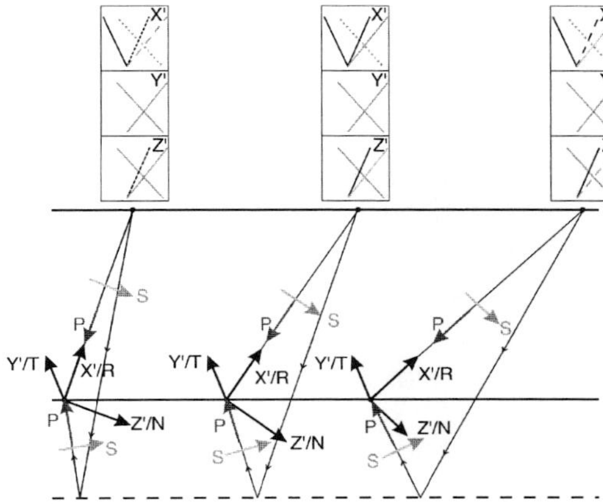

Figure 7.3. The dynamic axis rotation is appropriate for direct waves, but a bad approximation for reflected waves. In this example, the energy from the upgoing P-waves moves from the radial (*R*) to the normal (*N*) axis. Other reflectors also exhibit a different apportionment. This shows the need for a more sophisticated procedure in VSPs with a range of offsets.

7.1.1 Stochastic polarization filtering

These techniques use only the polarization information at each receiver, and require separation into up/downgoing waves prior to, or following, their application.

7.1.1.1 Rotation-based procedures

With a window of *N* samples around the direct P wave, a rotation about the vertical

$$\mathbf{d}'(t) = \begin{pmatrix} \cos\phi & \sin\phi & 0 \\ -\sin\phi & \cos\phi & 0 \\ 0 & 0 & 1 \end{pmatrix} \mathbf{d}(t) \tag{7.1}$$

followed by a rotation about a horizontal axis

$$\mathbf{d}''(t) = \begin{pmatrix} \cos\theta & 0 & \sin\theta \\ 0 & 1 & 0 \\ -\sin\theta & 0 & \cos\theta \end{pmatrix} \mathbf{d}'(t) \tag{7.2}$$

aligns the radial receiver axis with the ray path (Figure 7.2). Techniques used for obtaining an estimate of the rotation angles are described in Section 4.3. After receiver rotation for each source-receiver pair, the downgoing P-waves and upgoing S-waves now lie mainly on the radial component, while the downgoing S-waves and upgoing P-waves lie on the normal component. This rotated coordinate frame is referred to as the "dynamic coordinate frame". The method provides a simple and robust procedure for wavefield separation in walk-away VSP (Ahmed et al. 1986; Noble et al. 1987). The separation is possible irrespective of shear wave birefringence. Ellipticity of the P-wave reduces the effectiveness of this technique, as does the offset-dependent nature of the reflected wavefield (Figure 7.3). A field example of the procedure applied to walk-away VSP data is shown in Figures 7.4 to 7.6.

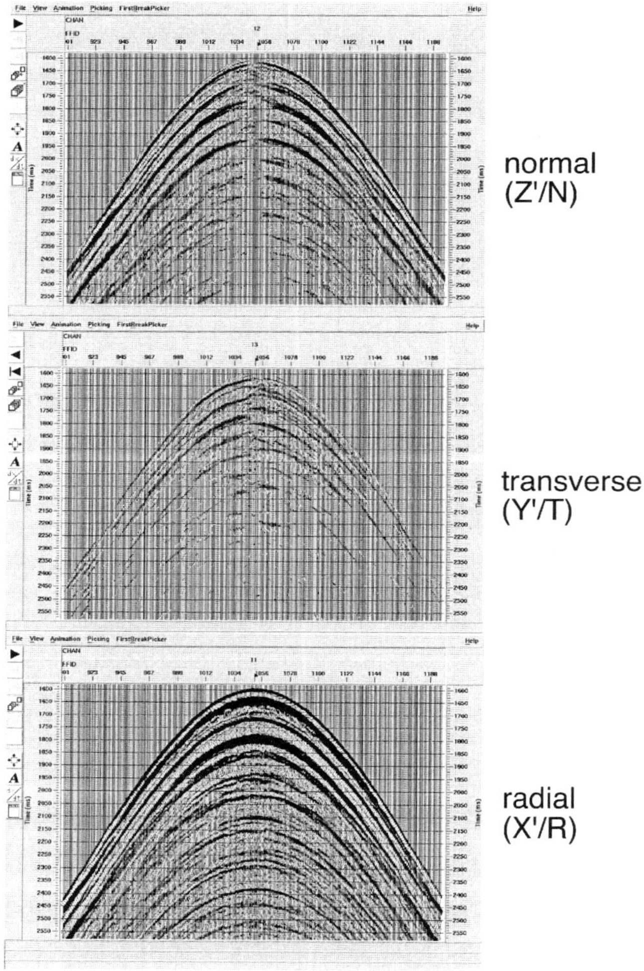

Figure 7.4. Downgoing wavefield from a North Sea walk-away VSP after rotation into the dynamic axes using the initial P-wave onset. The receiver depth is 3017 m and the maximum offset 3500 m. The rotation procedure has effectively separated the P- and S-waves on the transverse component, and to some extent on the normal component. Note the polarity change of the S-waves between positive and negative offsets – a characteristic discussed in Chapter 8. The three-component traces are normalized to a common vector amplitude. Data courtesy of the British Geological Survey and Norsk Hydro.

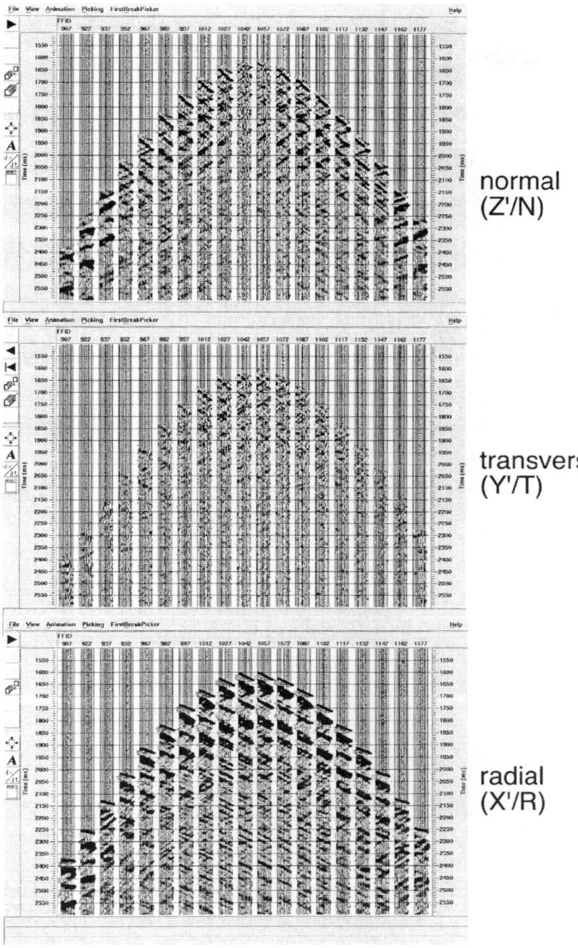

Figure 7.5. Common-shot gathers for the rotated walk-away data in Figure 7.4, but before separation of the up- and downgoing wavefields. Strong upgoing shear waves are visible on the far-offset traces. Data courtesy of the British Geological Survey and Norsk Hydro.

7.1.1.2 *Polarization filtering in the time-domain*

An extension of rotation analysis is the design of a filter function using the P-wave polarizations. This provides better control over which particle motion directions and ellipticities to accept. The method differs from rotation analysis – which is based solely on cosine and sine functions arising from the projection geometry. The first step in filter design is eigenanalysis of the trace covariance matrix

$$\mathbf{P}_j = \sum_{i=j}^{N+j-1} \mathbf{d}(t_i)\mathbf{d}^T(t_i) \tag{7.3}$$

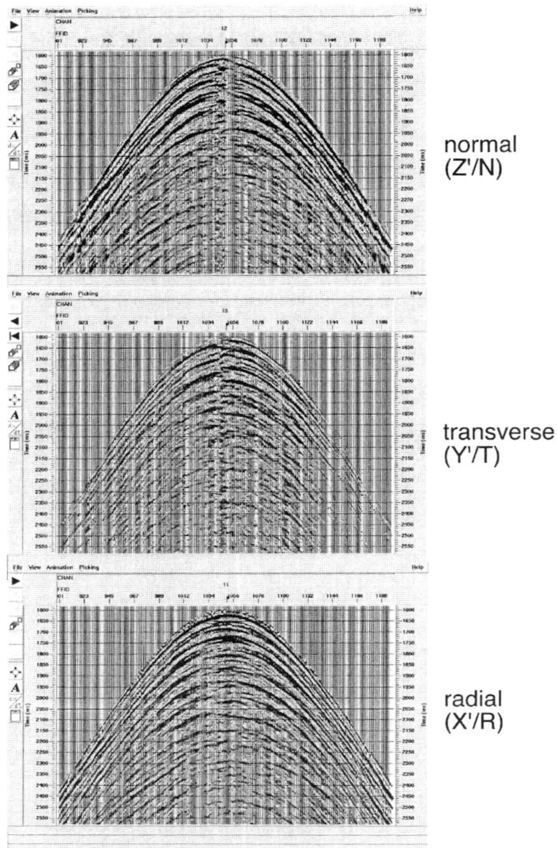

Figure 7.6. Upgoing wavefield from the North Sea walk-away data of Figure 7.4, again after rotation into the dynamic axes using the initial P-wave onset. In this case the separation into S-waves on the transverse and normal component is not as complete as for the downgoing wavefield. The three-component traces are normalized to a common vector amplitude. Data courtesy of the British Geological Survey and Norsk Hydro.

for a moving window of N samples (see Section 4.3 and A.2). For each window position, the relative magnitude of the eigenvalues (λ_1^j, λ_2^j and λ_3^j) reveals the nature of the average polarization and the normalized eigenvectors ($\hat{\mathbf{e}}_1^j$, $\hat{\mathbf{e}}_2^j$ and $\hat{\mathbf{e}}_3^j$) determine its direction (section A.2). Suitable combinations of these eigenvalues can be used to obtain a statistical measure of the polarization state, and how it evolves with time. The degree of linearity can be assessed using (Flinn 1965)

$$F^L(t) = \frac{\lambda_1(t) - \lambda_3(t)}{\lambda_1(t) + \lambda_3(t)}, \tag{7.4}$$

which lies between 0 and 1, or (Kanasewich 1981)

$$F^L(t) = \left(1 - \left(\frac{\lambda_3}{\lambda_1}\right)^d\right)^c, \tag{7.5}$$

where c and d are numbers to be evaluated for the particular application. Alternatives include the Mth order Butterworth functions (Esmersoy 1984)

$$F^L(t) = \frac{1}{\sqrt{1 + \left(p_0 \frac{\lambda_3}{\lambda_1}\right)^{2M}}} \,, \tag{7.6}$$

where p_0 is a design cut-off parameter, or the Gaussian weighted functions discussed by Perlberg and Hornbostel (1994). The polarization filter acts by defining a time domain reject filter that operates on the recorded trace $\mathbf{d}(t)$, using the polarization direction $\mathbf{a}$ evaluated for the window around most of the direct P-wave. This gives

$$\mathbf{d}'(t) = F^L(t) D(t) \mathbf{d}(t) \,, \tag{7.7}$$

where

$$D(t) = \cos^2 \psi \tag{7.8}$$

and

$$\cos^2 \psi = \left(\mathbf{a}^T \hat{\mathbf{e}}_1^j(t)\right)^K \,. \tag{7.9}$$

The number K lies between 0 and infinity, and when set to 2 gives a power-related directional operator. When K becomes smaller than unity, a greater range of directions are included in the resultant trace. Another directional function is the *Watson axial distribution* of Aster, Shearer and Berger (1990)

$$D(t) = \cos^2 \psi \cdot e^{\kappa\left(\cos^2 \psi - 1\right)} \,, \tag{7.10}$$

where $\kappa > 0$ is determined by a normalization procedure. The choice of the function to use for $F(t)$ and $D(t)$ depends largely on the data.

7.1.1.3 Polarization filtering in the frequency domain

The frequency domain is useful because linearly and elliptically polarized arrivals are separately represented by distinct eigenvalues. Thus it can be used when the P- or S-wave motion is consistently elliptical. For application in this domain, the spectral matrix (Samson 1983; Mars et al. 1987) is defined as

$$\mathbf{S} = \sum_{receivers} \sum_{\omega} g(\omega) \mathbf{d}^H(\omega) \mathbf{d}(\omega) \,, \tag{7.11}$$

where the superscript H indicates the Hermitian complex operation (complex conjugate transpose). This summation can be extended over the entire frequency band or just over selected frequency intervals. It incorporates an optional weighting element $g(\omega)$ and should be smoothed or averaged over a number of receiver levels, since otherwise the matrix is singular (Spitz, personal communication). The spectral matrix is Hermitian ($\mathbf{S}^H\mathbf{S} = \mathbf{I}$) and can be decomposed into the eigenproduct

$$\mathbf{S}_j = \begin{pmatrix}\hat{\mathbf{e}}_1^j & \hat{\mathbf{e}}_2^j & \hat{\mathbf{e}}_3^j\end{pmatrix}^H \begin{pmatrix}\lambda_1^j & 0 & 0\\ 0 & \lambda_2^j & 0\\ 0 & 0 & \lambda_3^j\end{pmatrix} \begin{pmatrix}\hat{\mathbf{e}}_1^j & \hat{\mathbf{e}}_2^j & \hat{\mathbf{e}}_3^j\end{pmatrix}, \tag{7.12}$$

where the complex eigenvectors satisfy the relations $\left(\mathbf{e}_m^j\right)^H\left(\mathbf{e}_n^j\right) = \delta_{mn}$ and the eigenvalues are real. The eigenvalues reflect the degree of dominance, or purity, of a particular polarization state (linear, elliptical or otherwise) in the window of interest.

As before, separate P- and S-waves are recovered by the filter

$$\mathbf{d}'(\omega, \tau_j) = \left(\mathbf{a}(\omega)^H.\hat{\mathbf{e}}_1^j(\omega)\right)^K \mathbf{d}(\omega), \tag{7.13}$$

where each $\hat{\mathbf{e}}_1^j$ is determined, as before, for the sliding window positioned at time τ_j, but also for a specific frequency band. If the contribution of the principal eigenvector is required, then **a** is replaced by **d**. To enhance the efficiency of the filtering, several three-component traces can be used simultaneously by grouping them in the input vector (Glangeaud and Cliet, 1990; Glangeaud and Mari 1994). As intermixing of the different polarization states makes little sense in the context of this analysis, there is no need to include the function $F(\omega)$. The technique has benefit in that specific frequency ranges can be targeted: this is particularly useful when the bandwidth of the P- and S-waves is known to be different, or a band of high signal-to-noise ratio is to be eliminated. Indeed, an advantage of this method is that it can deal with the frequency-dependent polarization signature from shear wave birefringence. The technique attempts to unravel the dominant arrivals in order of energy without knowledge of either medium properties or the number of wave components. No prior model assumptions are needed, but the formulation depends on the maximum number of arrivals.

7.1.2 Treating polarization and moveout together

7.1.2.1 Downgoing P and S waves

A multi-component transfer matrix connecting two receivers, or connecting each receiver pair of a group, can be estimated by following the birefringence techniques described in Section 4.2. Here, the incoming multi-component wavefield is a mixture of P- and S-waves, with complex polarizations $\mathbf{d}_P(\omega)$ and $\mathbf{d}_S(\omega)$, and corresponding waveforms $w_P(\omega)$ and $w_S(\omega)$. The displacement is

$$\mathbf{d}_k(\omega) = \begin{pmatrix}\mathbf{d}_P(\omega) & \mathbf{d}_S(\omega)\end{pmatrix} \begin{pmatrix} w_P(\omega)\\ w_S(\omega)\end{pmatrix}, \tag{7.14}$$

or

$$\mathbf{d}_k(\omega) = \mathbf{A}(\omega)\mathbf{w}(\omega) \; . \tag{7.15}$$

The objective is to estimate $\mathbf{A}(\omega)$ and thus obtain $\mathbf{w}(\omega)$. The result for an adjacent receiver, below the first, is

$$\mathbf{d}_{k+1}(\omega) = \mathbf{A}(\omega)\begin{pmatrix} \lambda_P(\omega) & 0 \\ 0 & \lambda_S(\omega) \end{pmatrix} \mathbf{w}(\omega) \ , \tag{7.16}$$

or

$$\mathbf{d}_{k+1}(\omega) = \mathbf{A}(\omega)\Lambda(\omega)\mathbf{w}(\omega) \ , \tag{7.17}$$

where $\lambda_P(\omega)$ and $\lambda_S(\omega)$ are time-shift operators. With an array of equal depth increments Δz and the slownesses q_P and q_S,

$$\lambda_P = \mathrm{e}^{\mathrm{i}\omega q_P \Delta z} \tag{7.18}$$

and

$$\lambda_S = \mathrm{e}^{\mathrm{i}\omega q_S \Delta z} \ . \tag{7.19}$$

The transfer matrix (propagator matrix) $\mathbf{Y}(\omega)$ is now given by

$$\mathbf{d}_{k+1}(\omega) = \left(\mathbf{A}(\omega)\Lambda(\omega)\mathbf{A}(\omega)^{-1}\right)\mathbf{d}_k(\omega) \ , \tag{7.20}$$

or

$$\mathbf{d}_{k+1}(\omega) = \mathbf{Y}(\omega)\mathbf{d}_k(\omega) \ . \tag{7.21}$$

This fundamental relation can be generalized for a receiver array (equations 4.28 and 6.28), from which a least-squares estimate of $\mathbf{Y}(\omega)$ can be obtained. Inversion of the equation is guaranteed provided the P- and S-wave polarizations are not collinear, which in turn restricts the technique to either downgoing or upgoing waves. The separate wavefields can be recovered through eigenanalysis of $\mathbf{Y}(\omega)$ to obtain an estimate of $\mathbf{A}(\omega)$. If necessary, further analysis can also give the moveout and amplitude decay across the array (Cho and Spencer 1992). The eigenvalues of $\mathbf{Y}(\omega)$ can be determined by solving the characteristic equation

$$\left|\mathbf{Y}(\omega) - \mathbf{I}\lambda\right| = 0 \ , \tag{7.22}$$

and the eigenvectors $\mathbf{v}$ are estimated by

$$\mathbf{Y}(\omega)\mathbf{v}(\omega) = \lambda\mathbf{v}(\omega) \ . \tag{7.23}$$

Routines are available from standard mathematical libraries such as NAG or MATLAB to solve these equations. As $\mathbf{Y}(\omega)$ is generally complex, both eigenvalues and eigenvectors are also complex. Wavefield separation is finally achieved by applying an estimate of the inverse eigenvector matrix $\hat{\mathbf{A}}^{-1}$ to each recording $\mathbf{d}_k(\omega)$

$$\mathbf{d}'_k(\omega) = \hat{\mathbf{A}}^{-1}(\omega)\mathbf{d}_k(\omega) \; , \tag{7.24}$$

from which the desired time series can be obtained by inverse Fourier transform. If only three-component data are recorded, the minimum requirement for P- and S-wave separation is a three-level tool, since for $\mathbf{Y}(\omega)$ the number of arrivals (rank of $\mathbf{Y}(\omega)$) must be less than its dimension.

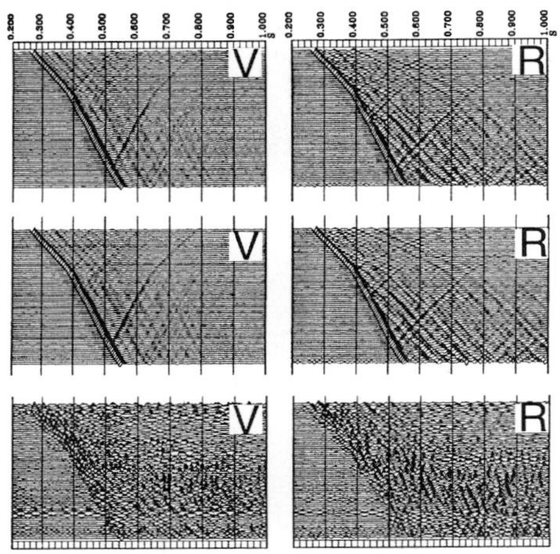

Figure 7.7. Application of the local plane wave expansion to two-component near-offset VSP data. The original (top) is re-assembled after parameter estimation, and then the difference (normalized) is plotted for comparison. Adapted from Leaney (1990).

7.1.2.2 *Up- and downgoing P and S waves*

The formulation is now

$$\begin{pmatrix} \mathbf{d}_k(\omega) \\ \mathbf{d}_{k+1}(\omega) \end{pmatrix} = \begin{pmatrix} \mathbf{d}_P^D(\omega) & \mathbf{d}_S^D(\omega) & \mathbf{d}_P^U(\omega) & \mathbf{d}_S^U(\omega) \\ \lambda_P^D \mathbf{d}_P^D(\omega) & \lambda_P^D \mathbf{d}_S^D(\omega) & \lambda_P^U \mathbf{d}_P^U(\omega) & \lambda_P^U \mathbf{d}_S^U(\omega) \end{pmatrix} \begin{pmatrix} w_P^D(\omega) \\ w_S^D(\omega) \\ w_P^U(\omega) \\ w_S^U(\omega) \end{pmatrix} , \tag{7.25}$$

where U and D refer to the up- and downgoing components, respectively. For isotropic, TIH and TIV media the vertical slownesses for up- and downgoing propagation are related:

$$\lambda_P^D = \frac{1}{\lambda_P^U} \tag{7.26}$$

and

$$\lambda_S^D = \frac{1}{\lambda_S^U} \; . \tag{7.27}$$

Equation (7.25) leads to a 4×4 transfer matrix and the minimum requirement of a three-level array (Cho and Spencer 1992)

$$\begin{pmatrix} \mathbf{d}_k(\omega) \\ \mathbf{d}_{k+1}(\omega) \end{pmatrix} = \mathbf{Y}(\omega) \begin{pmatrix} \mathbf{d}_{k+1}(\omega) \\ \mathbf{d}_{k+2}(\omega) \end{pmatrix} , \tag{7.28}$$

while for a five level tool it is possible to combine the information for robustness

$$\begin{pmatrix} \mathbf{d}_k(\omega) & \mathbf{d}_{k+1}(\omega) & \mathbf{d}_{k+2}(\omega) \\ \mathbf{d}_{k+1}(\omega) & \mathbf{d}_{k+2}(\omega) & \mathbf{d}_{k+3}(\omega) \end{pmatrix} = \mathbf{Y}(\omega) \begin{pmatrix} \mathbf{d}_{k+1}(\omega) & \mathbf{d}_{k+2}(\omega) & \mathbf{d}_{k+3}(\omega) \\ \mathbf{d}_{k+2}(\omega) & \mathbf{d}_{k+3}(\omega) & \mathbf{d}_{k+4}(\omega) \end{pmatrix} . \tag{7.29}$$

7.1.2.2.1 Birefringent shear waves

For large birefringence it is necessary to contend with fast and slow shear waves, and hence with three separate arrivals. With small birefringence, the similar moveout velocity reduces the stability of the inversion and necessitates the use of more receiver levels. If the number of receiver levels is insufficient, the shear waves can be treated as a single arrival with a frequency-dependent polarization state. This composite polarization is elliptical, but varies in ellipticity and orientation (for example, a 20 ms lag time introduces a phase shift of π at 25 Hz, but only $2\pi/5$ at 10 Hz). Thus, for (almost) orthogonal shear wave polarizations (see Chapter 8 for the reason)

$$\mathbf{d}_{S1} = \begin{pmatrix} \sin\theta_S \\ \cos\theta_S \end{pmatrix} , \tag{7.30}$$

and

$$\mathbf{d}_{S2} = \begin{pmatrix} \cos(\theta_S + \Delta\theta_S) \\ -\sin(\theta_S + \Delta\theta_S) \end{pmatrix} . \tag{7.31}$$

The combined signature is

$$\mathbf{d}_S = \begin{pmatrix} \sin\theta_S \\ \cos\theta_S \end{pmatrix} + \mathrm{e}^{\mathrm{i}\omega\Delta\tau} \begin{pmatrix} \cos(\theta_S + \Delta\theta_S) \\ -\sin(\theta_S + \Delta\theta_S) \end{pmatrix} , \tag{7.32}$$

giving a normalized polarization of

$$\mathbf{d}_S = \begin{pmatrix} \sin\theta'_S \,\mathrm{e}^{\mathrm{i}\eta(\omega)} \\ \cos\theta'_S \end{pmatrix} , \tag{7.33}$$

where

$$\tan\theta'_S \approx \tan\theta_S (1 + \Delta\theta_S \cot\theta_S) , \tag{7.34}$$

and

$$\eta(\omega) \approx \frac{\Delta\theta_S}{2\sin 2\theta_S} \frac{\sin(\omega\Delta\tau)}{(1+\cos(\omega\Delta\tau))^2} . \tag{7.35}$$

It is therefore possible to apply the above technique assuming only P- and S-waves as two arrivals, provided the transfer matrix is treated separately for each frequency.

7.1.3 P- and S separation using the parametric inversion

The idea of inverse modeling of array recordings as a small number of direction-dependent arrivals was developed by Seeman and Horowicz (1983) and later by Panziera and Arens (1985) for single-component applications. It was further extended by Esmersoy (1990b) to downgoing waves in multi-component VSP. The technique is a parametric inversion, as it establishes a model of the wavefield components, and then seeks to estimate each individual parameter. It is assumed at the outset that the number of plane waves is known.

7.1.3.1 Downgoing P- and S waves

The P- and S-waves are assumed to possess fixed, frequency-independent vertical slownesses (q_P, q_S) and polarization states. In an isotropic or TIV medium, these polarizations are

$$\mathbf{d}_P = \begin{pmatrix} -\sin\theta_P \\ \cos\theta_P \end{pmatrix}, \tag{7.36}$$

and

$$\mathbf{d}_S = \begin{pmatrix} \cos\theta_S \\ \sin\theta_S \end{pmatrix}, \tag{7.37}$$

where θ_P and θ_S are the angles of incidence. They also possess frequency-dependent waveforms $w_P(\omega)$ and $w_S(\omega)$; all these properties are constant across the array. The two-component displacements at each receiver can be written

$$\mathbf{u}(\omega) = \mathbf{d}_P w_P(\omega) + \mathbf{d}_S w_S(\omega) \ . \tag{7.38}$$

For example, generalizing (7.38) to a five-level tool , this becomes

$$\begin{pmatrix} \mathbf{u}_1(\omega) \\ \mathbf{u}_2(\omega) \\ \mathbf{u}_3(\omega) \\ \mathbf{u}_4(\omega) \\ \mathbf{u}_5(\omega) \end{pmatrix} = \begin{pmatrix} \mathbf{d}_P\lambda_P^2 & \mathbf{d}_S\lambda_S^2 \\ \mathbf{d}_P\lambda_P & \mathbf{d}_S\lambda_S \\ \mathbf{d}_P & \mathbf{d}_S \\ \mathbf{d}_P\lambda_P^{-1} & \mathbf{d}_S\lambda_S^{-1} \\ \mathbf{d}_P\lambda_P^{-2} & \mathbf{d}_S\lambda_S^{-2} \end{pmatrix} \begin{pmatrix} w_P(\omega) \\ w_S(\omega) \end{pmatrix}, \tag{7.39}$$

where the phase shift factors $\lambda_P(\omega)$ and $\lambda_S(\omega)$ are applied to the waves as they progress across the array and away from the central reference level. A similar formula holds for arrays with other numbers of receiver elements. Equation (7.39) can be written in the more compact form

$$\mathbf{u}(\omega) = \mathbf{G}(\omega)\mathbf{w}(\omega) \ . \tag{7.40}$$

The least-squares error E between the observed and the model-generated data is

$$E = \sum_{\omega} \left\| \mathbf{G}(\omega)\mathbf{w}(\omega) - \mathbf{u}(\omega) \right\|^2. \tag{7.41}$$

In this expression, each Fourier component of the unknown waveforms $w_P(\omega)$ and $w_S(\omega)$ enters linearly at each frequency, but the dependence on the slownesses q_P and q_S, and on the polarizations $\mathbf{d}_P$ and $\mathbf{d}_S$ is non-linear. Thus for each specific set of slownesses and polarizations in solution space, the waveform values required to minimize the fit are fixed by

$$\hat{\mathbf{w}}(\omega) = \left(\mathbf{G}^{*T}\mathbf{G}\right)^{-1}\mathbf{G}^{*T}\mathbf{u}(\omega) \ . \tag{7.42}$$

Use of the singular value decomposition

$$\mathbf{G} = \mathbf{U}\boldsymbol{\Sigma}\mathbf{V}^{*T} \tag{7.43}$$

provides an efficient form for the smallest squared difference

$$E = \sum_{\omega} \left\| \mathbf{u}(\omega) - \mathbf{U}(\omega)\mathbf{U}(\omega)^{*T}\mathbf{u}(\omega) \right\|^2 . \tag{7.44}$$

Thus, it is possible to optimize the objective function E without considering the waveforms. At each step, $\mathbf{G}(\omega)$ is evaluated, an SVD applied, and then $\mathbf{U}(\omega)$ inserted into (7.44). A non-linear search of E using standard gradient-based search techniques yields the desired parameter estimates. Once slownesses and polarizations (and hence $\mathbf{G}$) are determined, the waveform estimates can be recovered at each frequency using

$$\hat{\mathbf{w}}(\omega) = \hat{\mathbf{G}}^{-1}\mathbf{v}(\omega) = \hat{\mathbf{V}}\hat{\boldsymbol{\Sigma}}^{-1}\mathbf{U}^{*T}\mathbf{v}(\omega) \ . \tag{7.45}$$

Use of the SVD decomposition helps to monitor – and cope with – instability in the inversion $\mathbf{G}^{*T}\mathbf{G}$. Instability occurs at notches in the frequency spectrum for which the matrix becomes singular. Stability can be regained by "damping" the inverse by some suitable small quantity. Damping must be cautiously traded against undesirable trace mixing. A quality control for this procedure is provided by the eigenvalue spectrum and descriptors such as the condition number (ratio of maximum to minimum eigenvalue).

This technique is suited to near and moderate offset VSP, and has been used on marine walk-away VSPs in the North Sea (see Figure 7.13). It has also been applied for the separation of fast and slow waves in a near-offset VSP (Esmersoy 1990b). However, application to large offsets and complex structures is limited by the arrival angles becoming increasingly time-invariant. One adaptation in this case is to employ a sliding time window. Another is the convolutional operator technique proposed by Herrmann and Wapenaar (1992), where in particular the concern for the greater range of plane waves is addressed.

7.1.3.2 Up- and downgoing P and S waves

Leaney (1990) extended the method above to include upgoing waves. The general two-component displacement is now

$$\mathbf{u}(\omega)=\mathbf{d}_P^U w_P^U(\omega)+\mathbf{d}_S^U w_S^U(\omega)+\mathbf{d}_P^D w_P^D(\omega)+\mathbf{d}_S^D w_S^D(\omega)\ , \tag{7.46}$$

where the polarizations $\mathbf{d}_P^D$ and $\mathbf{d}_S^D$ remain as in (7.36) and (7.37), but in the upgoing waves the sign of the vertical component reverses

$$\mathbf{d}_P=\begin{pmatrix}-\sin\theta_P\\-\cos\theta_P\end{pmatrix}, \tag{7.47}$$

and

$$\mathbf{d}_S=\begin{pmatrix}\cos\theta_S\\-\sin\theta_S\end{pmatrix}, \tag{7.48}$$

The complete equation for a five-level tool now becomes

$$\begin{pmatrix}\mathbf{u}_1(\omega)\\\mathbf{u}_2(\omega)\\\mathbf{u}_3(\omega)\\\mathbf{u}_4(\omega)\\\mathbf{u}_5(\omega)\end{pmatrix}=\begin{pmatrix}\mathbf{d}_P^D\lambda_P^2 & \mathbf{d}_S^D\lambda_S^2 & \mathbf{d}_P^U\lambda_P^{-2} & \mathbf{d}_S^U\lambda_S^{-2}\\ \mathbf{d}_P^D\lambda_P & \mathbf{d}_S^D\lambda_S & \mathbf{d}_P^U\lambda_P^{-1} & \mathbf{d}_S^U\lambda_S^{-1}\\ \mathbf{d}_P^D & \mathbf{d}_S^D & \mathbf{d}_P^U & \mathbf{d}_S^U\\ \mathbf{d}_P^D\lambda_P^{-1} & \mathbf{d}_S^D\lambda_S^{-1} & \mathbf{d}_P^U\lambda_P & \mathbf{d}_S^U\lambda_S\\ \mathbf{d}_P^D\lambda_P^{-2} & \mathbf{d}_S^D\lambda_S^{-2} & \mathbf{d}_P^U\lambda_P^2 & \mathbf{d}_S^U\lambda_S^2\end{pmatrix}\begin{pmatrix}w_P^D(\omega)\\w_S^D(\omega)\\w_P^U(\omega)\\w_S^U(\omega)\end{pmatrix}, \tag{7.49}$$

where it can be assumed that the moveout for the upgoing and downgoing waves has equal magnitude but opposite sign. An example of the results from this procedure is shown in Figures 7.8 and 7.9.

It should be noted that the larger matrix in (7.49) creates more opportunity for collinear polarizations. This is caused by P- and S-waves with different moveouts and thus necessitates careful control of the matrix inverse. An example is shown in Section 7.3.2 where this technique is used as a precursory step in a phase-slowness extraction described by Miller, Leaney and Borland (1994). Note that although the above is formulated in the vertical slowness domain, it works equally well in the horizontal slowness domain. Thus, for a walk-away VSP one can transform the common receiver gathers into the horizontal slowness domain and then perform the filtering. Indeed, this domain has a wider spatial aperture and thus more potential resolution.

7.1.4 Deterministic polarization filtering

Deterministic polarization filtering is applied to only down- or upgoing waves. Each common shot gather is transformed into the frequency (ω) and vertical slowness (q) domain using a 2D Fourier transformation over time and depth

$$\mathbf{u}(\omega,q)=\frac{1}{4\pi^2}\int_{-\infty}^{\infty}\int_{-\infty}^{\infty}\mathrm{U}(t,z)\mathrm{e}^{-\mathrm{i}\omega qz}\,\mathrm{e}^{\mathrm{i}\omega t}\,dt\ dz\ . \tag{7.50}$$

This is a plane wave expansion of the data into the range of directions specified by each ω–q point. The transformation results in a generalization of (7.38) to incorporate the directional plane-wave behavior

$$\mathbf{u}(\omega,q)=\mathbf{d}_P(\omega,q)w_P(\omega,q)+\mathbf{d}_S(\omega,q)w_S(\omega,q)\ , \tag{7.51}$$

or

$$\mathbf{u}(\omega,q)=\begin{pmatrix}\mathbf{d}_P(\omega,q) & \mathbf{d}_S(\omega,q)\end{pmatrix}\begin{pmatrix}w_P(\omega,q)\\ w_S(\omega,q)\end{pmatrix}\ . \tag{7.52}$$

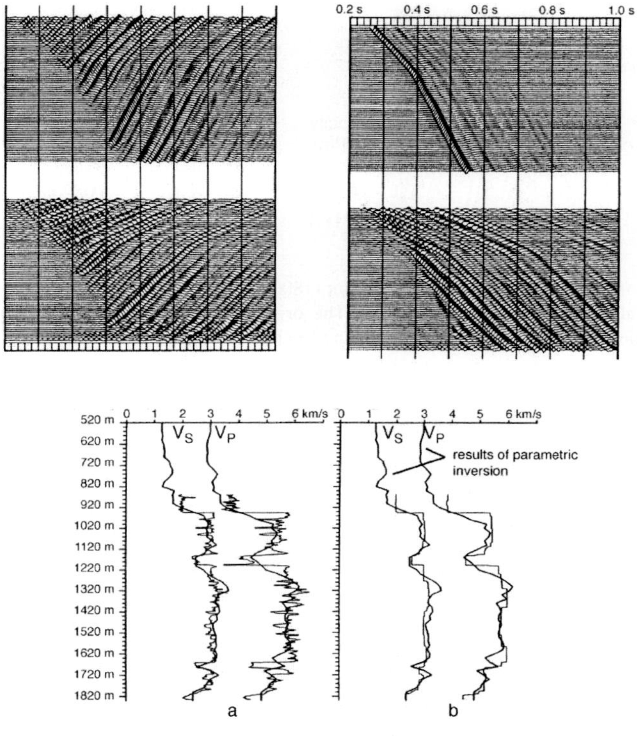

Figure 7.8. Top: separated P and S-wave fields after the decomposition of Figure 7.7.
Bottom: results of best-fit velocity parameters superimposed on the sonic logs with two levels of blocking. Adapted from Leaney (1990).

The essence of the technique is a prediction of these polarizations for each vertical slowness q and frequency ω by guessing or evaluation from the logs. Once calculated, the polarizations can be used to construct an inverse filter to give the requisite separation into $w_P(\omega,q)$ and $w_S(\omega,q)$

$$\begin{pmatrix}\hat{w}_P(\omega,q)\\ \hat{w}_S(\omega,q)\end{pmatrix}=\begin{pmatrix}\mathbf{d}_P(\omega,q) & \mathbf{d}_S(\omega,q)\end{pmatrix}^{-1}\mathbf{u}(\omega,q)\ , \tag{7.53}$$

from which the P- and S-waves can be recovered by the inverse transform

$$\mathbf{u}(t,z)=\int_{-\infty}^{\infty}\int_{-\infty}^{\infty}\mathbf{U}(t,z)\mathrm{e}^{\mathrm{i}\omega qz}\,\mathrm{e}^{-\mathrm{i}\omega t}\,dq\ d\omega\ . \tag{7.54}$$

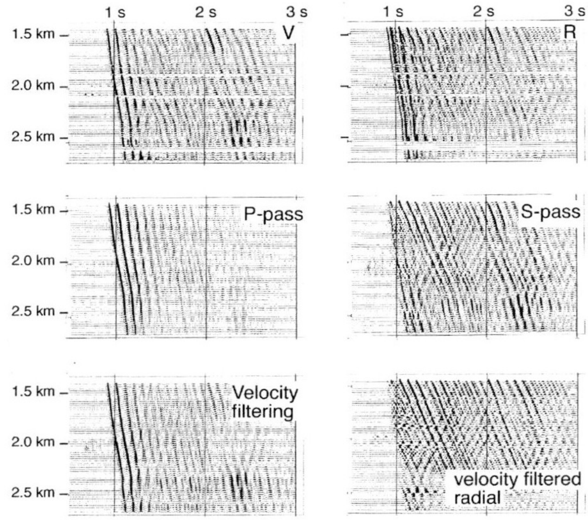

Figure 7.9. Common source gather from a fixed offset (1800m) VSP used to test the deterministic wavefield separation algorithm of Dankbaar (1987). The original data are shown at the top. The results of the separation (middle) are compared against a velocity filter (bottom).

The inverse of the polarization matrix is a multi-component filter weighting the individual contributions to the P- and S-wavefield. Writing

$$\mathbf{d}_P = \begin{pmatrix} \sin\phi_P(q) \\ \cos\phi_P(q) \end{pmatrix} \tag{7.55}$$

and

$$\mathbf{d}_S = \begin{pmatrix} \sin\phi_S(q) \\ \cos\phi_S(q) \end{pmatrix}, \tag{7.56}$$

the two-component sagittal-plane version of equation (7.53), for which $\mathbf{u} = (u_r, u_z)^T$, gives

$$\hat{w}_P(\omega) = \frac{\cos\phi_P(q)u_r(\omega,q) + \sin\phi_P(q)u_z(\omega,q)}{D} \tag{7.57}$$

$$\hat{w}_S(\omega) = \frac{-\sin\phi_S(q)u_r(\omega,q) + \cos\phi_S(q)u_z(\omega,q)}{D}, \tag{7.58}$$

where

$$D = \cos\phi_P(q)\cos\phi_S(q) - \sin\phi_P(q)\sin\phi_S(q) \; . \tag{7.59}$$

The first equation (7.57) passes the P-wave only, and the second (7.58) passes the S-wave. The equations correspond to fan filters that spread along radial lines in the ω-q space. In practical implementation of this procedure, the following points should be noted:

(a) the upper limit of the integral is in both cases the maximum slowness, i.e., the limits are different for the P-pass and S-pass equations;
(b) the separation is exact for a line source or single plane wave excitation, but not for a point source – this can produce artifacts (tests on synthetic seismograms by Dankbaar (1987) show that this approximation appears adequate for most applications);
(c) the technique cannot treat out-of-plane arrivals;
(d) the technique requires sufficient aperture;
(e) separation of the up and downgoing waves must be done afterwards;
(f) the geophones must be well calibrated and evenly coupled.

Construction of these filters for the isotropic elastic case is based on the known orthogonality of $\mathbf{d}_P$ and $\mathbf{d}_S$ for any given propagation direction. Thus, from Aki and Richards (1981)

$$\mathbf{d}_P = V_P \begin{pmatrix} \sqrt{1-q^2V_P^2} \\ q \end{pmatrix} \tag{7.60}$$

and

$$\mathbf{d}_S = V_S \begin{pmatrix} -q \\ \sqrt{1-q^2V_S^2} \end{pmatrix} . \tag{7.61}$$

This isotropic version was implemented by Devaney and Oristaglio (1986). It has been applied in the frequency domain by Dankbaar (1987) (Figure 7.9) and by Dillon, Ahmed and Roberts (1988). There have also been applications in the slant-stack domain by Chiburos et al. (1988).

For the general anisotropy case, the polarizations are orthogonal for a common direction of the wave normal, but not for a common vertical slowness q. However, the vertical slowness components q corresponding to a particular phase angle can be computed directly using the Kelvin-Christoffel equation, evaluated either numerically or analytically. For the TIV case with known elastic constants $a_{11} = c_{11}/\rho$, $a_{33} = c_{33}/\rho$, $a_{13} = c_{13}/\rho$, $a_{55} = c_{55}/\rho$ and $a_{66} = c_{66}/\rho$, the horizontal slowness must firstly be calculated for the P-waves (p_P) and the S-waves (p_S) using Chapter 3:

$$p_P = -B - \frac{\sqrt{B^2 - 4AC}}{2A} \tag{7.62}$$

and

$$p_S = -B + \frac{\sqrt{B^2 - 4AC}}{2A} , \tag{7.63}$$

where

$$A = a_{11}a_{55} , \tag{7.64}$$

$$B = \left(a_{11}a_{33} + a_{55}^2\right)q^2 - \left(a_{11} + a_{55}\right) - \left(a_{13} + a_{55}\right)^2 q^2 \ , \tag{7.65}$$

and

$$C = 1 - \left(a_{33} + a_{55}\right)q^2 + a_{33}a_{55}q^4 \ . \tag{7.66 }$$

These horizontal slownesses then determine the phase angles for the P- and S-waves, $\theta_P = \tan^{-1}(p_P/q)$ and $\theta_S = \tan^{-1}(p_S/q)$, respectively, from which the polarizations and $\mathbf{d}_S$ can be determined. The polarizations lie in the vertical plane, and are obtained from

$$\mathbf{d}_P = \begin{pmatrix} \sqrt{Q+R} \\ \sqrt{Q-R} \end{pmatrix} \frac{1}{\sqrt{2Q}} \ , \tag{7.67}$$

and

$$\mathbf{d}_S = \begin{pmatrix} -\sqrt{Q-R} \\ \sqrt{Q+R} \end{pmatrix} \frac{1}{\sqrt{2Q}} \ , \tag{7.68}$$

where

$$Q = \sqrt{R^2 + 4S^2} \ , \tag{7.69}$$

$$R = \left(a_{11} - a_{55}\right)\sin^2\theta - \left(a_{33} - a_{55}\right)\cos^2\theta \ , \tag{7.70}$$

and

$$S = \left(a_{13} + a_{55}\right)\sin\theta\cos\theta \ . \tag{7.71}$$

Alternative expressions for the polarization angles $\mathbf{d} = (\cos\phi, \sin\phi)^T$, valid for any strength of anisotropy, are given by Musgrave (1970):

$$\tan\phi = \frac{\left(a_{13} + a_{55}\right)\sin 2\theta}{\rho V^2 - a_{11}\sin^2\theta + a_{55}\cos^2\theta} \ , \tag{7.72}$$

where V is the phase velocity ($V^{-2} = p^2 + q^2$), and by de Parscau (1991) (see eq. (7.101)). Other approximations exist for weak anisotropy. The TIV case has been considered by Cho and Spencer (1990) and discussed by Dellinger and Etgen (1990).

7.2 ANISOTROPY ESTIMATION USING LOCAL PHASE SLOWNESS

The method of local phase slowness (LPS) has a long history of development in borehole-related seismic techniques. In fact, it has been discussed and applied as far back as the 1930s as the "method of wavefronts" and was the basis for the classic wavefront chart (for example, Ansel 1931, Thornburgh 1930). The most recent form of this method has evolved from the work of White et al. (1983), Gaiser (1990) and Miller et al. (1994), and is applied to wide aperture walk-away VSPs shot with multi-level tools.

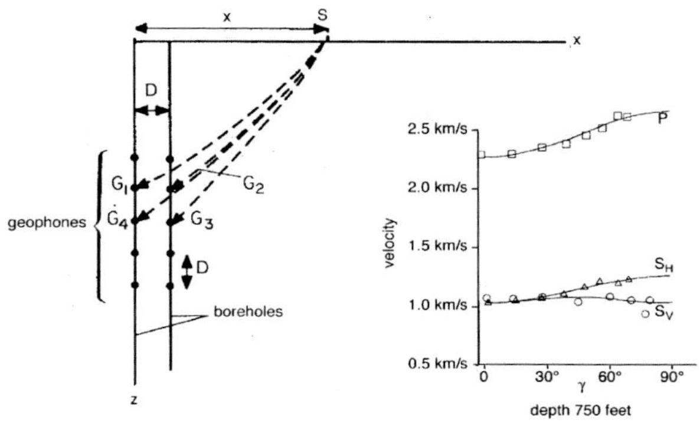

Figure 7.10. Acquisition configuration used by White et al. (1983) to estimate both horizontal and vertical phase slownesses in the subsurface. The VSPs are shot into two shallow boreholes 30 m apart, with the receivers positioned at 240 m. It is rare in a producing field to have two such closely spaced wells. The resultant velocities are shown as a function of phase angle γ at the depth where the horizontal birefringence of this rock (Pierre shale, TIV) is observed.

The approach obtains estimates of phase slownesses from a receiver array situated within the target zone and lines of shot points. From these estimates, knowledge of the functional relationship between these locally determined phase slownesses and the elastic stiffnesses allow the construction of an inversion procedure. The method is attractive as it operates directly on the phase slownesses, with no need to determine phase or propagation angles. The estimates relate to a rock volume of about a wavelength in diameter around the array. For a TIV medium, a single walk-away line and a vertical receiver array ensures that the first breaks of the qP arrivals are sufficient to determine four of the five elastic stiffnesses: a_{11}, a_{33}, a_{13}, and a_{55}. Application to SH waves provides information on the remaining elastic modulus a_{66}, and an additional estimate of a_{55}. For extracting details of vertical fractures set in a TIV background, several lines at different azimuths are required. Recent advances in long multi-level receiver tools (Chapter 9) have the potential to extend the method to variations in the anisotropy with depth.

A major condition for the general validity of this technique is a flat overburden (dip less than a few degrees). A vertical borehole is a further preferred condition for simplification of the slowness estimation procedure and more accurate results. For vertical fractures an appropriate set of line azimuths is also required. The LPS method is sensitive to acquisition errors arising from source elevation, navigation geometry, tool mis-orientation or mis-identification of wave type.

7.2.1 Phase slowness estimation in walk-away VSPs

In theory, the spatial gradient of the travel-time function determines the local phase slowness **p**

$$\mathbf{p} = \begin{pmatrix} p_x \\ p_y \\ q \end{pmatrix} = \begin{pmatrix} \partial t/\partial x \\ \partial t/\partial y \\ \partial t/\partial z \end{pmatrix} . \tag{7.73}$$

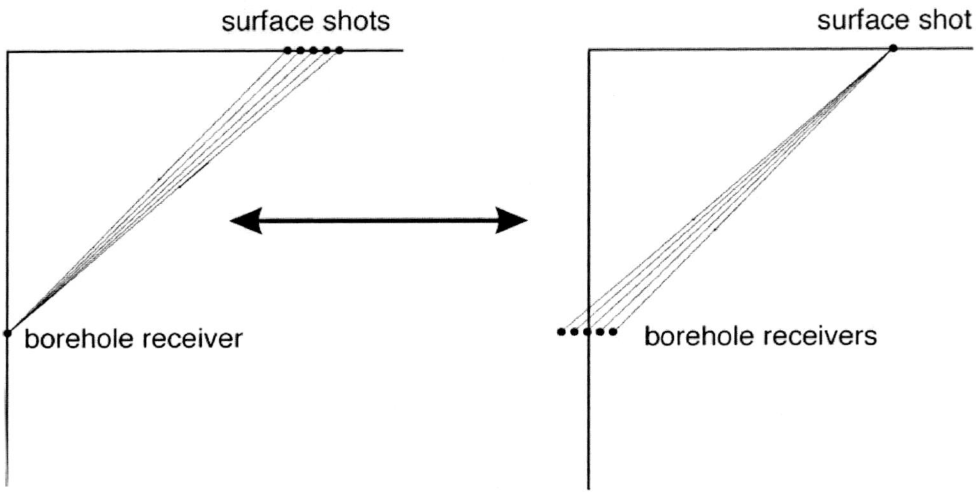

Figure 7.11. Horizontal phase slowness at the target is evaluated on the assumption that a closely spaced group of shots into a single receiver is equivalent to a single shot into a short horizontal array of receivers.

This relationship is valid for a single point in an arbitrary anisotropic medium, and for medium fluctuations that are small over the seismic wavelength (Bubuska and Cara 1991). The essence of the LPS technique is that these derivatives can be approximated by using discrete measurements from downhole receivers. The approach is satisfactory provided the array length is shorter than a wavelength; with a typical receiver spacing between 10 m and 25 m this is just fulfilled by most downhole tools. Unfortunately, determination of all derivatives would require a three-dimensional subsurface array. In practice, the best that can be expected is two closely spaced wells (see for example White et al. 1983, or Figure 7.10), thus determining the vertical slowness and a component of the horizontal slowness. For most applications only one well is available, providing p_a, the projection of the phase slowness on the array axis. A way to get around this difficulty was demonstrated by Gaiser (1990) and Miller et al. (1994): they showed that an independent estimate of horizontal slowness can be provided using multiple offset VSP. The basis for this is the assumption that traveltimes in a receiver gather from a short array of source points (typical spacing 25 m) correspond to a horizontal array of receivers at the target (interchange of source and receiver positions, see Figure 7.11). For this to be valid, the horizontal slowness must be conserved with depth, a condition rendered invalid if dipping subsurface interfaces are present. This approach is applied to each receiver gather, yielding many separate estimates of horizontal slowness with offset.

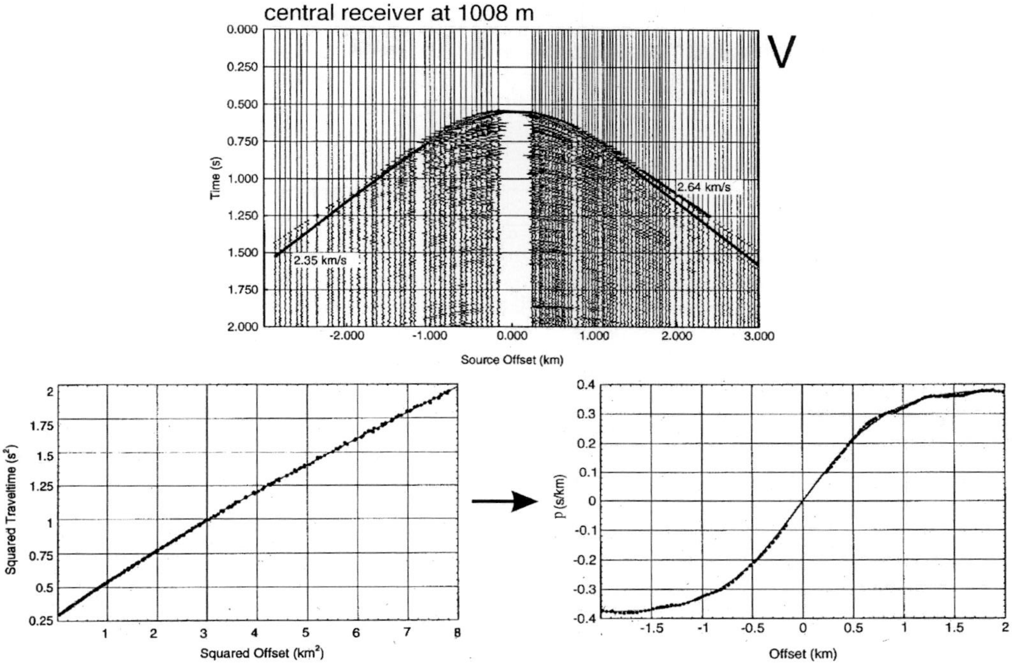

Figure 7.12. In the local phase slowness approach for a walk-away VSP. Travel-times are picked in common receiver gathers and used to determine the horizontal slowness p as a function of offset. Diagrams extracted from Miller et al. (1994).

7.2.1.1 Horizontal slowness estimation using the common receiver gathers

There are typically several hundred shot points, and several numerical schemes for obtaining the travel-time gradient:

(a) Assuming a normal (hyperbolic) moveout equation, a regression analysis fits a straight line to the squared travel-times versus squared offsets (Gaiser 1990); the resulting slope is p^2.

(b) To retain a greater generality in the moveout, fit a smooth, symmetric global travel-time curve in the squared domain to all points (Figure 7.12):

$$t^2\left(x^2\right) = t_0^2 + bx^2 + cx^4 + dx^6 \quad . \tag{7.74}$$

Once the coefficients of the polynomial are estimated, calculate the slowness analytically using

$$p = \frac{\partial t}{\partial x} = \frac{b + 2cx^2 + 3dx^3}{\sqrt{t_0^2 + bx^2 + cx^4 + dx^6}} \quad . \tag{7.75}$$

(c) A third approach by Ohlsen et al. (1998) evaluates $\partial t/\partial x$ by fitting a cubic in x to the points in the t-x domain, but for a running window. The window can be as large as 45 points.

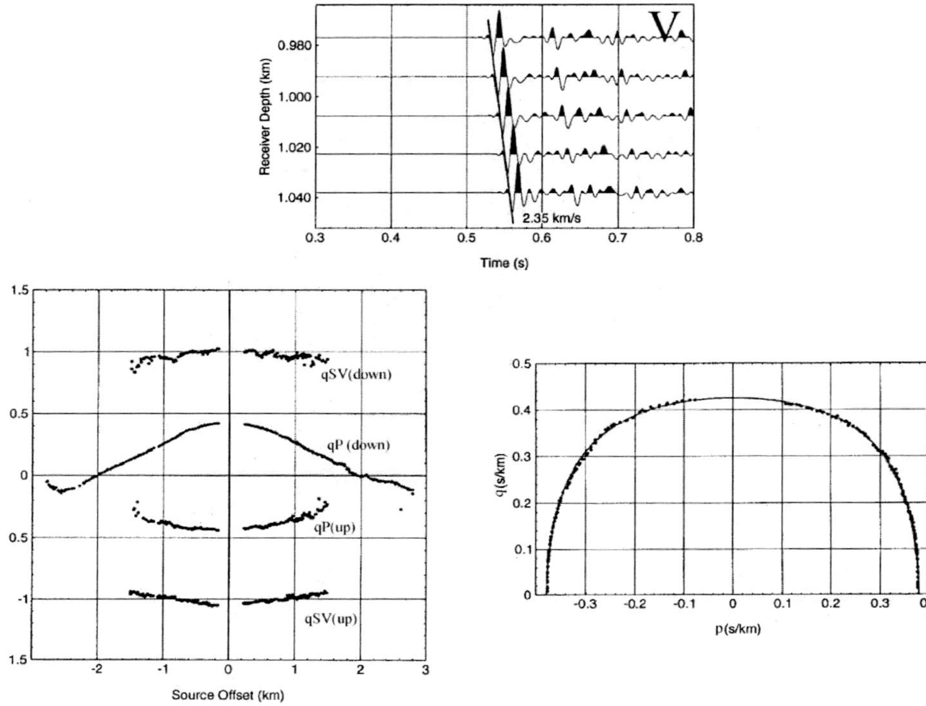

Figure 7.13. The vertical slowness q is calculated by determining the moveout across the receiver array. If there are interfering up- and downgoing P and S waves, the wavefield expansion method of Esmersoy (1990) can be used instead. This yields the vertical slownesses directly as a function of offset. Plotting p against q for each offset gives the curve in the lower right, which can be used to invert for the elastic stiffnesses. Diagrams extracted from Miller et al. (1994).

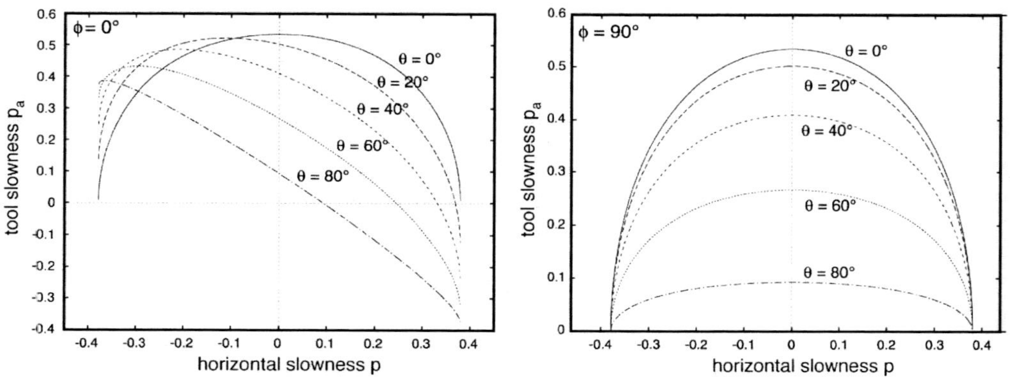

Figure 7.14. The effect of receiver tool orientation on the relationship between horizontal slowness and array tool slowness. Due to errors in predicting tool orientation some of this behavior remains in the plots of vertical slowness versus horizontal slowness. Curves are plotted for different inclinations θ along common azimuths ϕ of 0° and 90°.

7.2.1.2 *Vertical slowness estimation using the common shot gathers*

Evaluating the gradient of the travel-times gives the apparent slowness p_a along the axis of the array tool (Figure 7.13). There are typically 3 to 12 elements in the receiver tool, and the slownesses can be obtained numerically

(a) by a local linear regression of the travel-time picks. It is advisable for the picks used in each regression to come from the same lock position of the multi-level tool, so that errors involved with source drift are avoided. The slope from this regression gives a direct indication of the array slowness, with this value then being assigned to the central depth. For a long receiver array with more than 5 elements a running window can be used, and slowness assigned to different depths.

(b) by fitting a second order polynomial to the travel-times and calculate the gradient as in (7.75).

(c) with a wavefield separation approach (Section 7.1) such as the parametric inversion of Esmersoy (1990a); results from this technique are shown in Figure 7.13.

A vertical borehole is desirable, as q is then directly output in the above procedures. However, in a deviated borehole (common in the North Sea and the Gulf of Mexico), the apparent slowness p_a is now measured along the tool axis; it is a combination of both q and the horizontal slowness p. This means that the directly measured slownesses by themselves cannot be inverted for elastic moduli without further knowledge of the tool orientation. This transformation can be accomplished by projecting the slowness vector **s** onto the direction of the tool axis. Consider an *XYZ* coordinate frame; the horizontal projection of the tool deviation along the *X*-axis defines a unit vector along the tool axis

$$\mathbf{t} = \left(\sin\theta, \quad 0, \quad \cos\theta\right)^T \quad . \tag{7.76}$$

The acquisition line at an angle ϕ to the *X*-axis, points along the vector

$$\mathbf{l} = \left(\cos\phi, \quad \sin\phi, \quad 0\right)^T \quad , \tag{7.77}$$

giving the projected slowness along the array

$$p_a = \mathbf{s}\cdot\mathbf{t} = -p\sin\theta\cos\phi + q\cos\theta \quad , \tag{7.78}$$

in terms of the horizontal slowness p measured along the line. This has the inverse

$$q = p\tan\theta\cos\phi + \frac{p_a}{\cos\theta} \quad . \tag{7.79}$$

The effect of (7.23) on the observed slownesses is shown in Figures 7.14, 7.15 and 7.16, for different tool azimuths and inclinations. The slownesses are very sensitive to orientation: a deviation of a few degrees produces a 10 % change in the slownesses at an inclination of 20°. Indeed, these errors have a more significant effect on the slownesses than the medium itself, inducing an azimuthally varying slowness response, and an asymmetry or flattened shape. In view of this sensitivity, accurate well surveying is necessary. The inverse transformation becomes increasingly "ill conditioned" as the tool orientation approaches horizontal (in this case no q projection exists), and sensitivity to errors increases.

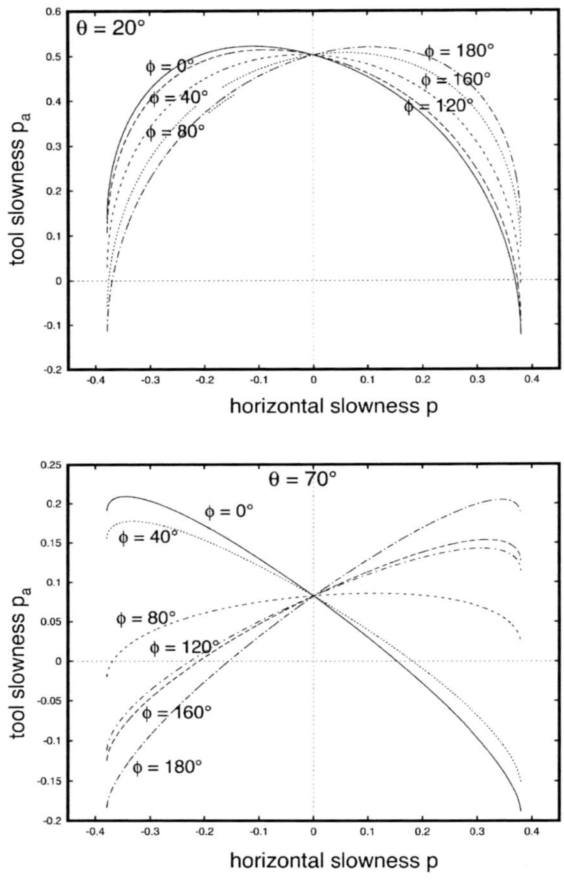

Figure 7.15. The effect of tool orientation on the relation between horizontal slowness and tool slowness, for changes in inclination azimuth ϕ at a fixed inclination θ. There is a dramatic azimuthal dependence on sub-horizontal inclinations.

Errors in the well trajectory exist (Wolff and Wardt 1981). Well surveys using the earth's magnetic field can give a deviation accuracy of 1°–3°, while a gyroscopic tool gives an accuracy of 1°. The well position, on the other hand, is typically accurate in the vertical direction to 15 m, and azimuthally to 60 m (Dodds, personal communication). This uncertainty in the receiver location is removed by transforming the receiver position to the minimum travel-time location at the surface (Sayers 1997). Near this location, the travel times all fall on a hyperboloid, and fitting to this surface for a range of small offsets can be used as a way to find the minimum. The availability of a number of different line azimuths is an advantage (Horne et al. 1998). Interestingly, the seismic response represents the best method for positioning well, as it can be achieved to within 1°.

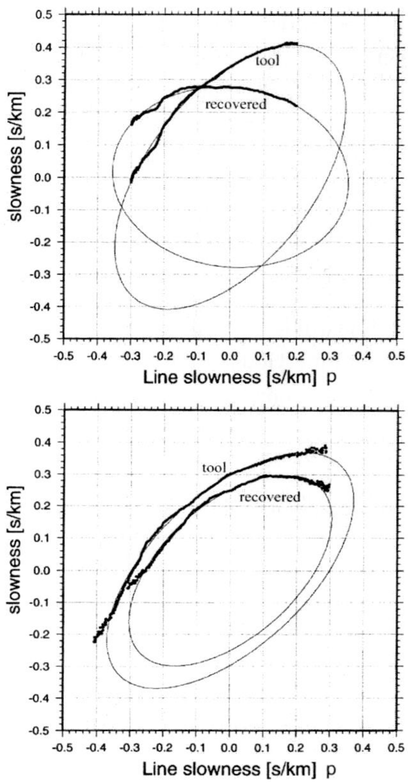

Figure 7.16. Example of tool slowness transformation to vertical slowness, and its effect on the plot against horizontal slowness for a North Sea VSP.

7.2.2 Inversion of qP phase slownesses for a TIV medium

One of the major advantages of the slowness method is the ability to estimate the TIV anisotropy, particularly in shales. Generally, for qP-qSV propagation four density-weighted elastic stiffnesses ($a_{ij} = c_{ij}/\rho$) are involved in this slowness estimate inversions: a_{11}, a_{13}, a_{33} and a_{55}. As before, an independent measure of the density ρ from the logs is required to determine the actual elastic stiffnesses. The phase dispersion relation can be determined directly from the Kelvin-Christoffel equation (Helbig 1994), and re-arranged from Chapter 3 as (Miller and Spencer 1994)

$$a_{11}\left(a_{55}p^4 - p\right) + a_{33}\left(a_{55}q^4 - q^2\right) + A\left(p^2q^2\right) = \left(a_{55}\left(p^2 + q^2\right) - 1\right) , \tag{7.80}$$

where

$$A = a_{11}a_{33} + a_{55}^2 - \left(a_{13} + a_{55}\right)^2 . \tag{7.81}$$

Thus, given the elastic stiffnesses, the expression in 7.13 can be used to generate a curve predicting the relationship between p and q for comparison with the slowness data. For each q, the corresponding p can be obtained from the quadratic equation in p^2

$$\left(a_{11}a_{55}\right)p^4+\left(-a_{11}-a_{55}+Aq^2\right)p^2+\left(a_{33}\left(a_{55}q^4-q^2\right)+1-a_{55}q^2\right)=0 \ . \tag{7.82}$$

The square root of the quadratic solution gives two possible signs, one relates to the qP solution and the other to the qSV solution. This expression is non-linear in the elastic stiffnesses and thus some form of inversion is required to estimate the elastic stiffnesses for a given set of slowness estimates; a direct solution is not possible. One approach is to use an efficient global optimization procedure, such as a Genetic Algorithm (Godin 1998) or "simulated annealing" (Horne et al. 1998). There are several other possibilities for simplification.

7.2.2.1 *Linearization of the qP dispersion relation*

Gaiser (1990) introduced a "weak anisotropy" approximation to linearize equation (7.80)

$$C_0+C_2\left(q^2-p^2\right)+C_4\left[\left(q^2-p^2\right)^2-4p^2q^2\right]=\left(p^2+q^2\right) \ , \tag{7.83}$$

where

$$C_0=\tfrac{1}{2}\left(a_{11}+a_{33}\right)-\tfrac{1}{16}\left(a_{11}+a_{33}-2a_{55}\right)e \ , \tag{7.84}$$

$$C_2=-\tfrac{1}{2}\left(a_{11}-a_{33}\right) \ , \tag{7.85}$$

$$C_4=\tfrac{1}{16}\left(a_{11}+a_{33}-2a_{55}\right)e \ , \tag{7.86}$$

and the anellipticity term e is

$$e=4\left(1-\frac{\left(a_{13}+a_{55}\right)^2}{\left(a_{11}-a_{55}\right)\left(a_{33}-a_{55}\right)}\right) . \tag{7.87}$$

Here, phase angle terms of higher order than $\cos 4\theta$ ($\tan\theta=p/q$) are neglected. The expressions break down for phase angles of 45° and anisotropies above 15 %. After linear inversion, the elastic stiffnesses are obtained from the coefficients using the relations

$$a_{11}=C_0+C_4-C_2 \ , \tag{7.88}$$

$$a_{33}=C_0+C_4+C_2 \ . \tag{7.89}$$

Determination of a_{13} again requires knowledge of a_{55}, with which e can be determined from either (7.84) or (7.86), and hence from the quadratic relation in (7.87).

$$a_{13} = -a_{44} - \sqrt{(a_{11} - a_{44})(a_{33} - a_{44})\left(1 - \frac{e}{4}\right)} \ . \tag{7.90}$$

An alternative weak-anisotropy expression for linearized inversion is given by Thomsen (1986) (see Chapter 3):

$$2\varepsilon p^4 + 2\delta p^2 q^2 = \frac{1}{\alpha_0^2} - 1 \ , \tag{7.91}$$

where α_0 is the vertical velocity.

Miller and Spencer (1994) suggest an approach based on a good guess of a_{55}. This then linearizes (7.80), giving

$$a_{11}u_i + a_{33}v_i + Aw_i = d_i \ , \tag{7.92}$$

where i runs from 1 to N, the total number of points. An estimate of the coefficients is now obtained by a single inversion step

$$\begin{pmatrix} \hat{a}_{11} \\ \hat{a}_{33} \\ \hat{A} \end{pmatrix} = \begin{pmatrix} u_1 & u_2 & \dots & u_N \\ v_1 & v_2 & \dots & v_N \\ w_1 & w_2 & \dots & w_N \end{pmatrix} \begin{pmatrix} d_1 \\ d_2 \\ \cdot \\ \cdot \\ d_N \end{pmatrix} . \tag{7.93}$$

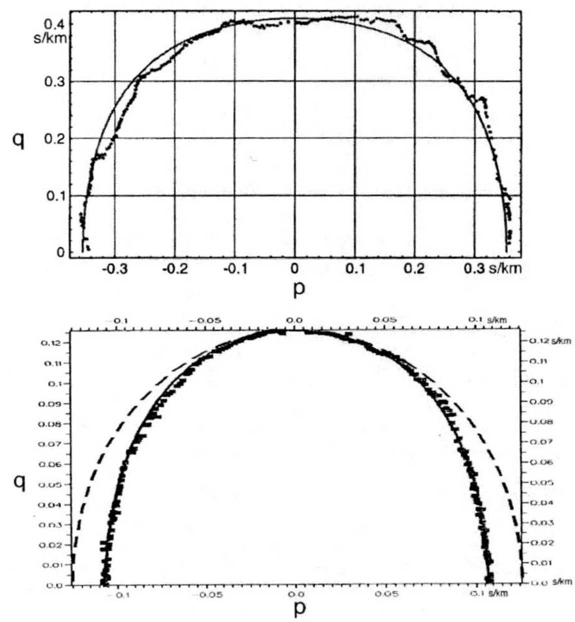

Figure 7.17. Examples of plotting vertical against horizontal slowness for walk-away VSPs. Top: North Sea VSP data by Hornby et al. (1995); bottom: Java sea example by Leaney et al. (1994).

After determining the coefficients, a_{13} can be calculated using (7.81). This particular approach is justified if the slowness solutions are not sensitive to a_{55}, which is generally the case. A good estimate of this parameter can be obtained by measuring the fast shear waves near vertical incidence, or using dipole logging measurements. If an accurate estimate is not possible, a sequence of solutions for different trial values could be used instead.

Examples of using this and similar techniques are shown in Figures 7.17, with further results from other studies shown in Table 7.1.

Table 1. Estimates of Thomsen parameters from the local phase slowness technique applied to multi-offset VSP.

Refer.	ε	δ	ε–δ	γ	Origin	Rock type
1	0.110	0.090	0.020	0.165	Colorado, 450 m	Pierre shale
1	0.195	0.175	0.020	0.300	Colorado, 650 m	Pierre shale
1	0.015	0.060	–0.045	0.030	Colorado, 950 m	Pierre shale
2	0.05–0.10	–	0.10–0.15	–	East Texas	sand+chalk
2	0.10–0.15	–	0.12–0.30	–	East Texas	shales
2	0.15–0.20	–	0.40–0.55	–	East Texas	Calcareous marls
3	0.132	–0.160	0.292	–	South China Sea	Petronas shale
4	0.170	–0.030	0.200	–	North Sea	shale
6	0.021	0.231	–0.210	–	North Sea	chalk
6	0.059	0.050	0.009	–	North Sea	chalk
6	–0.007	0.101	–0.108	–	North Sea	chalk
5	0.231	0.008	0.223	–	Java Sea	shale
5	0.196	0.029	0.167	–	Java Sea	shale
5	0.063	–0.068	0.131	–	Java Sea	shale
5	0.226	0.048	0.178	–	West Africa	–
7	0.350	–0.180	0.530	–	West Africa	–

1 - White et al. (1983);
2 - Gaiser (1990) (estimated from qP and qSV);
3 - Miller et al. (1994);
4 - Hornby et al. (1995);
5 - Leaney (1994);
6 - Ohlsen et al. (1998);
7 - Godin (1998).

Almost without exception these shales show a positive anellipticity. This was also demonstrated in the selection by Thomsen (1986) shown in Chapter 3.

7.2.2.2 *Why not fit an ellipse to the slowness data?*

Examples of some typical plots of vertical slowness against horizontal slowness are shown in Figures 7.17 to 7.19, together with the best fits of the predicted behavior. Figures 7.18 and 7.19 show application of the LPS technique to a 3D-VSP. With these, as with examples from the other geographical locations shown in Figure 7.20, Ohlsen and MacBeth (1999) found that it was possible to adequately fit a number of slowness results using an ellipse ($\varepsilon = \delta$), instead of equation (7.80). An ellipse appears to be the simplest geometric figure that can fit these observations, but it should not be considered as the final solution. In fact, it was found that a small region of ε and δ solutions existed with similar RMS deviations, and the ability to provide a satisfactory fit. The width of this region increases with decreasing angular aperture, and can support a variation in ellipticity of between 0.02 and 0.05, for ellipticities of 0.05–0.30. Due to these uncertainties, it is recommended to fit an ellipse as the most stable geometric figure, and then use this shape as a template for evaluating further best-fit solutions. This strategy should be pursued for data residuals greater than 2 %.

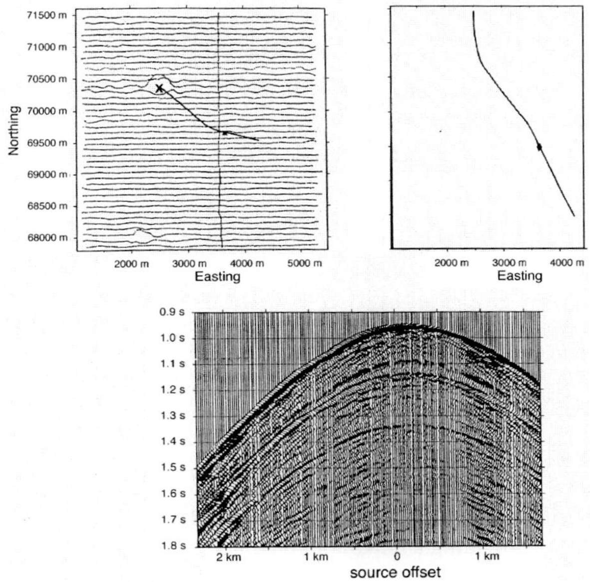

Figure 7.18. Example of the local phase slowness technique applied to a 3D-VSP acquired in the North Sea. Above: plan view of the source spread, horizontal trajectory of the well and predicted tool location, together with a vertical section showing the extent of the tool deviation. Below: typical receiver gather. The slowness technique can be applied to interpolated travel times lying within selected vertical planes at a range of azimuths. After Kühnel and MacBeth (1999). Data courtesy of the British Geological Survey.

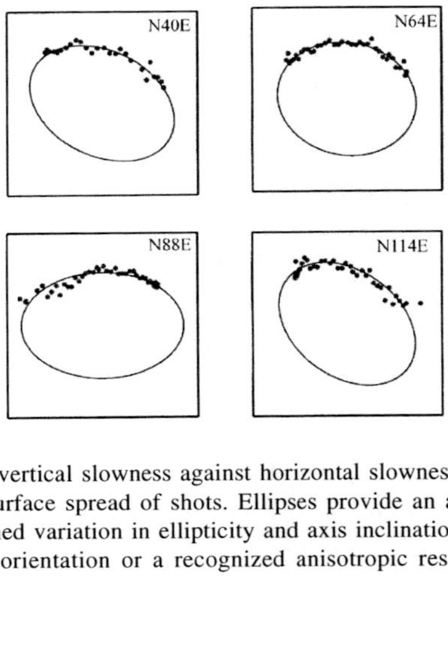

Figure 7.19. Variation in plots of vertical slowness against horizontal slowness for four of the selected azimuthal slices from the 3D surface spread of shots. Ellipses provide an adequate fit to the slowness points. There is an unexplained variation in ellipticity and axis inclination with azimuth that does not correspond to either tool orientation or a recognized anisotropic response. After Kühnel and MacBeth (1999).

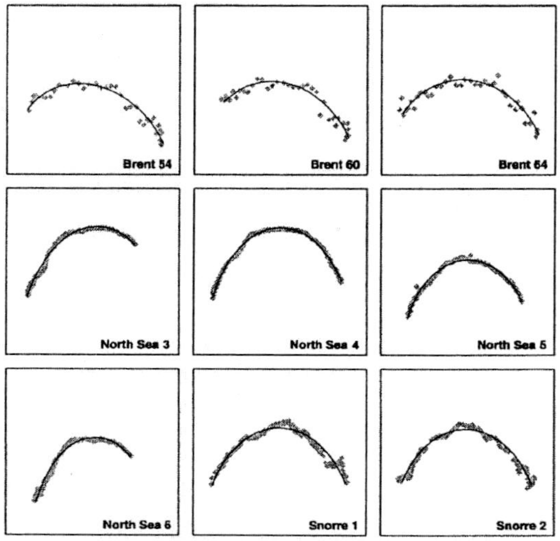

Figure 7.20. A selection of examples of ellipses fitting plots of vertical slowness versus horizontal slowness. After Ohlsen and MacBeth (1999).

7.2.2.3 Picking specific directions

Picking vertical and horizontal propagation directions gives a_{33} and a_{11}, respectively, with one additional direction at 45° providing a quadratic (obtained by setting $p = q$ in (7.80)) to solve for the remaining coefficient. In practice, this is possible in walk-away VSP as the ray paths can quickly reach their critical angle at the target.

7.2.3 Using shear wave phase slownesses and polarizations

For a TIV medium, shear waves are necessary for providing a_{55} for the qP inversion, but also to obtain a_{66}. Shear wave slowness estimation is generally less reliable, as the travel-times are not well defined. Inversion is often restricted by the range of possible slownesses. For offshore applications, the main energy is concentrated in the qSV waves. In this case, use is again made of (7.80). Gaiser (1990) offers another linearization:

$$(p^2 + q^2) = B_0 + B_4\left((p^2 - q^2)^2 - 4p^2q^2\right) , \tag{7.94}$$

where

$$B_0 = a_{55} + \tfrac{1}{16}(a_{11} + a_{33} - 2a_{55})e , \tag{7.95}$$

and

$$B_4 = \tfrac{1}{32}(a_{11} - a_{33})e , \tag{7.96}$$

and recovering

$$a_{55} = B_0 + B_4 \ . \tag{7.97}$$

An alternative is the form of Thomsen (1986)

$$\beta_0^2\left(p^4 + q^4\right) + p^2 q^2\left(\beta_0^2 + 2\alpha_0^2(\varepsilon - \delta)\right) - p^2 - q^2 = 0 \ . \tag{7.98}$$

Land surveys offer the potential of recording SH waves also, with a simpler and better defined slowness surface given by

$$a_{55}q^2 + a_{66}p^2 = 1 \ . \tag{7.99}$$

If the subsurface can be assumed to be strictly TIV, cross-line- and in-line sources excite only SH or qSV waves, respectively, and there is no complication in identification. Complications arise, however, in fracture-related TIH media, for which the fast shear waves at near-offsets are polarized perpendicular to the fracture plane. Thus the source could excite both fast and slow shear waves, which – unless properly identified and separated – will confuse the results. Another potential problem is the cross-over of the slowness sheets at the line singularity (Chapter 3). Another approach using the waves excited by shot lines at different azimuths is discussed in Chapter 8.

7.2.4 Factors leading to inaccurate slowness estimation

7.2.4.1 Dipping subsurface layers

If the overburden exhibits dip, then horizontal slowness p is not conserved, but instead the slowness component parallel to the interfaces (according to Snell's law). Lateral heterogeneities near the surface have a stronger effect than similar deeper heterogeneities (Gaiser 1990), and can be significant for dips as small as 5°. The overall effect is to move each observed p-q point towards smaller values of p in the up-dip direction (Figure 7.21). This effect depends on the direction of the acquisition line direction relative to the dip. A ray tracing study by Sayers (1997) showed that slownesses remain unaffected in the strike direction. In addition, it was shown that the travel times for opposite line azimuths (positive and negative offsets x at either side of the well) can be averaged

$$t(x) = \tfrac{1}{2}\left(t(x,\phi) + t(-x,\phi)\right) = \tfrac{1}{2}\left(t(x,\phi) + t(x,\phi+\pi)\right) \tag{7.100}$$

to give a measure that is independent of dip and line azimuth to first order in dip. This quantity is equal to that expected along the strike direction.

7.2.4.2 Finite array length

Heterogeneity and wave front curvature affect the stability of the numerical approximations for the travel-time gradients. However, White et al. (1983) found that the travel-time error due to ray bending amounted to no more than 0.26 % for path lengths of a few hundred meters and spatial gradients evaluated over 30 m.

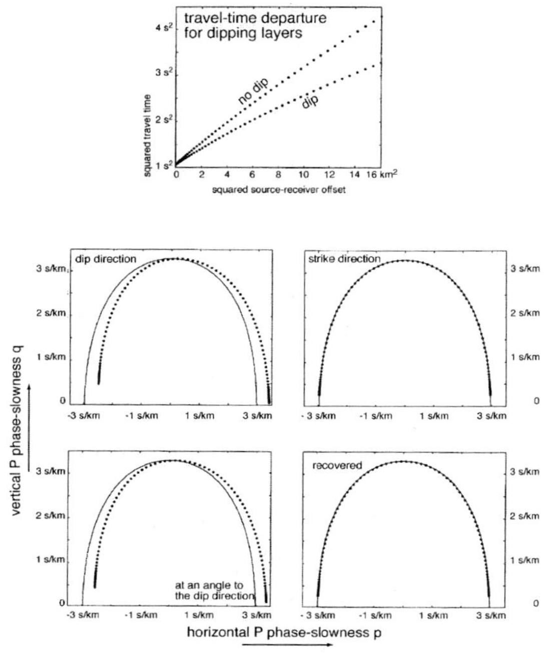

Figure 7.21. The effect of a dipping overburden on travel-time curves and plots of vertical slowness against horizontal slowness. The bottom right diagram is recovered by averaging travel-times for positive and negative offsets. After Sayers (1997).

7.2.4.3 Deviation of gropu velocity from phase velocity

For general anisotropy, the phase velocity vector deviates from the source-receiver plane. Thus, this deviation must be included in the projection onto the array orientation, and the overall inversion scheme. In this case it is best to invert using a numerical procedure such as that of Horne et al. (1998).

7.2.5 Using qP polarizations

Additional information for the inversion procedure can be provided by the polarization of the qP waves, as these deviate from both the group and the phase direction. For a TIV medium, for which the qP and qSV polarizations lie in the vertical symmetry plane, de Parscau (1991) showed that the direction of the qP polarization vector $(\cos\phi, 0, \sin\phi)^T$ with respect to the vertical axis is

$$\tan 2\phi = \frac{2(a_{13}+a_{55})\sin 2\theta}{(a_{33}-a_{11})+((a_{11}+a_{33})-2a_{55})\cos 2\theta}, \tag{7.101}$$

where θ is the phase angle. The expression is valid for mild anisotropy, a_{55} less than the minimum of a_{11} and a_{33}, and $a_{13} + a_{55} > 0$. The phase angle can be estimated directly from the estimates of p and q as in Section 7.2.3, thus leading to a further constraint on the elastic stiffnesses. This is an alternative procedure when the horizontal slowness cannot be reliably used due to significant overburden dip. The approach makes use of an additional relationship for q determined from the magnitude of the slowness vector s and θ. s is calculated by re-arranging (7.80) using the substitution $q = s \cos \theta$ and $p = s \sin \theta$. This yields the quadratic

$$\left(s^{-2}\right)^2 - a\left(s^{-2}\right) - b = 0 \ , \tag{7.102}$$

where

$$a = \frac{a_{11} + a_{33}}{2} + a_{55} - \frac{a_{11} - a_{33}}{2} \cos 2\theta \ , \tag{7.103}$$

and

$$b = a_{55}\left(\frac{a_{11} + a_{33}}{2} - \frac{a_{11} - a_{33}}{2} \cos 2\theta\right) + \tfrac{1}{4}\left(\left(a_{11} - a_{55}\right)\left(a_{33} - a_{55}\right) - \left(a_{55} + a_{13}\right)^2\right)\sin^2 2\theta \ . \tag{7.104}$$

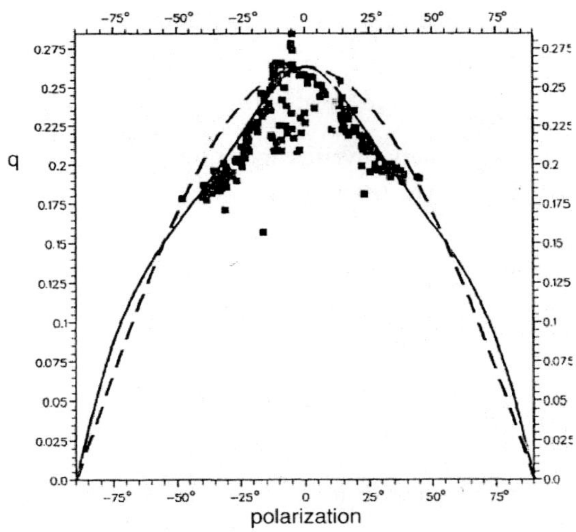

Figure 7.22. Vertical slowness versus polarization angle for downgoing qSV waves extracted from a walk-away VSP in the Java sea. The dashed line is the isotropic result, while the solid line is the best-fitting TIV model. After Leaney et al. (1994).

With both sets of observations for q and $\tan 2\phi$ now described as functions of the independent parameters θ, a_{11}, a_{33}, a_{13}, and a_{55}, estimation of these parameters is possible using a Gauss-Newton inversion scheme. An advantage of this particular approach is that explicit expressions for each partial derivative can be obtained (Hsu et al. 1991). An example of this technique on qSV waves is shown in Figure 7.22.

7.2.6 Vertical fractures in a TIV background

Vertical fractures in a TIV background form an orthorhombic medium (Chapter 2), characterized by the nine elastic stiffnesses a_{11}, a_{22}, a_{33}, a_{23}, a_{13}, a_{12}, a_{44}, a_{55}, and a_{66}. The stiffnesses are directly related to the three Thomsen-parameters and the two fracture indices.

7.2.6.1 *Known fracture strike*

An extension of the local phase slowness is offered by Miller and Spencer (1994). Here, the technique must be applied to two walk-away lines parallel and perpendicular to a known fracture strike. The fracture strike is obtained from borehole information or shear wave polarizations. For the line perpendicular to the fracture strike, the dispersion relation is of a form similar to (7.80):

$$a_{11}\left(a_{55}p^4 - p^2\right) + a_{33}\left(a_{55}p^4 - p^2\right) + A\left(p^2q^2\right) = a_{55}\left(p^2 + q^2\right) - 1 \ , \tag{7.105}$$

where

$$A = a_{11}a_{33} + a_{55}^2 - \left(a_{13} + a_{55}\right)^2 \ .$$

Similarly, for acquisition along the fracture strike a_{22} replaces a_{11}, and a_{23} replaces a_{13}:

$$a_{22}\left(a_{44}p^4 - p^2\right) + a_{33}\left(a_{44}p^4 - p^2\right) + A\left(p^2q^2\right) = a_{44}\left(p^2 + q^2\right) - 1 \ , \tag{7.106}$$

where

$$A = a_{22}a_{33} + a_{44}^2 - \left(a_{23} + a_{44}\right)^2 \ .$$

Thus, each line can be separately inverted to estimate $\hat{a}_{11}$, $\hat{a}_{22}$, $\hat{a}_{33}$, $\hat{a}_{13}$, $\hat{a}_{23}$, $\hat{a}_{44}$ and $\hat{a}_{55}$. From these, it is possible to recover the fracture parameters

$$e_t = \frac{\hat{a}_{44} - \hat{a}_{55}}{\hat{a}_{44}} \ , \tag{7.107}$$

and

$$e_n = \frac{\hat{a}_{11}\left(\hat{a}_{23} - \hat{a}_{13}\right)^2}{\hat{a}_{23}^2\hat{a}_{11} - \hat{a}_{13}^2\hat{a}_{22}} \ , \tag{7.108}$$

from which the TIV background medium can also be recovered (Chapter 2).

7.2.6.2 *Approximation for weak fracturing but arbitrary TIV background*

If the fracture orientation is unknown, then the approximation of MacBeth (1999a) for the Kelvin-Christoffel equation – treating the fracture indices as a first order perturbation – can be used. It yields a TIV-like equation

$$a'_{11}\left(a'_{55}p^4 - p^2\right) + a_{33}\left(a'_{55}p^4 - p^2\right) + A'\left(p^2q^2\right) = a'_{55}\left(p^2 + q^2\right) - 1 \ , \tag{7.109}$$

for any azimuth ϕ with respect to the fracture normal, where the individual stiffnesses a'_{11}, a'_{13} and a'_{44} have been adapted to absorb the additional azimuthal variations. The elastic stiffness a_{33} is independent of azimuth. The azimuth-dependent stiffnesses are given by

$$a'_{55}(\phi) \approx a_{55}\left(1 - e_t \sin^2 \phi\right) \tag{7.110a}$$

or

$$a'_{55} = a_{55}\left(1 - \alpha_t(\phi)\right) ,$$

$$a'_{11}(\phi) \approx a_{11}\left(1 - e_t \frac{a_{66}}{a_{11}} \sin^2 2\phi - e_n + e_n \frac{a_{12}}{a_{11} - a_{66}} \sin^2 2\phi + e_n\left(1 - \zeta^2\right)\cos^4 \phi\right) , \tag{7.110b}$$

or

$$a'_{11} = a_{11}\left(1 - \beta_t(\phi)\right) ,$$

and

$$a'_{13}(\phi) \approx a_{13}\left(1 - e_n + e_n(1 - \zeta)\cos^2 \phi\right) , \tag{7.110c}$$

or

$$a'_{13} = a_{13}\left(1 - \gamma_t(\phi)\right) .$$

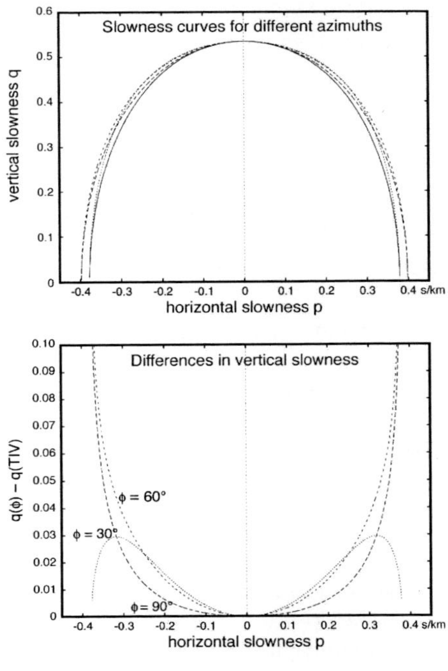

Figure 7.23. Expected azimuthal variation in slownesses for an isotropic medium containing vertical fractures with $e_t = 0.3$ and $e_n = 0.0$. Line azimuths are with respect to the fracture strike.

These lead to an azimuth-dependent ε and δ:

$$\varepsilon'(\phi) \approx \varepsilon - \frac{\varepsilon + \frac{1}{2}}{\beta_t} , \tag{7.111}$$

and

$$\delta'(\phi) \approx \delta + \zeta\gamma_t + \frac{\eta}{1-\eta}\zeta\alpha_t . \tag{7.112}$$

With observations parallel to the fracture strike the TIV background parameters are recovered, with observations normal to it $\varepsilon' = -\frac{3}{8}$ and $\delta' = -2\eta(e_t + \zeta e_n)$. As partial affirmation of this result, it should be noted that the numerical study of Miller and Spencer (1994) found that a TIV equation could be fit to any vertical plane of observations in an orthorhombic medium generated by weak fracturing. The variation in the slownesses in a vertical plane with azimuth are shown in Figures 7.23 and 7.24 for different e_n and e_t.

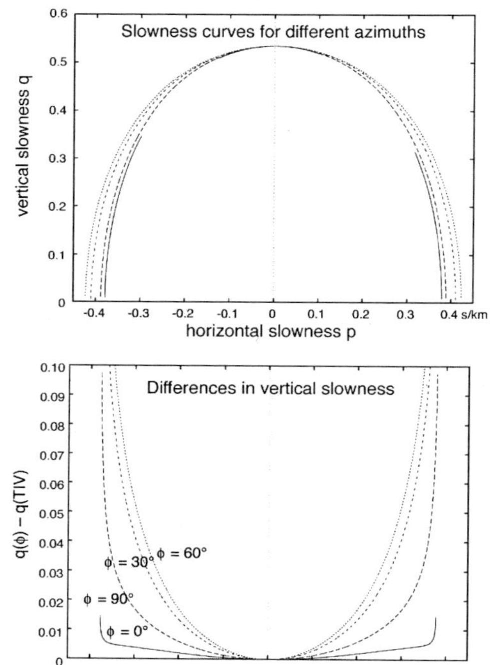

Figure 7.24. Azimuthal variation in slownesses for vertical fractures with $e_n = 0.3$ and $e_t = 0.0$. The behavior for this case is more regular and predictable.

The above equations do not circumvent the need for several walk-away lines, but suggest a practical strategy for fracture evaluation using a_{55} and appropriate line azimuths. Given the regular azimuthal variation of the stiffnesses, separate inversion results for each line can be combined to give details of the fracturing. For example, four walkaway lines consisting of two orthogonal sets *A*-*C* and *B*-*D* separated by 45° (Figure 7.25) give the fracture strike ϕ_f with

$$\tan 2\phi_f = \frac{a'^{C}_{55} - a'^{A}_{55}}{a'^{B}_{55} - a'^{D}_{55}} \,. \tag{7.113}$$

If four lines are impractical, three lines separated by 45° give

$$\tan\left(2\phi_f - \frac{\pi}{4}\right) = \frac{a'^{C}_{55} - a'^{B}_{55}}{a'^{C}_{55} - a'^{D}_{55}} \,. \tag{7.114}$$

For three lines of arbitrary orientation, separated by angles of ψ_1 and ψ_2 give

$$\eta = \frac{a'^{B}_{55} - a'^{C}_{55}}{a'^{B}_{55} - a'^{A}_{55}} \tag{7.115}$$

and

$$\tan 2\phi_f = \frac{\eta \sin(2\psi_2 + \psi_1)\sin\psi_1 + \sin^2\psi_2}{\eta\cos(2\psi_2 + \psi_1)\sin\psi_1 - \sin\psi_2\cos\psi_2} \,. \tag{7.116}$$

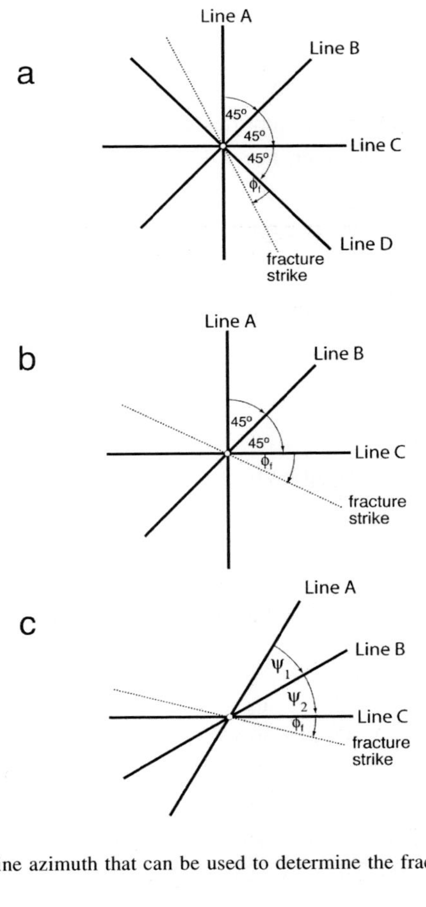

Figure 7.25. Configurations of line azimuth that can be used to determine the fracture strike and other fracture details.

Equation (7.116) is accurate provided line azimuths are 30°–60° apart. After estimation of the fracture strike, e_t and the background value a_{55} can be recovered. It should be noted that a similar strategy cannot be followed for $a'_{11}(\phi)$, since the term with $\sin^2 2\phi$ varies more rapidly with azimuth, and the stiffness also includes an inter-mixing of e_t and e_n terms. However, e_n can be recovered from the estimates of $a'_{13}(\phi)$ using formulae similar to the above. A disadvantage of this approach is the general insensitivity of the TIV dispersion equations to a_{55}. This leaves a_{13} as the only possibility for the determination of fracture strike and fracture index.

7.3 MOVEOUT CORRECTION AND VELOCITY ANALYSIS

The most appropriate geometry for anisotropy analysis is the walk-away VSP, with the combination of rig and boat-source (normal incidence) also being particularly useful for deviated wells. For the direct waves a line of multiple shot offsets recorded on a common downhole receiver generates one half of a surface CDP (CMP) gather. As a consequence, the moveout can be treated as if a surface seismic experiment had been performed with a fictitious horizontal reflector at the depth of the receiver, by substitution of two-way times for one-way times. The advantage of anisotropy analysis using moveout in VSP data is the accuracy of the vertical velocity estimates, due to the receiver depth being known. For TIV media, normal moveout velocity on short-spread gathers provides δ as a function of depth from the P-waves, $\varepsilon - \delta$ from the qSV waves, and γ from SH waves. Wider offsets induce a non-hyperbolic term that supplies an additional parameter η_{eff}, that is related to $(\varepsilon - \delta)/(1 - 2\delta)$ for a homogeneous layer. For a TIH or orthorhombic medium, the fracture strike can be determined by shooting several different line azimuths. If moveout from P and S waves can be determined together, then the fracture parameters e_t and e_n can also be resolved. Reflected waves are to be used when anisotropy parameters are required for targets below the recording interval, with the analysis accessing similar parameters to those in the direct waves.

Isotropic moveout follows the familiar hyperbolic form

$$t^2 = t_0^2 + \frac{X^2}{V_{iso}^2} , \tag{7.117}$$

where t_0 is the (known) one-way time for vertical incidence

$$t_0 = \sum_{i=1}^{N} \Delta t_0(i) , \tag{7.118}$$

and the normal moveout velocity is defined as the RMS-average of the vertical interval velocities $\alpha_0(i)$

$$V_{iso}^2 = \frac{1}{t_0} \sum_{i=1}^{N} \alpha_0^2(i) \Delta t_0(i) . \tag{7.119}$$

where $\Delta t_0(i)$ is the vertical interval time.

Seismic anisotropy influences this moveout in two ways. Firstly, it changes the value of the (short-spread) normal moveout to some new anisotropy-dependent quantity V. Secondly, it creates a deviation from the familiar hyperbolic form, adding to the non-hyperbolic term already expected for long offsets ($2X/Z > 1$) (Taner and Koehler 1969; Al-Chalabi 1974)

$$t^2 = t_0^2 + \frac{X^2}{V^2} + A_4^{iso} X^4 . \tag{7.120}$$

However, the non-hyperbolic term arising from anisotropy has a different character. It is present even for a single homogeneous layer (Hake et al. 1984), and loses accuracy far quicker with offset than the isotropic case. To account for all long offset effects Alkalifah and Tsvankin (1995) suggest an expression that includes a truncation error

$$t^2 = t_0^2 + \frac{X^2}{V^2} + \frac{A_4 X^4}{B + A_5 X^2} , \tag{7.121}$$

for which it is possible to relate the coefficients by the asymptotic behavior, with the moveout velocity tending towards the horizontal velocity at large offsets ($A_4/A_5 = V^2(1/V_h^2 - 1/V^2)$). Equation (7.120) is valid for offsets $X \leq Z_R$.

7.3.1 Normal moveout and velocity analysis for direct waves (TIV)

Moveout remains close to hyperbolic for short spreads ($2X/Z < 1$), even in the presence of anisotropy, and isotropic velocity analysis tools can still be used (Figure 7.26). Thomsen (1986) gave the moveout velocities for short spreads for a single horizontal layer:

$$V^2(qP) = \alpha_0^2(1 + 2\delta) , \tag{7.122}$$

$$V^2(qSV) = \beta_0^2\left(1 + 2\frac{\alpha_0^2}{\beta_0^2}(\varepsilon - \delta)\right) = \beta_0^2(1 + 2\sigma) , \tag{7.123}$$

and

$$V^2(SH) = \beta_0^2(1 + 2\gamma) . \tag{7.124}$$

These equations are exact, and hold for TIV-anisotropy of any strength. There are several points worth mentioning:

(a) for VSP application the vertical velocities α_0 and β_0 are generally known, so that the determination of V gives immediate access to the anisotropy coefficients ε, δ and γ.

(b) offset variations for the qSV-wave are more pronounced than those for the P-wave (since $\alpha_0^2/\beta_0^2 > 2.5$).

(c) for elliptical anisotropy ($\varepsilon = \delta$), $V(qP)$ is equal to the horizontal velocity and $V(qSV)$ is equal to the vertical velocity.

(d) for the SH waves, V is equal to the horizontal velocity $\beta_0(1 + \gamma)$.

(e) the influence of β_0 on the qP travel-times is negligible, even for strong anisotropy (Tsvankin and Thomsen 1994, Alkalifah and Larner 1994).

For qP waves in a stack of TIV layers (Hake et al. 1984 and Tsvankin and Thomsen 1994), the moveout velocity is as a function of the vertical velocities $\alpha_0(i)$ and anisotropies $\delta(i)$ for each interval i:

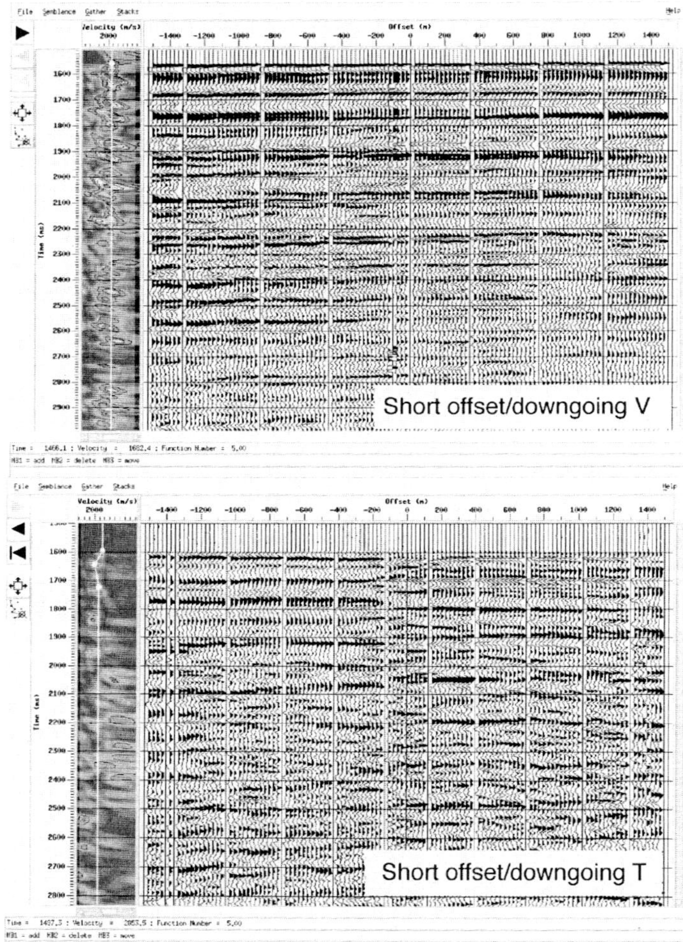

Figure 7.26. Walk-away VSP data after wavefield separation by rotation and normal moveout correction. The P-waves are predominantly on the vertical component and shear waves on the transverse. The short offset amplitudes align, demonstrating that the hyperbolic moveout correction is adequate for these data. Data courtesy of the British Geological Survey and Norsk Hydro.

$$V^2 = \frac{1}{t_0} \sum_{i=1}^{N} \alpha_0^2(i) \Delta t_0(i) \left(1 + 2\delta(i)\right) , \tag{7.125}$$

or

$$V^2 = V_{iso}^2 \left(1 + 2\zeta\right) . \tag{7.126}$$

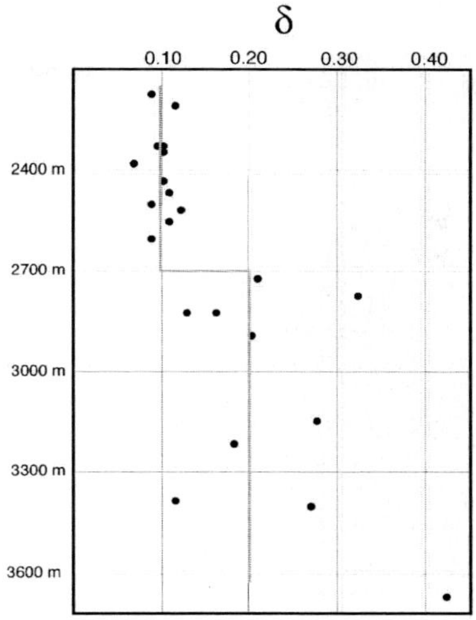

Figure 7.27. This figure, re-drawn from Banik (1984), shows the distribution of δ for predominantly shaly sections in the North Sea as a function of mean depth. The anisotropy is constant down through the Cretaceous section until the Jurassic, where there is a large jump. Banik (1984) found a strong correlation between anisotropy and the presence of shales, but little anisotropy in the limestones/chalks. The shales also correspond to significant depth errors in the surface seismic.

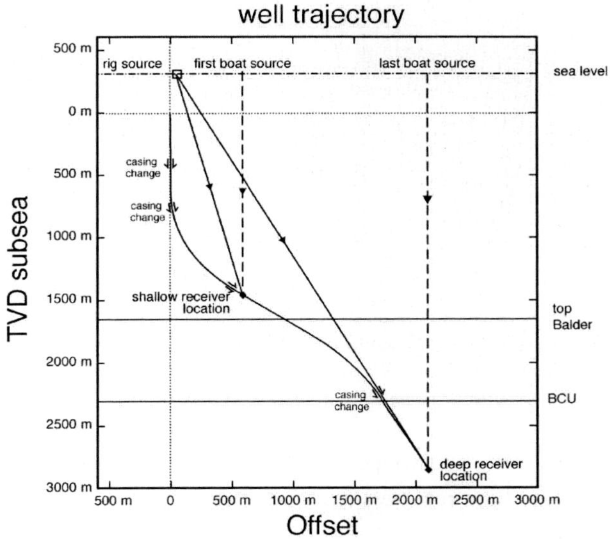

Figure 7.28. A combination of rig- and boat source VSPs for a deviated well permits the computation of δ as a function of depth using normal moveout velocity. Data courtesy of the British Geological Survey and Norsk Hydro.

Thus the isotropic term is now modified by an additional term

$$\zeta = \frac{1}{V_{iso}^2 t_0} \sum_{i=1}^{N} \alpha_0^2(i) 2\delta(i) \Delta t_0(i) \ . \tag{7.127}$$

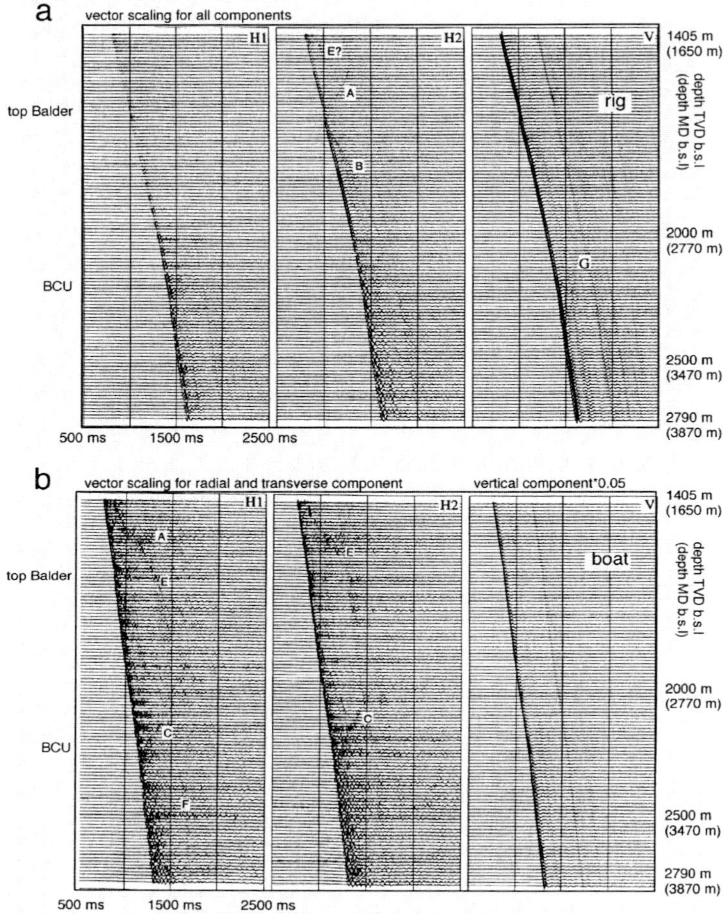

Figure 7.29. Three-component data from rig- and boat source VSPs shot in the Norwegian Sea using the acquisition geometry of Figure 7.28. Hyperbolic moveout is noticeable in the rig data on both the P and S-waves. Data courtesy of the British Geological Survey and Norsk Hydro.

A similar relation exists for the other arrivals. From these equations, the interval velocity can be immediately recovered using a Dix-type formula (Alkalifah and Tsvankin 1995) applied to the NMO velocities determined from receivers at the top ($v(n-1)$) and bottom ($v(n)$) of the region of interest

$$\alpha_0^2(n)(1+2\delta(n)) = \frac{t_0(n)(v(n))^2 - t_0(n-1)(v(n-1))^2}{t_0(n)-t_0(n-1)}, \tag{7.128}$$

where $t_0(n-1)$ and $t_0(n)$ are the vertical travel-times to the top and bottom of the layer of interest. With an $\alpha_0(n)$ from the near-offset VSP, the anisotropy parameters $\delta(n)$ can be determined as a function of depth.

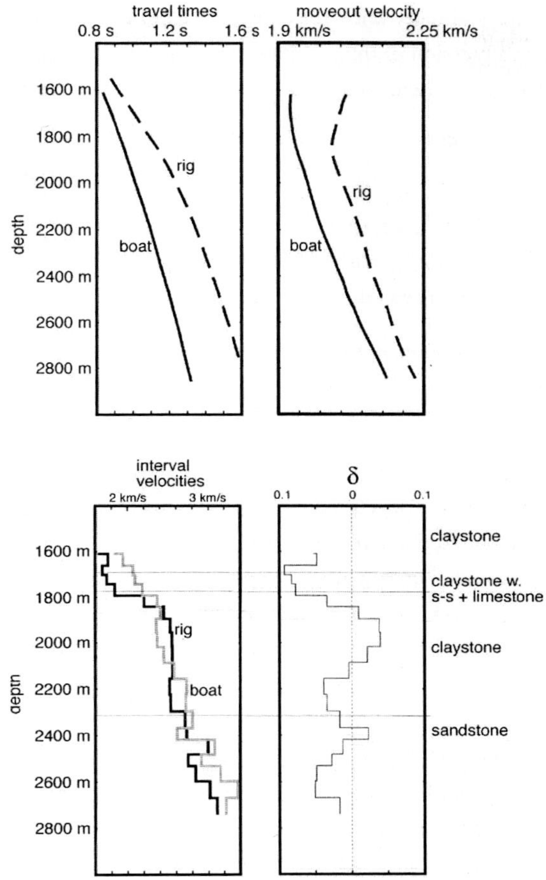

Figure 7.30. First break picks, together with corresponding moveout and interval velocities for both rig- and boat source VSP. The final δ values show some lithology correlation.

The expected magnitude of some typical δ measurements is shown in Figure 7.27 for shales, chalks and sandstones from a surface seismic study of Banik (1984) covering 21 wells, where the vertical velocity is supplied by the well logs. A second example is provided by the combination of fixed-offset (rig) and normal-incidence (boat) VSP shot for a deviated well (Berg et al. 1995, Figures 7.28 and 7.29). The normal-incidence VSP provides the vertical velocity, while the fixed-offset VSP defines the moveout behavior. In particular, the fixed-offset VSP generates a sequence of time and offset pairs $\{t_k^2, X_k^2\}$ (Figure 7.30). These do not directly lie along a hyperbola because of the borehole deviation, but the moveout velocity corresponding to any particular pair (and hence receiver depth k) can be accurately determined using

$$V_k^2 = \frac{X_k^2}{t_k^2 - t_{0k}^2}, \tag{7.129}$$

as the vertical traveltime t_{0k} is already known in advance from the normal incidence VSP shooting into the same receiver. Dix inversion of the resultant moveout velocities combined with estimates of the vertical interval velocity α_{0k} – again supplied by the normal incidence VSP –

$$V_{0k} = \frac{1}{t_{0k}} \sum_k \alpha_{0k} \Delta t_{0k} \tag{7.130}$$

provides an interval value of δ_k for each moveout velocity. An example of a δ-profile for the North Sea is shown in Figure 7.30. This procedure is adequate if the subsurface is laterally uniform, otherwise inversion ray tracing is required (Berg et al. 1995).

7.3.2 Non-hyperbolic correction for the direct waves (TIV)

7.3.2.1 P-waves

The need for a non-hyperbolic moveout expression is apparent in most walk-away VSPs (Figure 7.31). One of the first corrections for VSP data was based on the skewed hyperbolic formula of Byun et al. (1990). This accounted for the long-spread moveout due to layering and anisotropy, approximated for weak and nearly elliptical anisotropy. The formula is parameterized by the average vertical ($V_v = \alpha_0$), horizontal ($V_h = \alpha_0(1 + \varepsilon)$) and skew moveout ($V_\gamma = \alpha_0(1 + \delta)$) velocities:

$$t^2 = t_0^2 + \left(t_0^2 \frac{V_v^2}{V_\gamma^2} + \frac{X^2}{V_h^2} \right) \frac{X^2}{t_0^2 V_v^2 + X^2} , \tag{7.131}$$

where at very large offsets or shallow depths the moveout velocity again becomes the horizontal velocity V_h. Velocity analysis can be performed for each t_0 by computing the semblance along trajectories defined by (7.131) for particular values of V_v, V_h and V_γ. With knowledge of V_v from a near-offset VSP, a two-parameter scan is required (Figure 7.32). All velocities can be expressed as cumulative average properties, leading to Dix-type formulae for the interval properties.

For a layered isotropic or anisotropic medium (and $X/Z < 2$) the non-hyperbolic term in (7.121) can also be written according to the chain of development by Tsvankin and Thomsen (1994), Alkalifah and Tsvankin (1995) and Alkalifah (1997):

$$\frac{A_4 X^2}{B + A_5 X^2} = \frac{-2\eta_{eff} X^4}{V^2 \left(t_0^2 V^2 + \left(1 + 2\eta_{eff} \right) X^2 \right)} , \tag{7.132}$$

where

$$\eta_{eff} = \frac{1}{8} \frac{\sum_{i=1}^{N} V^4(i)(1 + 8\eta(i))\Delta t_i - t_0 V^4}{t_0 V^4} , \tag{7.133}$$

and $\eta(i)$ is the interval value of the anisotropy combination

$$\eta(i) = \frac{\varepsilon(i) - \delta(i)}{1 + 2\delta(i)} . \tag{7.134}$$

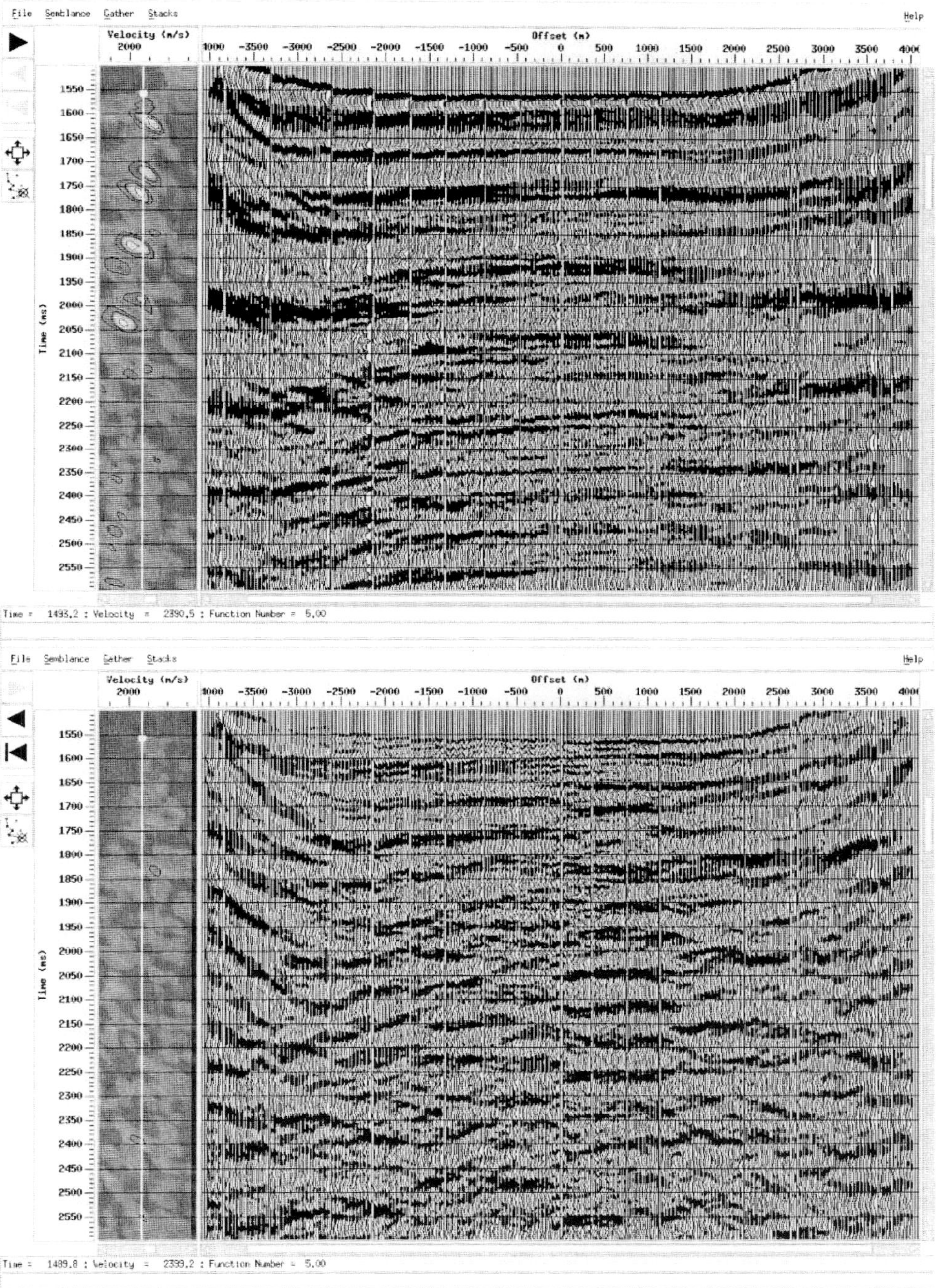

Figure 7.31. The data of Figure 7.25, except now for the full range of offsets. This demonstrates the need for some form of long-spread correction. Data courtesy of the British Geological Survey and Norsk Hydro.

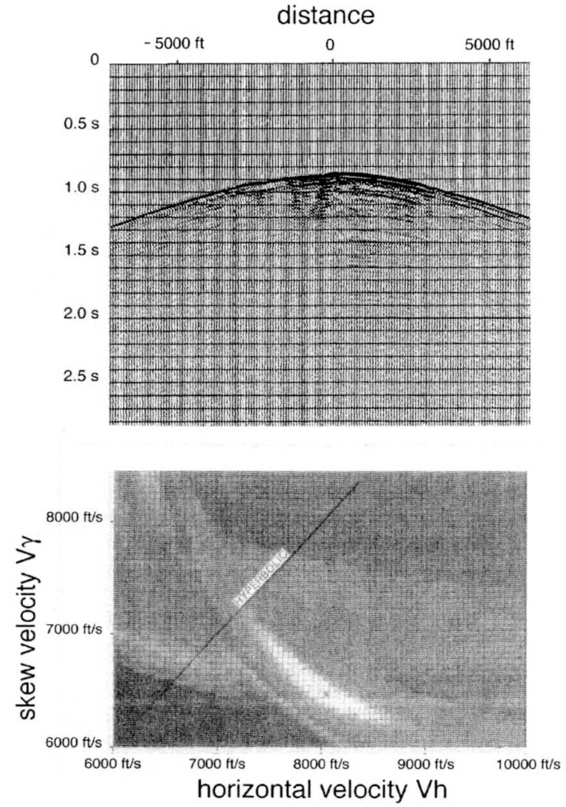

Figure 7.32. Two-parameter semblance scan for VSP field data from the Gulf of Mexico. light gray: peak; dark gray: background. Adapted from Byun, Corrigan and Gaiser (1990). The hyperbolic line corresponds to $V_\gamma = V_h$. The position of the peak clearly shows a strong non-hyperbolicity.

Values for $\eta(i)$ are normally positive for shales, but negative if fracturing is present (Chapter 2). The parameter η_{eff} can also be written as

$$\eta_{eff} = \frac{1}{2}\left(\frac{V_h^2}{V^2} - 1\right), \tag{7.135}$$

where V_h is the horizontal velocity. The increased accuracy of this expression is demonstrated in Figure 7.33. Velocity analysis now consists of determining two controlling parameters, the normal moveout velocity V and η_{eff}. For application, a 2D semblance search over V and V_h (or η_{eff}) for the full range of offsets is suggested. An alternative scheme determines V using the short-spread data in the conventional way, corrects the data, and then calculates η_{eff} from the residual moveout between the hyperbolic results and the observed data Δt over the entire range of offsets

$$\eta_{eff} = \frac{\Delta t^2 V^2 \left(t_0^2 V^2 + X^2\right)}{2X^2\left(X^2 - \Delta t^2 V^2\right)} . \tag{7.136}$$

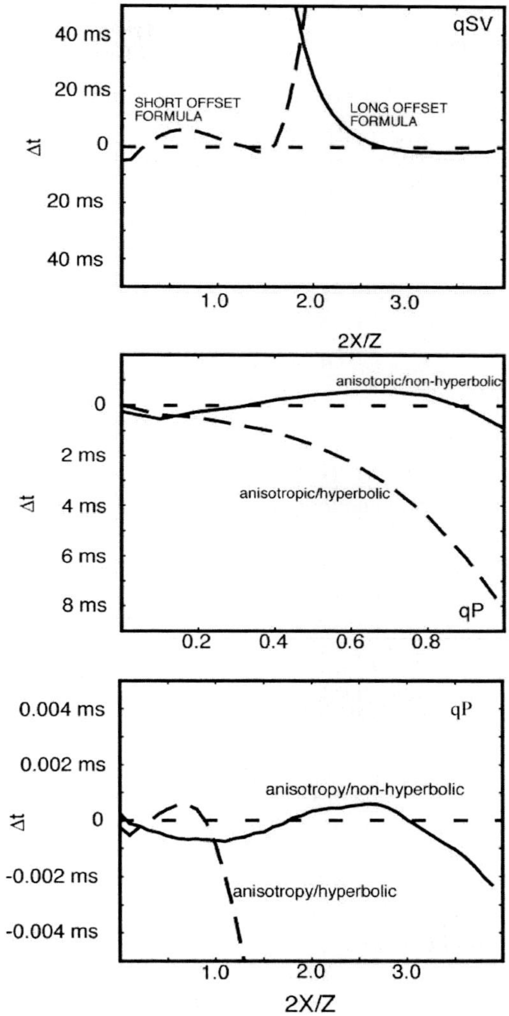

Figure 7.33. Difference between travel times predicted from synthetic seismograms and the best-fitting moveout curve for a single homogeneous TIV layer ($\varepsilon = 0.101$, $\delta = -0.035$, Taylor sandstone, Thomsen 1986). The wide offset qP results demonstrate the stability of the non-hyperbolic equation, while the S-wave results (top) reveal a difficult region that cannot be adequately fitted.

One final approach determines V using the short spread data, but then scans over V_h (or η_{eff}) in semblance using the long offset spread. The scan over V is repeated, and the sequence continued until convergence is reached. Finally, although it is possible to evaluate Dix-type formulae for the coefficients in the above scheme, Alkalifah (1997) has observed that the complexity of the equation leads to a progressive accumulation of errors. Non-linear inversion is suggested as an alternative, where an "interval velocity- and η_{eff} model" are fit to the measured V and V_h, with some regularization to account for instabilities. This inversion approach can also mix measurements from neighboring gathers to further stabilize results. An example from such an inversion has been shown by Alkalifah et al. (1996), where η_{eff} values of up to 0.25 were found below a Tertiary section.

7.3.2.2 S-waves

Comparison of the above with Tsvankin and Thomsen (1994) gives

$$\frac{A_4 X^2}{B + A_5 X^2} = \frac{\eta' X^4}{V^2 \left(t_0^2 V^2 + \eta' X^2\right)} \tag{7.137}$$

for a single layer, where

$$\eta' = \frac{\alpha_0^2}{\beta_0^2}(\varepsilon - \delta)\frac{1 + 2\delta}{(1 + 2\sigma)^2} \ . \tag{7.138}$$

which varies more strongly than the corresponding P-wave expression. The validity of this hyperbolic equation is limited ($2X/Z < 0.75$). This can cause difficulties, even for a single homogeneous layer. Despite this, good fits at short and long offsets separately are expected (Figure 7.33).

7.3.2.3 Instantaneous velocity

Some insight into the effects of non-hyperbolic moveout can be gained by re-organizing the equations above as (7.117), replacing V_{iso} by an offset-dependent moveout velocity $V'(X)$. The behavior of this velocity is a sensitive indicator of the performance of the moveout equation. For P-waves, this leads to

$$V'^2(X) = V_\gamma^2 \left(1 - \frac{\left(V_\gamma^2 / V_h^2 - 1\right) X^2}{t_0^2 V_v^2 + X^2}\right) \tag{7.139}$$

(Byun et al. 1989), and

$$V'^2(X) = V^2 \left(1 + \frac{2\eta_{eff} X^2}{t_0^2 V^2 + X^2}\right) = V^2 \left(1 + \frac{\left(V_h^2 / V^2 - 1\right) X^2}{t_0^2 V^2 + X^2}\right) \tag{7.140}$$

(Alkalifah 1997). Both of these functions predict a smooth transition from a short-spread value V to a long-spread value V_h with increasing spread length X. These functions adequately match the actual behavior (Figure 7.34). In the second case, the slope and position of this transition region is controlled by η_{eff}. For S-waves, the instantaneous velocity function is now

$$V'^2(X) = V^2 \left(1 - \frac{X^2}{C + 2X^2}\right) , \tag{7.141}$$

where

$$C = t_0^2 V^2 \frac{(1 + 2\sigma)^2}{1 + 2\delta} \ . \tag{7.142}$$

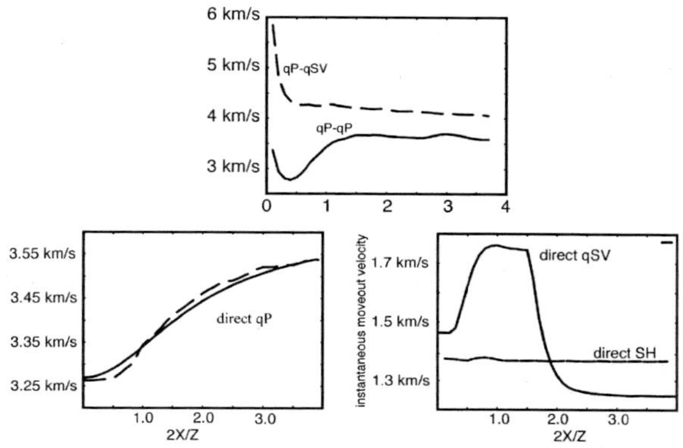

Figure 7.34. The instantaneous moveout velocity for the direct qSV waves highlights the difficulty in predicting the moveout for this wave. There is an abrupt jump around the position of a velocity maximum in the group velocity. By contrast, the qP behaves more regularly, and the reflected or converted waves are also much smoother. Fitting a single normal moveout velocity to the near and far offsets for qSV can circumvent, but not solve, this problem.

The example in Figure 7.34 reveals the problem in this case: there is now a rapid variation in the instantaneous velocity within a transition region corresponding to a propagation angle of 45°. Tests show that for negative σ this is strong even for small offsets. This rapid variation cannot be adequately simulated by a smooth function, and thus leads to inevitable breakdown. Fortunately, converted shear waves tend to have a smoother and more predictable behavior (Figure 7.34) due to the P-wave component in their path.

7.3.3 Azimuthal variations in moveout for the direct waves (TIH, ORTHO)

7.3.3.1 Hyperbolic moveout

Both direct and reflected waves in TIH or orthorhombic media possess azimuthally varying moveout signatures. Expressions for the normal (hyperbolic) moveout are given by Sena (1991) and Tsvankin (1995), with a simplified expression by Li (1999). Following the latter, on the assumption of an isotropic overburden and a single fractured layer at depth, an interval time is defined between events at the top and bottom of the fractured layer. The difference Δt in the interval times for two orthogonal directions is considered as a function of azimuth. For a TIH medium, weak anisotropy assumptions permit the separation of azimuth- and offset dependent terms:

$$\Delta t(\phi, X) = X^2\left(\frac{1}{V_{\perp Bot}^2} - \frac{1}{V_{\perp Top}^2} - \frac{1}{V_{\| Bot}^2} + \frac{1}{V_{\| Top}^2}\right) = A(X, e_t, e_n)\cos 2\phi \ , \tag{7.143}$$

where ϕ is the azimuth of the fracture strike with respect to the line (offset) direction. Figures 7.35 and 7.36 show this variation in interval time as a function of fracture strike, and confirm the applicability of this approximation to media with different contrasts. Due to the $\cos 2\phi$ behavior, a practical scheme to evaluate fracture strike can use four-line configurations, similar to the phase-slowness method in Section 7.2. In particular, for two orthogonal lines acquired with azimuths differing by 45° one has

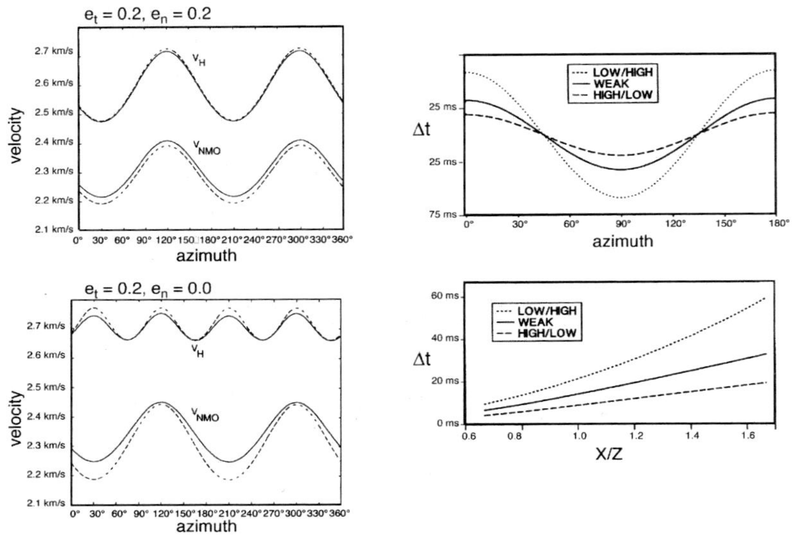

Figure 7.35. Left hand side diagrams: predicted variation in short-spread moveout and horizontal velocity against an accurate numerical computation for two vertical fracture models – from Sayers and Ebrom (1997). Right hand side diagrams: predicted azimuthal variations for three reservoir contrasts (2.4/3.5, 3.6/4.0 and 3.0/2.1). Also shown is the increase in interval time with offset for a thickness of 300m. After Li (1999).

$$\Delta t(\phi, X) = A(X, e_t, e_n)\cos 2\phi \tag{7.144}$$

and

$$\Delta t(\phi, X) = \Delta t(\tfrac{\pi}{4} - \phi, X) = A(X, e_t, e_n)\sin 2\phi \ , \tag{7.145}$$

from which the fracture strike can be determined by least-squares methods. As an alternative, two sets of orthogonal lines oriented at an angle ϕ_0 can be used. In this case

$$\tan 2\phi = \frac{\Delta t(\phi_0 - \phi, X) - \cos 2\phi_0 \Delta t(\phi, X)}{\sin 2\phi_0 \Delta t(\phi, X)} \ . \tag{7.146}$$

Li (1999) further suggests that a robust measure of fracture azimuth can be obtained by cross plotting the quantities in the numerator and denominator. The technique is believed to be accurate for $2X/Z < 2$. The sensitivity of the technique increases with increasing offset.

Tsvankin (1995) and Grechka and Tsvankin (1998) have shown that the azimuthal variation of the normal moveout velocity V for qP and qS waves can be represented by an ellipse, with the orientation of its axes determining the fracture strike. This is true regardless of whether the medium has TIH, orthorhombic or an arbitrary inhomogeneous anisotropy. It is also valid for layered media. For an orthorhombic medium the qP-NMO velocity at some azimuth ϕ can thus be generalized to

$$V^{-2}(\phi) = V_2^{-2}\cos^2\phi + V_1^{-2}\sin^2\phi \ , \tag{7.147}$$

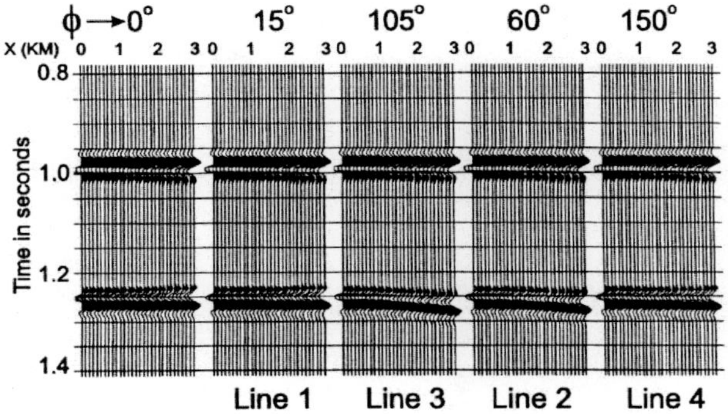

Figure 7.36. The evolution of normal moveout as a function of line azimuth in five surface gathers calculated for a TIH medium. The top events are all aligned to highlight the differential moveout. After Li (1999).

where the NMO velocities V_1 and V_2 are determined in the two vertical symmetry planes as

$$V_1^2 = V_0^2(1+2\delta_1) , \tag{7.148}$$

and

$$V_2^2 = V_0^2(1+2\delta_2) . \tag{7.149}$$

The relation in (7.147) can also be written as

$$V^{-2}(\phi) = \tfrac{1}{2}(V_1^{-2} + V_2^{-2}) + \tfrac{1}{2}(V_1^{-2} - V_2^{-2})\cos 2\phi , \tag{7.150}$$

which is similar to (7.143). The parameter V_0 is the vertical velocity and δ_1 and δ_2 are the effective Thomsen δs in each of the two vertical symmetry planes defined as in Tsvankin (1997). For vertical fractures in a TIV background

$$\delta_1 = \delta \tag{7.151}$$

and

$$\delta_2 = \delta - 2\eta_0(1-\eta_0)e_n - 2\eta_0 e_t . \tag{7.152}$$

This suggests a lithological indicator based on the differential moveout velocity. As the vertical velocities are known in VSP surveys, this indicates that at least three azimuthal measurements are necessary to reconstruct the best-fitting ellipse and evaluate the e_n, e_t combination. Separation of e_n and e_t would require additional information from the shear wave moveout.

7.3.3.2 *Non-hyperbolic moveout*

Al-Dajani, Tsvankin and Toksöz (1998) have extended the concept of the elliptical relation in (7.147) to large offsets. In this case, the quartic term in (7.120) becomes

$$A_4 = c_1 \sin^4 \phi + c_2 \cos^4 \phi + c_3 \sin^2 \phi \cos^2 \phi \ , \tag{7.153}$$

with

$$c_1 = \frac{-2\eta_1}{t_0 V_1^4} \ , \tag{7.154}$$

$$c_2 = \frac{-2\eta_2}{t_0 V_2^4} \ , \tag{7.155}$$

and

$$c_3 = \frac{2}{t_0 V_1^2 V_2^2}(1 - E) \ , \tag{7.156}$$

where η_1 and η_2 are defined as in (7.134), and E is a cross term in these anisotropy parameters which is equal to unity for isotropy.

An alternative insight is offered by Sayers (1995), who provides a non-hyperbolic equation based on an expansion of the group velocity in spherical harmonics. The flexibility of this expansion means that arbitrary, non-weak anisotropy can be accommodated, and expressions given for symmetries as low as monoclinic. This gives

$$t^2 \approx t_0^2 + \frac{X^2}{V^2(\phi)} - \frac{B(\phi)X^4}{X^2 + t_0^2 V_0^2} \ , \tag{7.157}$$

where $t_0 = Z/V_0$; V_0 is the vertical velocity,

$$V^{-2}(\phi) = b_1 + b_2 \cos 2\phi + b_3 \sin 2\phi \ , \tag{7.158}$$

and $B(\phi)$ is a measure of the anellipticity of the medium:

$$B(\phi) = \frac{1}{V^2(\phi)} - \frac{1}{V_h^2(\phi)} = b_4 + b_5 \cos 2\phi + b_6 \sin 2\phi + b_7 \cos 4\phi + b_8 \sin 4\phi \ . \tag{7.159}$$

The expansion coefficients and related to the elastic stiffnesses in a complicated way; only for weak anisotropy the relationship is linear. Equation (7.157) has the same form as that of Byun et al. (1989) for TIV media and that of Sena (1991) for TIH media, but is valid not only for weak anisotropy. Indeed, the equation fits a wide range of ray angles and strongly anisotropic media. For the TIH and orthorhombic case b_3, b_6 and b_8 are zero and the term with $\cos 2\phi$ remains in the normal moveout (Figure 7.35). The coefficients must be determined by fitting to the travel-time variations at a sufficient number of azimuths. Thus, for a monoclinic medium at least five azimuths are required to determine the necessary combinations. For azimuthal variations in V for a short spread in principle only three measurements are required. Given the complexity of the non-hyperbolic terms it appears to be sensible to use only short spreads and hence the lower order terms for fracture analysis.

7.3.4 Normal moveout of reflected events (TIV)

7.3.4.1 P-P reflections

Reflected arrivals have asymmetric ray paths; this must be taken into account in interpretation of their hyperbolic moveout at short offset (Figure 7.37). Thus, for example, for a horizontal reflector at depth Z and receiver at depth Z_R, t_0 is given by $(2Z–Z_R)/V$. The NMO velocity V is an RMS average over the asymmetric reflection path, involving individual downgoing (a) and upgoing (b) path segments

$$V^2 = V_a^2 \frac{t_0^a}{t_0} + V_b^2 \frac{t_0^b}{t_0}, \tag{7.160}$$

where $t_0 = t_0^a + t_0^b$, with

$$V_a^2 = V_a^2(iso)(1 + \zeta_a) \tag{7.161}$$

and

$$V_b^2 = V_b^2(iso)(1 + \zeta_b) \tag{7.162}$$

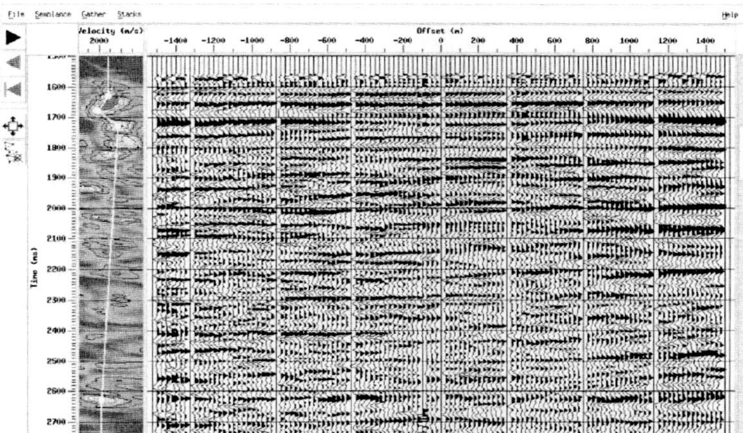

Figure 7.37. Upgoing P-waves for a walk-away VSP, after wavefield separation and normal moveout correction for a short spread of offsets. The amplitudes align, demonstrating normal moveout. Data courtesy of the British Geological Survey and Norsk Hydro.

for a receiver located on the Mth interface. The additional anisotropy terms are given by

$$\zeta_a = \frac{1}{t_0^a V_a^2(iso)} \sum_{i=1}^{N} \alpha_0^2(i) \Delta t(i) \delta(i) \tag{7.163}$$

for the downward leg to the receiver level, and

$$\zeta_b = \frac{1}{t_0^b V_b^2(iso)} \sum_{i=M}^{N} \alpha_0^2(i)\Delta t(i)\delta(i) \tag{7.164}$$

for the remaining ray path. For shallow receivers or deep events, the receiver gather approaches the CMP gather, so that the analysis from Section 7.3.3 is appropriate. However, for most acquisitions a special treatment is required, and the anisotropy between reflectors must be evaluated by separating the average velocity contributions V_{OB} associated with the overburden from the contributions V_{sym} associated with the symmetric portion of the ray path.

With a datum level through the common receiver level, the average velocity can be written

$$V^2 = \frac{t_{OB}}{t_0} V_{OB}^2 + \frac{t_{sym}}{t_0} V_{sym}^2 , \tag{7.165}$$

where the vertical times t_{OB} and t_{sym} are related to the total time $t_0 = t_{OB} + t_{sym}$. A direct insight from this expression is that the moveout velocity cannot be interpreted using generalized Dix-style interval velocity equations, as a contribution for t_{OB} would remain. Instead, evaluating the overburden quantities V^2_{OB} and t_{OB} from the direct wave provides a way of immediately correcting the moveout velocity for each reflected event in the common receiver gather:

$$V_b^2 = \frac{t_0 V^2 - t_{OB} V_{OB}^2}{t_0 - t_{OB}} , \tag{7.166}$$

with a corresponding adjusted time $t_0 – t_{OB}$. This adjusted time can now be used to determine interval velocity and anisotropy in the target.

7.3.4.2 P-S reflections

More asymmetry is introduced by P-waves that are converted to S-waves upon reflection (Thomsen 1998 has called these "C-waves"). However, overburden correction is possible as before, the result being C-wave moveout velocities referenced to receiver level. Each C-wave moveout velocity V_C can be written as a function of the individual P-wave (V_P) and S-wave (V_S) moveout velocities for the mixed path

$$V_C^2 = \frac{V_{Pb}^2 t_{0P}^b + V_{Sb}^2 t_{0S}^b}{t_{0P}^b + t_{0S}^b} . \tag{7.167}$$

Following Thomsen (1999) this reduces to

$$V_C^2 = V_P^2 \frac{\gamma_b^2 + \gamma_0}{\gamma_b^2(1+\gamma_0)} , \tag{7.168}$$

where the ratio of the vertical velocities is given by

$$\gamma_0 = \frac{t_{0S}}{t_{0P}} , \tag{7.169}$$

and the ratio of moveout velocities for the path segment b is

$$\gamma_b = \frac{V_P}{V_S} \ . \tag{7.170}$$

Thus, for a single layer $V_C^2 = V_P V_S$. Following wavefield separation, anisotropy evaluation proceeds using the following steps:

(a) analyse the reflected qP waves to establish the RMS velocity; obtain V_P and t_{0P} from the reflected P-waves, and V_{OB} and t_{OB} from the direct P-waves;

(b) determine a profile of δ with depth from each V_P;

(c) perform a C-wave moveout analysis to determine V_C and t_{0C};

(d) correlate major P-wave and C-wave events, and hence determine γ_0 and γ_b from (7.169) and (7.170);

(e) estimate $\sigma(i)$ and from the relation

$$\varepsilon(i) = \sigma(i)\gamma_0(i) + \delta(i) \ . \tag{7.171}$$

7.3.5 Non-hyperbolic moveout of the reflected arrivals (TIV)

Overburden correction for the wide offsets at which the P-P moveout is non-hyperbolic can utilize (7.166). Unfortunately, this approach leads to inaccuracy, and an alternative strategy is suggested. This employs an equation adapted from "straight ray" geometry

$$t_P = \frac{1}{V_P^a}\sqrt{X_P^2 + (Z - Z_R)^2} + \frac{1}{V_P^b}\sqrt{(X - X_P)^2 + Z^2} \ , \tag{7.172}$$

or for the converted waves

$$t_C = \frac{1}{V_P^a}\sqrt{X_C^2 + (Z - Z_R)^2} + \frac{\gamma_b}{V_P^b}\sqrt{(X - X_C)^2 + Z^2} \ , \tag{7.173}$$

where $Z = t_0 V_0$, and the velocities still obey the generalized Dix averaging equations for anisotropy. A key point in the application of these formulae is the evaluation of the reflection point (X_P) or conversion point (X_C). A suitable expression is the iterative equation of Zhang and Robinson (1992), that progresses from a best guess to converge on a final solution. Adapted to VSP application, the iterative procedure advances according to the rule

$$X_C^{new} = \frac{\sqrt{\gamma_b^2 + \left(X_C^{old}/Z\right)^2\left(\gamma_b^2 - 1\right)}}{1 + D\sqrt{\gamma_b^2 + \left(X_C^{old}/Z\right)^2\left(\gamma_b^2 - 1\right)}} \ , \tag{7.174}$$

where

$$D = \frac{Z_R \gamma_b X_C}{\sqrt{\left(X_c^{old}\right)^2 + Z^2}} \ , \tag{7.175}$$

and the depth to the reflector is $Z = t_{0C}\tilde{V}_C - Z_R$. This method has been shown to converge after only a few iterations (Zhang 1996). A starting point can be obtained by the single-layer solution for the P-P waves

$$X_P(start) = \frac{Z - Z_R}{2Z - Z_R} \,, \tag{7.176}$$

and for the C-waves

$$X_C(start) = X_{P0} + (\gamma_b - 1)\frac{(Z - Z_R)^2 + X_{P0}^2}{X - X_P} \,. \tag{7.177}$$

7.3.6 Dip moveout for the reflected events (TIV)

Here, the travel-times are now dependent upon the position relative to the dip direction of the reflector, with the moveout equation for a common receiver gather containing odd powers of offset even for a single layer. For a single homogeneous overburden with velocity V, Hardage (2000) gives an expression for a line shot along the dip direction

$$t^2 = \frac{(2Z - Z_R)^2}{V^2} + \frac{X^2}{V^2} - \frac{4(Z - Z_R)}{V^2}\left(Z\sin^2\alpha + X\sin\alpha\cos\alpha\right) \tag{7.178}$$

or

$$t^2 = A + \frac{(X - B)^2}{V^2} \,, \tag{7.179}$$

where

$$A = \frac{(2Z - Z_R)^2}{V^2} - \frac{4(Z - Z_R)\sin^2\alpha\left(Z\sin^2\alpha - Z_R\cos^2\alpha\right)}{V^2} \,, \tag{7.180}$$

and

$$B = \left(X - 2(Z - Z_R)\sin\alpha\cos\alpha\right)^2 \,, \tag{7.181}$$

which is again a hyperbola, but now described around a reduced offset and vertical travel time according to the dip angle α, and receiver and reflector depths. The formula depends both upon the sign of the offset and the line direction. For a line at an angle ϕ to the dip direction, α must be replaced by α', where $\sin\alpha' = \sin\alpha\cos\phi$. For converted waves, a general expression can be constructed following (7.172):

$$t_C = \frac{1}{V_P^a}\sqrt{X_P^2 + (Z - X_P\sin\alpha)^2} + \frac{\gamma_b}{V_P^b}\sqrt{(X - X_P)^2 + (Z - Z_R - X_P\sin\alpha)^2} \,. \tag{7.182}$$

CHAPTER 8 – MULTIPLE-OFFSET VSP – DYNAMICS

The vector convolutional model developed in Chapter 4 for near-offset multi-component VSP is now extended to include multi-component VSP using larger offsets and hence more oblique ray paths. This requires development of the existing tools for shear wave rotational analysis, and also lays the foundation for their adaptation to converted shear waves. Arising from this are two specialized converted shear wave techniques, one that treats specifically orthogonal offsets and another tailored principally to walk-away VSP acquisition. These converted-shear wave methods are ideally suited to offshore applications, where multi-component sources cannot be used. Finally, amplitude variations with offset from walkaway VSP provide information on e_n, e_t, ε, δ and γ that supplements the local phase slowness and moveout analysis of Chapter 7.

8.1 THE EXTENDED VECTOR CONVOLUTIONAL MODEL

The vector convolutional model for offset multi-component VSP can be formed by arguments analogous to those in Section 4.1. Thus, the source motions along the in-line (*X*), cross-line (*Y*), and vertical (*Z*) axes are firstly partitioned into the P- and S-waves, their polarization direction determining the relative energies involved. Separate propagation of these waves through the medium then results in differential lags between all three waves. Finally, the waves are projected onto receiver axes aligned with the *X-Y-Z* coordinate frame. This entire procedure can be simulated by a sequence of rotations and a time-shift operator in a fashion similar to (4.13)

$$\mathbf{D}^{(9)}(t) = \mathbf{C}_Y^T(\theta)\mathbf{C}_{Z'}(\phi_N)\Lambda^{(9)}(t) * \left(\mathbf{C}_{Z'}^T(\phi_N)\mathbf{C}_Y(\theta)\mathbf{S}_0^{(9)}(t)\right), \tag{8.1}$$

where

$$\mathbf{C}_Y(\theta) = \begin{pmatrix} \cos\theta & 0 & \sin\theta \\ 0 & 1 & 0 \\ -\sin\theta & 0 & \cos\theta \end{pmatrix}, \tag{8.2}$$

with the familiar operators

$$\mathbf{C}_{Z'}(\phi_N) = \begin{pmatrix} \cos\phi_N & \sin\phi_N & 0 \\ -\sin\phi_N & \cos\phi_N & 0 \\ 0 & 0 & 1 \end{pmatrix}, \tag{8.3}$$

$$\Lambda^{(9)}(t) = \begin{pmatrix} \lambda_1(t) & 0 & 0 \\ 0 & \lambda_2(t) & 0 \\ 0 & 0 & \lambda_3(t) \end{pmatrix}, \tag{8.4}$$

and

$$\mathbf{S}_0^{(9)}(t) = \begin{pmatrix} s(t) & 0 & 0 \\ 0 & s(t) & 0 \\ 0 & 0 & s(t) \end{pmatrix} . \tag{8.5}$$

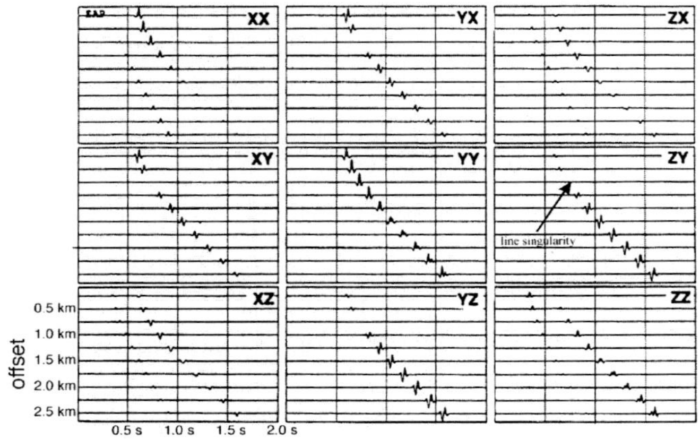

Figure 8.1. Nine-component data matrix generated by source motions along the in-line (*X*), cross-line (*Y*) and vertical (*Z*) directions. These synthetic seismograms are simulated for a fracture-related TIH half-space and a range of offsets. The fracture strike makes an acute angle with the in-line direction. Note that the P-waves appear only on the *XX*, *XZ*, *ZX* and *ZZ* components as predicted by equation (8.6), while the S waves are on all components. The P-wave energy is also down-weighted as the contribution arising from the point force (representing the source) is proportional to β^2/α^2 (see Chapter 4). The S-wave amplitudes dim on the *XY* and *YX* components as the ray paths pass through the line singularity (MacBeth and Li 1996).

The main effect of offset has been the introduction of additional rotations about the *Y*-axis. These are inserted outside the *Z*′-axis rotations. This order is critical to the interpretation of the sequence: indeed, if the order is reversed, it is only possible to simulate SV and SH waves but not the more general polarizations for the TIH medium. If the sources are balanced and $\mathbf{S}_0^{(9)}(t) = \mathbf{I}$, equation (8.1) can be written

$$\mathbf{D}^{(9)}(t) = \begin{pmatrix} \cos^2\theta A(t) & \cos\theta C(t) & \sin\theta\cos\theta A(t) \\ \cos\theta C(t) & B(t) & \sin\theta C(t) \\ \sin\theta\cos\theta A(t) & \sin\theta C(t) & \sin^2\theta A(t) \end{pmatrix} + \begin{pmatrix} \sin^2\theta\lambda_3(t) & 0 & \sin\theta\cos\theta\lambda_3(t) \\ 0 & 0 & 0 \\ \sin\theta\cos\theta\lambda_3(t) & 0 & \cos^2\theta\lambda_3(t) \end{pmatrix} , \tag{8.6}$$

where

$$A(t) = \cos^2\phi_N \lambda_1(t) + \sin^2\phi_N \lambda_2(t) , \tag{8.7a}$$

$$B(t) = \sin^2\phi_N \lambda_1(t) + \cos^2\phi_N \lambda_2(t) , \tag{8.7b}$$

and

$$C(t) = \sin\phi_N \cos\phi_N \left(\lambda_2(t) - \lambda_1(t)\right) . \tag{8.7c}$$

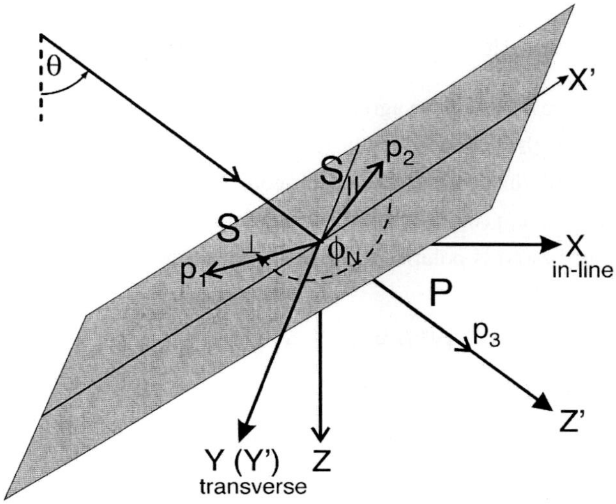

Figure 8.2. The offset VSP rotation model assumes that the S-wave polarizations $\mathbf{p}_1$ and $\mathbf{p}_2$ are confined to the plane normal to the ray path direction (Z'). The polarization $\mathbf{p}_1$ is defined by the rotation angle ϕ_N measured from the projection (X') of the in-line direction in this plane, with $\mathbf{p}_2$ orthogonal to $\mathbf{p}_1$. The P-wave polarization $\mathbf{p}_3$ is directed along the Z' axis in a uniform medium, but in a multi-layered medium with depth variant α/β it is, in general, differently oriented. In the model, the dynamic coordinate system X'- Y'- Z' is transformed into the acquisition X-Y-Z coordinates by a sequence of rotations depending on the incidence angle θ and the polarization angle ϕ_N.

The pattern of behavior predicted in (8.1) can be seen in the synthetic example of Figure 8.1. Unlike the equation for vertical incidence, this particular arrangement of operators now includes a simplifying assumption for the wave behavior. The assumption is that both shear wave polarizations lie in the plane normal to the ray path. This is justifiable if the anisotropy is weak (see Section 3.2.3). The assumption allows the expression in this convenient rotational form, but slightly detracts from the physical interpretation. In particular, the $S_\perp$ polarization is now defined by the rotation angle ϕ_N about the Z' axis, measured in the normal plane, with the $S_\parallel$ polarization perpendicular (Figure 8.2). However, for a fracture-related TIH medium, these polarizations must also agree with the solutions of the Kelvin-Christoffel equation (equations 3.48, 3.49 and 3.50). Thus, from these solutions (Chapter 3) it is known that the $S_\perp$ wave has a fixed polarization perpendicular to the symmetry plane containing the propagation direction, and always points along the fracture strike (that lies at an angle ψ to the X-axis). This means that its horizontal polarization $\phi_\perp$ constrains ϕ_N according to

$$\tan\phi_\perp = \frac{1}{\tan\phi_N \cos\theta} = \tan\psi \ . \tag{8.8}$$

The inevitable consequence is that the angle ϕ_N is expected to rotate in the normal plane as θ (offset) varies, in order to satisfy this relation (Figure 8.3). The conclusion is that a group of VSP offsets cannot be analysed using one rotation angle throughout. Additionally, once determined, ϕ_N should be immediately translated into $\phi_\perp$. For the $S_\parallel$ wave, the horizontal polarization angle $\phi_\parallel$ is given by

$$\tan\phi_{\parallel} = \frac{-\tan\phi_N}{\cos\theta} \quad . \tag{8.9}$$

When $\theta = 0$ the expressions above agree with the results from Chapter 3. However, for oblique angles the projections of the $S_{\perp}$ and $S_{\parallel}$ polarizations on the horizontal plane are not orthogonal ($\tan\phi_{\perp} \neq 1/\tan\phi_{\parallel}$). Indeed, this apparent non-orthogonality varies with incidence angle θ and fracture strike (and hence with ϕ_N) (Figure 8.4). Another assumption inherent in (8.1) is that the P-wave (with time-shift operator $\lambda_3(t)$) is polarized parallel to the shear wave ray path direction defined by θ. Thus, (8.6) can be written

$$\mathbf{D}^{(9)}(t,\theta,\phi_N) = \mathbf{D}_S^{(9)}(t,\theta,\phi_N) + \mathbf{D}_P^{(9)}(t,\theta) \quad . \tag{8.10}$$

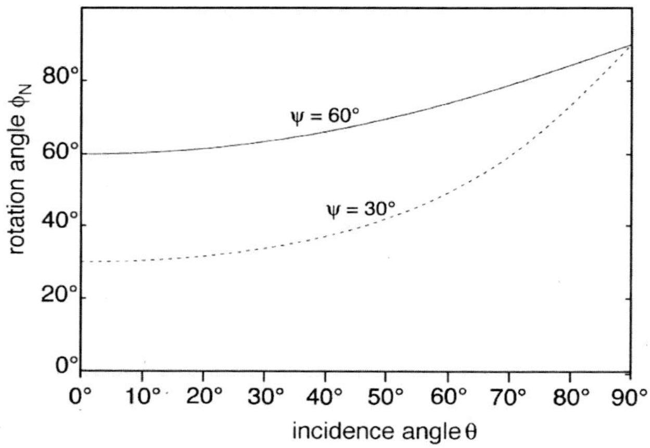

Figure 8.3. The $S_{\parallel}$ polarization is defined by the rotation angle ϕ_N in the normal plane. Because of this definition, for a fracture-related TIH medium ϕ_N must vary with incidence angle in order for its horizontal projection to remain fixed. The curves above show this variation for fracture strikes ψ of 30° and 60°.

This assumption breaks down in a multi-layered medium, especially when α/β€varies from layer to layer. When the deviation is significant, (8.8) must be re-written

$$\mathbf{D}^{(9)}(t,\theta_S,\theta_P,\phi_N) = \mathbf{D}_S^{(9)}(t,\theta_S,\phi_N) + \mathbf{D}_P^{(9)}(t,\theta_P) \quad . \tag{8.11}$$

In this case, it is not possible to treat both the P- and S-waves using the same rotational operator.

There are several practical schemes for applying the model in (8.1) to field data, to yield the separate shear waves and hence their lag time and polarizations:

(a) For the first pass, treat the entire wavefield without windowing, assuming $\theta_S \approx \theta_P$ and perform two separate multi-component (Alford) rotations, first about the Y-axis and then about the Z'-axis. Optimize by reducing all off-diagonal energy to a minimum as in Section 5.2. After this, take the resultant trace set and now window the shear waves, and repeat the procedure in a residual analysis.

(b) Use only the windowed shear waves, and cross-plot in-phase combinations of the matrix components in (8.6). Thus, any pair from the triplet $\{d_{XX}(t), d_{XZ}(t), d_{ZZ}(t)\}$ can yield an estimate of θ; and so can any pair from the set $\{d_{XY}(t)\ , d_{ZY}(t)\ , d_{YX}(t), d_{YZ}(t)\}$. Once this angle – and hence A(t), $B(t)$, and $C(t)$ – is known, the four-component shear wave matrix can be derived:

$$\mathbf{D}^{(4)}(t) = \begin{pmatrix} A(t) & C(t) \\ C(t) & B(t) \end{pmatrix}, \tag{8.12}$$

which can be solved using the multi-component analysis techniques of Section 5.3.

(c) perform an eigenanalysis of the (singular) covariance matrix

$$\mathbf{M} = \sum_{i=1}^{N} \left(\mathbf{D}_S^{(9)}(t_i) (\mathbf{D}_S^{(9)}(t_i))^T \right), \tag{8.13}$$

where each eigenvector must be interpreted to yield θ_S and ϕ_N, and hence $\phi_\perp$.

(d) rotate into the dynamic axes by windowing the P-wave and using the methods in Section 4.3, on the assumption that $\theta_P \approx \theta_S$. This yields the four-component shear wave matrix, that can be treated using the techniques in Section 5.1.

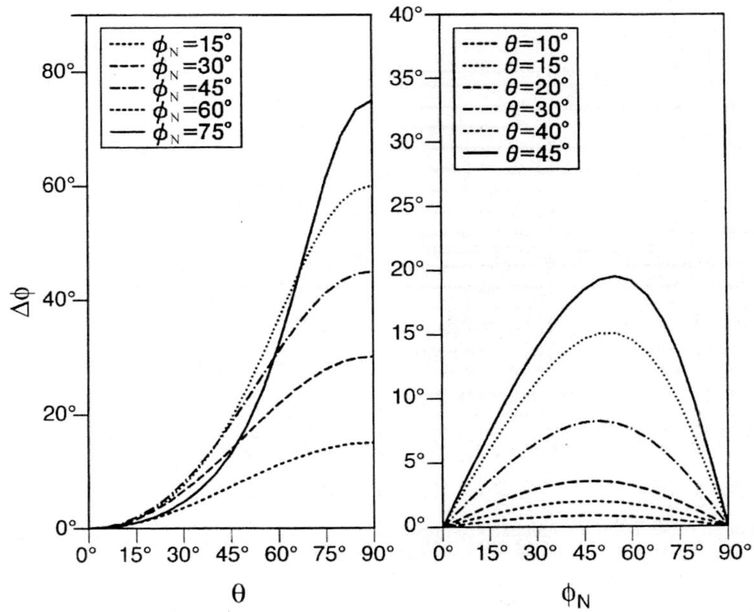

Figure 8.4. The horizontal projections of the S-wave polarizations are non-orthogonal due to the oblique incidence. The magnitude $\Delta\phi$ of this non-orthogonality depends on the rotation angle ϕ_N and the incidence angle θ, and can be quite significant. Since ϕ_N varies with offset, the non-orthogonality varies too (Li, MacBeth and Crampin 1998).

It is important to underline that the polarization directions and time-shift operators recovered above are appropriate to the oblique ray paths.

8.1.1 Alford rotation

Difficulties arise with offset VSP when the full nine-component data are not available. In this situation, the shear waves cannot be adequately separated by the multi-component rotation algorithms outlined above. There is an unavoidable breakdown in applying the Alford rotation of Chapter 5, and the resulting birefringence estimates are inaccurate (Figure 8.5). With only two horizontal sources (in-line and cross-line) the subset matrix is

$$\mathbf{D}_S^{(4)}(t) = \begin{pmatrix} \cos^2\theta_S A(t) & \cos\theta_S C(t) \\ \cos\theta_S C(t) & B(t) \end{pmatrix} \tag{8.14}$$

or

$$\mathbf{D}_S^{(4)}(t) = \begin{pmatrix} \cos\theta\cos\phi_N & \cos\theta\sin\phi_N \\ -\sin\phi_N & \cos\phi_N \end{pmatrix} \Lambda_S^{(4)}(t) \begin{pmatrix} \cos\theta\cos\phi_N & -\sin\phi_N \\ \cos\theta\sin\phi_N & \cos\phi_N \end{pmatrix}, \tag{8.15a}$$

which can be written

$$\mathbf{D}_S^{(4)}(t) = \begin{pmatrix} \cos\phi_{\parallel} & \cos\phi_{\perp} \\ \sin\phi_{\parallel} & \sin\phi_{\perp} \end{pmatrix} \Lambda_S^{(4)}(t) \begin{pmatrix} \cos\phi_{\parallel} & \sin\phi_{\parallel} \\ \cos\phi_{\perp} & \sin\phi_{\perp} \end{pmatrix}. \tag{8.15b}$$

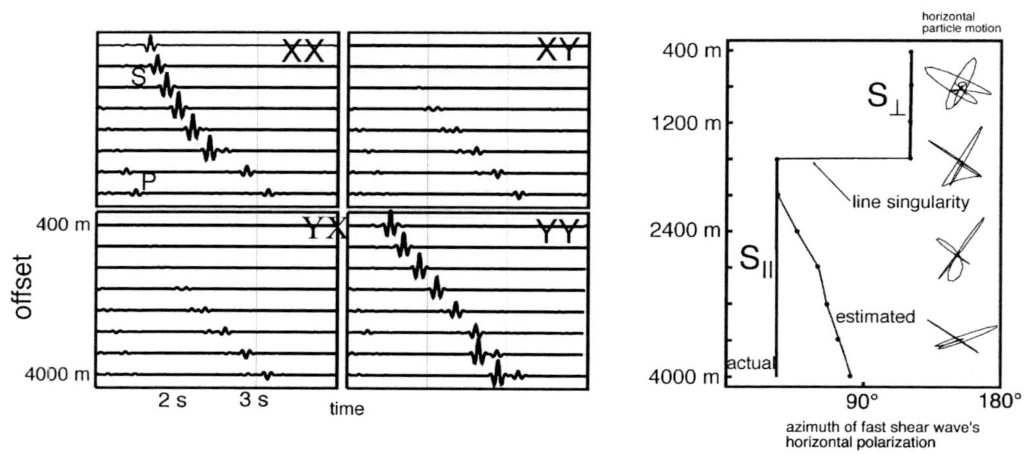

Figure 8.5. For an offset multi-component VSP, residual energy remains on the off-diagonals after Alford rotation. This is related to the oblique ray path and can be corrected by reversing the incidence-angle dependent weighting functions in equation (8.6). While the $S_{\perp}$ polarization in a TIH medium remains fixed with offset, the $S_{\parallel}$ polarization changes direction; the projection on the horizontal plane amplifies the variation. In the case above (receiver at 2000 m), indications of this projection problem only occur beyond the line singularity (at offsets of 2000 m). The true polarization estimates follow the fracture strike of 120° up to the line singularity. Beyond the line singularity they jump to the other shear mode and give an angle of 30°.

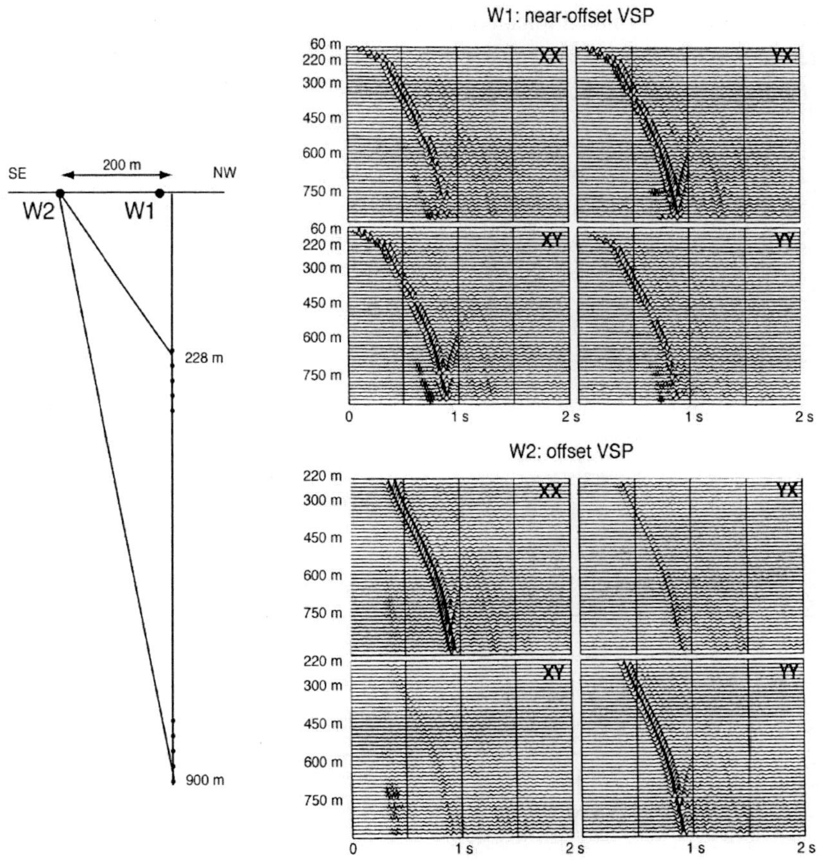

Figure 8.6. Four-component data matrices from a near-offset VSP (W1) and an offset (200 m) VSP (W2) at the former BP test site in Devine, Texas (Raikes 1991). The oblique paths to the shallowest receivers in the offset VSP give rise to apparent non-orthogonality. This non-orthogonality leads to inaccurate polarization and lag estimates. Extracted from the study by Li, MacBeth and Crampin (1998).

Thus, although the data matrix is symmetric, offset creates an imbalance in the components. Incomplete separation can be rectified by corrections using the appropriate trigonometric functions (Figures 8.6 and 8.7). This requires an estimate of the incidence angle from the P-waves or ray tracing. For simple geological structures with weak anisotropy and weak lateral heterogeneity this is adequate. Alternatively, as VSP data are always recorded with three-components, the vertical components can be included. Thus, the six-component matrix

$$\mathbf{D}^{(6)}(t) = \begin{pmatrix} \cos^2\theta_S A(t) & \cos\theta_S C(t) \\ \cos\theta_S C(t) & B(t) \\ \sin\theta_S \cos\theta_S A(t) & \sin\theta_S C(t) \end{pmatrix}, \tag{8.16}$$

yields several possibilities for cross-plotting: $d_{XZ}(t)$ versus $d_{XX}(t)$ or $d_{YX}(t)$ versus $d_{YZ}(t)$. An educated guess at θ_S is sufficient to produce an accuracy of 5 % or better in the polarization determination.

It should be noted that an additional complication enters through the deviation Δ of the $S_{||}$- and P-wave polarization (Section 3.2.3). However, the impact of this effect on the non-orthogonality in the horizontal plane is generally small for weak anisotropy.

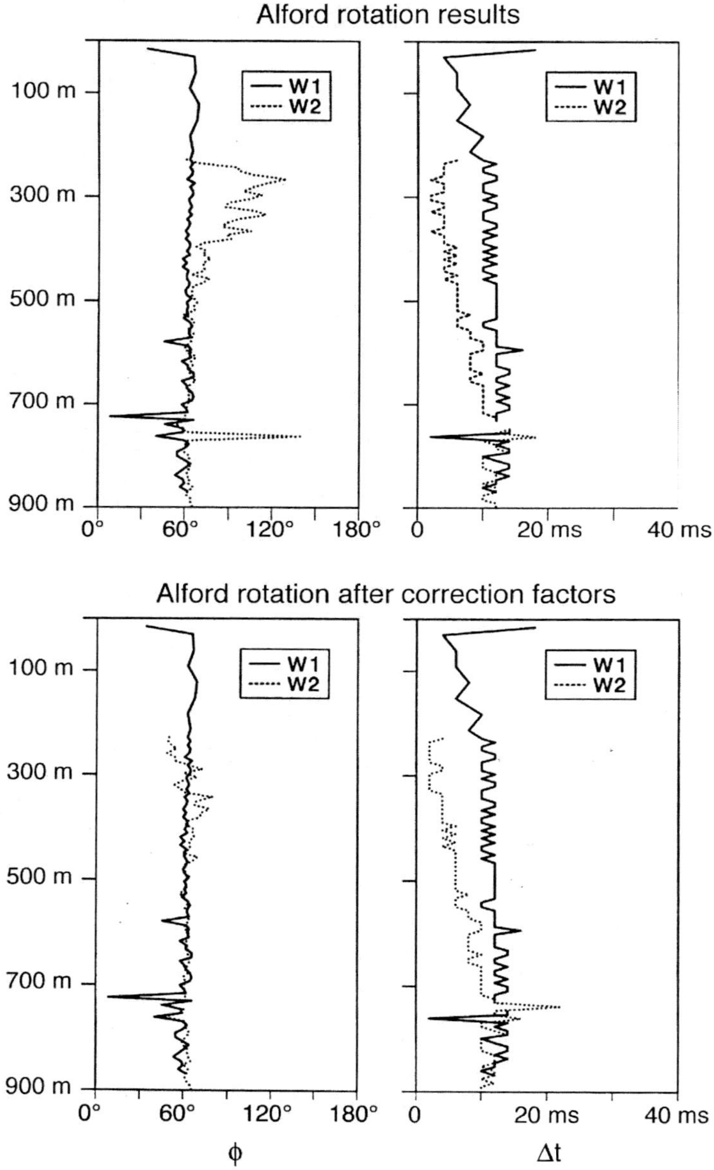

Figure 8.7. Non-orthogonal correction factors applied to the data matrix for VSP W2 create a closer agreement with the results of the near-offset VSP W1. In this particular example it appears that non-orthogonality has very little overall influence on the time-delay measurements. After Li, MacBeth and Crampin (1998).

8.2 NEAR-SURFACE CORRECTION

The adverse influence of source effects and near-surface effects discussed in Section 4.2 gets larger at oblique angles. However, the degree to which these effects influence the multi-component data can be directly tested using the anticipated linear relationships between the components in equation (8.6). A comparison should also be made with cross-plots of those components predicted to be out of phase (Figures 8.8 and 8.9). Thus, $d_{XX}(t)$ and $d_{XZ}(t)$ should be in phase, while the matrix components $d_{XX}(t)$, $d_{YY}(t)$ and $d_{XY}(t)$ are out of phase. Mismatch with the model prediction can be caused by a change in the orientation of the anisotropy or by near-surface effects.

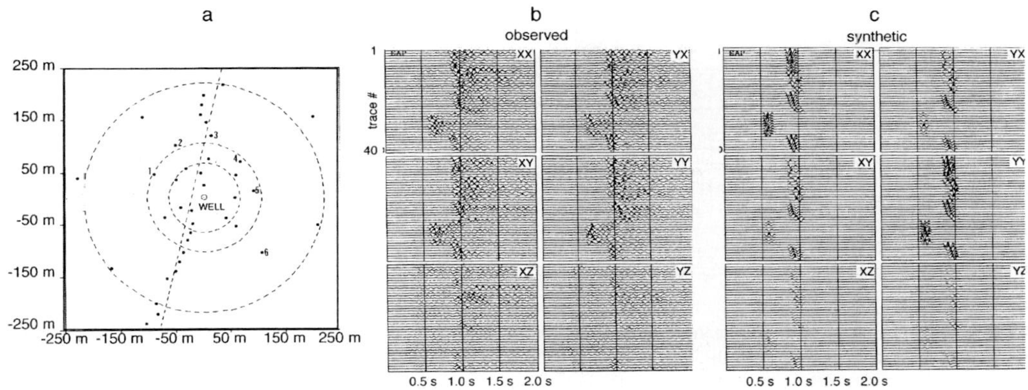

Figure 8.8. (a) Plan view of source positions and wellhead for offset VSPs acquired at the Cymric field, California. A single three-component receiver is positioned at 472 m depth, with a walkaround and walk-away spread using the ARIS source.
(b) The six-component data matrix showing the distribution of shear-wave energy.
(c) Synthetic seismograms corresponding to panel (b), computed for an anisotropic half space.
From MacBeth, Li and Horne (1994).

8.2.1 Anisotropic layer stripping

On the assumption that the near surface is anisotropic, but with an orientation different from that of the subsurface, it is possible to strip off a thin anisotropic layer from each separate source-receiver pair by applying an individually tailored Alford rotation (Section 6.2.2). From this rotation, shear wave polarizations and lags can be calculated that can then be subtracted from the data (Figure 8.10). Alford rotation is re-applied to the stripped data (Winterstein and Meadows 1991a). The technique is time-consuming but effective. The stripped lags are interpreted as originating from the shallow ray-path segments as the ray paths converge at depth.

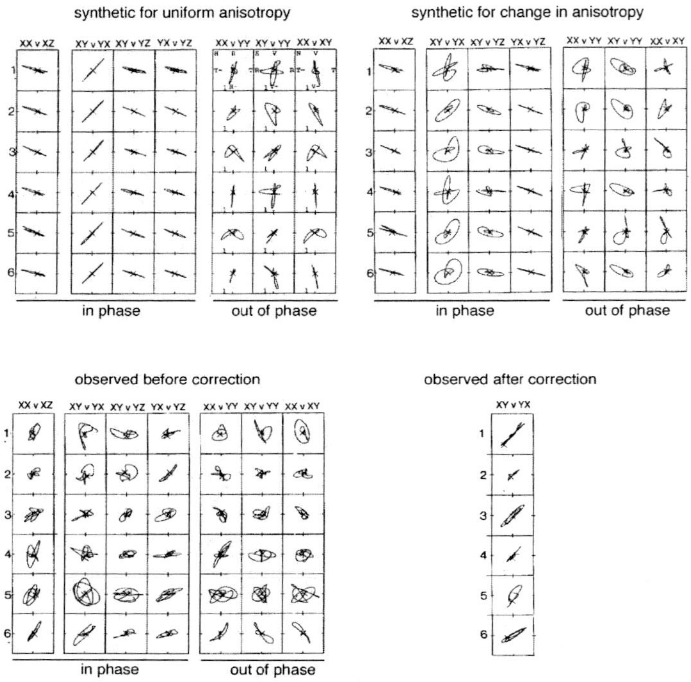

Figure 8.9 Cross plots of data matrix components for six different receivers in a six-component offset VSP simulated for (i) a uniform TIH medium, (ii) a change in anisotropy orientation, and (iii) the field data of Figure 8.8. Cross plots indicate the components that should be in phase and those that should be out of phase. The panel at lower right shows the result of near-surface corrections using equation (8.20). From MacBeth, Li and Horne (1994).

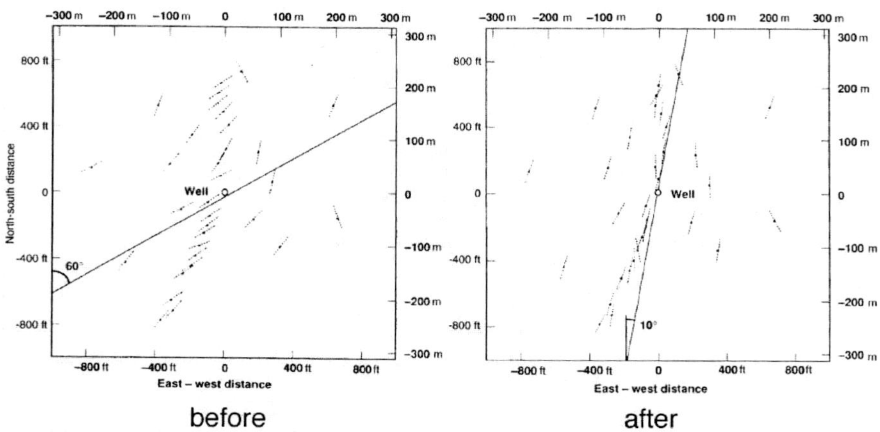

Figure 8.10. Polarization azimuths for the fast shear wave, calculated by applying Alford rotation to the data in Figure 8.8, before and after layer stripping by Winterstein and Meadows (1991b). Prior to stripping, most of the azimuths lie close to 60°, an indication that the upper zone dominates. Stripping a different shear wave lag for each source location revealed subsurface azimuths close to 10° in agreement with the near-offset VSP. The lag used varied strongly with location and was as much as 43 ms.

8.2.2 Downgoing deconvolution using a frequency dependent operator

8.2.2.1 Statistical correction

One approach to correction is to represent the near surface by the multi-component operator $\mathbf{N}(\omega)$, with the recording $\mathbf{D}(\omega)$ at each level j written in terms of the desired subsurface response $\mathbf{M}_A(\omega)$, i.e.,

$$\mathbf{D}(\omega) = \mathbf{M}_A(\omega)\mathbf{N}(\omega) \ , \tag{8.17}$$

where a datum level is drawn at some unspecified depth. Unlike the near-offset case in Section 4.2, the shallow recordings for large offsets generally no longer fully representative of the input for the deeper recordings, and a fully deterministic procedure is not possible. Instead, solution must involve a mathematical constraint for the medium or for the near-surface properties. This constraint can take one of two major forms, both of which result in a unique polar decomposition (Marcus and Minc 1992):

(a) Consider $\mathbf{M}_A(\omega)$ to be unitary (that is, $\mathbf{M}_A(\omega))^{*T}\ \mathbf{M}_A(\omega) = \mathbf{I}$), and $\mathbf{N}(\omega)$ to be Hermitian ($\mathbf{N}(\omega)^{*T} = \mathbf{N}(\omega)$) at each frequency. This unitarity condition is satisfied by the vector convolutional model (see Chapter 4 and Chapter 6) provided that no differential attenuation exists between the shear waves (for which $\lambda_1(\omega)$ differs from $\lambda_2(\omega)$). This particular decomposition does not, however, resolve two layers with different orientation of the anisotropy (the product of two unitary matrices is itself unitary). The Hermitian condition constrains the near-surface contribution to be independent or a symmetric function in time. These assumptions lead to the estimate

$$\hat{\mathbf{N}}(\omega) = \sqrt{\mathbf{D}^{*T}(\omega)\mathbf{D}(\omega)} \ , \tag{8.18}$$

from which a correction can be obtained for each recording separately. A correction operator common to all depth levels is given by

$$\overline{\mathbf{N}}(\omega) = \sqrt{\begin{pmatrix}\mathbf{D}_1^{*T} & \mathbf{D}_2^{*T} & \dots & \mathbf{D}_M^{*T}\end{pmatrix}\begin{pmatrix}\mathbf{D}_1 & \mathbf{D}_2 & \dots & \mathbf{D}_M\end{pmatrix}^T} \tag{8.19}$$

for the M recordings.

(b) Consider $\mathbf{M}_A(\omega)$ to be symmetric ($\mathbf{M}_A{}^T(\omega) = \mathbf{M}_A(\omega)$) and $\mathbf{N}(\omega)$ to be orthogonal ($\mathbf{N}^T(\omega)\mathbf{N}(\omega) = \mathbf{I}$). These alleviate the previous constraint on differential attenuation, as equation (8.6) now remains symmetric. The assumption of an orthogonal near-surface operator is adequate to define amplitude differences, polarity changes, time-delays, component rotations and small modifications of the wavelet. The condition leads to the solution at each receiver

$$\hat{\mathbf{N}}(\omega) = \sqrt{\mathbf{D}(\omega)^{*T}\mathbf{D}(\omega)} \ , \tag{8.20}$$

and the generalization for M recordings

$$\hat{\mathbf{N}}(\omega) = \sqrt{\left(\mathbf{D}_1^{*T} \quad \mathbf{D}_2^{*T} \quad \dots \quad \mathbf{D}_M^{*T}\right)\left(\mathbf{D}_1 \quad \mathbf{D}_2 \quad \dots \quad \mathbf{D}_M\right)} \tag{8.21}$$

Figure 8.19 shows the results of application of this procedure to a four-component dataset. This particular form of the near-surface operator has also been used by Karrenbach (1994), who minimized the sum of non-symmetric parts of the data matrix over all depth levels:

$$E = \sum_{j=1}^{M} \left\| \mathbf{D}_j(\omega)\mathbf{N}^{-1}(\omega) - \left(\mathbf{D}_j(\omega)\mathbf{N}^{-1}(\omega)\right)^T \right\| . \tag{8.22}$$

Here, a global minimum was sought to estimate the components of $\mathbf{N}(\omega)$ specified by a pre-defined minimum phase filter shape with a short time duration. This minimization can be achieved using a conjugate gradient method.

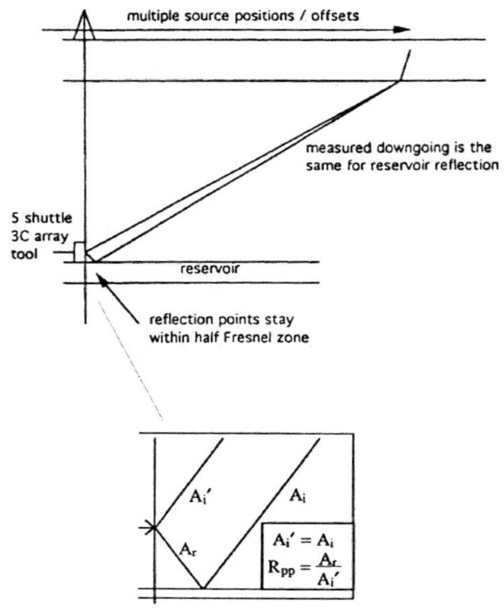

Figure 8.11. Schematic of a marine walkaway VSP designed for measuring anisotropy in the target zone, employing near-surface correction based on downgoing wavefield deconvolution.

8.2.2.2 *Deterministic correction for upgoing waves*

Deterministic deconvolution corrects for propagation effects such as multiples, geometric spreading, and source variation. An extension of Section 4.2 is possible for upgoing waves in a walk-away VSP. A particular requirement is that the receiver must be clamped just above the target interface (Figure 8.11). It is assumed that the downgoing wavefield of the direct waves is the same as the downgoing wavefield that generates the reflected arrivals (Coulombe, Stewart and Jones 1991). The assumption is true in the limit of the receivers being placed exactly on the target, and approximately true for small vertical distances above it. This method is adequate provided the reflection point at the target zone stays within half of a Fresnel zone from the borehole. This places a condition on the vertical distance, $h_{\min}$ above the target interface. For a particular maximum offset X_{max} and target depth H

$$h_{\min} \approx \frac{\lambda/4}{\sqrt{X_{\max}^2/H^2+1}-1} \quad . \tag{8.23}$$

where λ is the wavelength. For example, a maximum offset of 4 km, target depth of 2 km and a wavelength of 100 m gives the requirement for the receiver placement above the interface of only 20 m. This becomes over 200 m when the offset is reduced to 1000 m. An example of application to field data is shown in Figure 8.12.

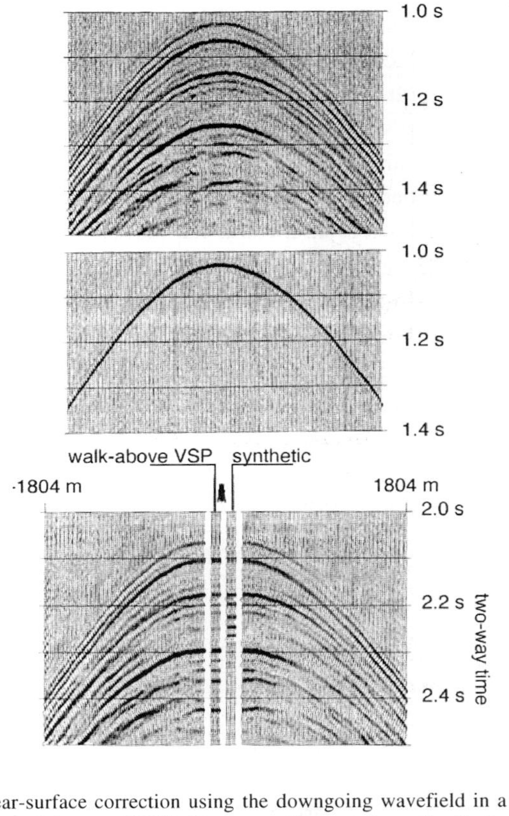

Figure 8.12. Example of near-surface correction using the downgoing wavefield in a walk-away VSP from Armstrong, Chmela and Leaney (1995). The lower diagram shows the final upgoing compressional wavefield after frequency domain deconvolution, based on the downgoing wavefield. The upper two show the up and downgoing wavefields after parametric wavefield separation (Section 7.1) and deterministic wave shaping deconvolution.

8.2.3 Downgoing deconvolution using a frequency- and slowness dependent operator

For wide offset VSP or the receiver placed high above the target zone, the recordings at the shallowest receiver can be used to correct for the deeper events if these depend on slowness or wavenumber. Thus, assuming the medium to be horizontally stratified and cylindrically symmetric, a single multi-component recording can be transformed directly into a function of horizontal slowness p and frequency. This transformation is accomplished using the technique described in Section 7.2. This wavefield $\mathbf{D}(\omega, p)$ can now be used to deconvolve the deeper layers, and an inverse transform can then be applied. If azimuthal anisotropy is present, the technique can proceed on a line-by-line basis. If full multi-component data are not available, the technique can also be applied on a component-by-component basis.

8.3 CONVERTED SHEAR WAVE ANALYSIS (TIH)

8.3.1 Pseudo-rotation using orthogonal shot lines

When multi-component sources are not available, or the VSP survey is shot offshore, a discrete set of offset azimuths can be used to generate converted shear waves to replace the components in the multi-component data matrix. It is then possible to apply the rotation analysis techniques described in Chapter 5 directly for anisotropy analysis. This approach relies on observing strong conversions along the transmission path. To understand this conversion process in the context of the fracture-related TIH model, consider a thin veneer of isotropic material at the converting interface. The incident P-wave is first converted to an SV wave at the top of the veneer, with amplitude T_{PS} depending on the incidence angle θ_S. This wave propagates through to the anisotropic medium and acts as an effective source loading the material below. By changing the offset azimuth, this material can be probed with a range of differently polarized 'sources'. Structural uniformity is required for this to be effective. A multi-component data matrix is generated by using orthogonal line azimuths, with receiver axes that have already been aligned along these line directions (Figure 8.13). By following the convention in (8.1), the three-component recording of source A, offset along the X-axis, as in Figure 8.13, is given by

$$\mathbf{d}_A(t) = T_{PS}(\theta_S)\mathbf{C}_Y^T(\theta_S)\mathbf{C}_{Z'}(\phi_N)\Lambda_S^{(9)}(t) * \mathbf{C}_{Z'}^T(\phi_N)\mathbf{C}_Y(\theta_S) * \begin{pmatrix} s_A(t) \\ 0 \\ 0 \end{pmatrix}. \tag{8.24}$$

For a source B, offset along the Y-axis, a new recording can be generated by

$$\mathbf{d}_B(t) = T_{PS}(\theta_S)\mathbf{C}_Y^T(\theta_S)\mathbf{C}_{Z'}(\phi_N)\Lambda_S^{(9)}(t) * \mathbf{C}_{Z'}^T(\phi_N)\mathbf{C}_Y(\theta_S) * \begin{pmatrix} 0 \\ s_B(t) \\ 0 \end{pmatrix}. \tag{8.25}$$

These are satisfied if $\Lambda_S^{(9)}$ and the polarizations remain the same. This limitation requires that the medium properties are invariant when probed at different offsets. This assumption is justified for ray paths at large incidence angles (see Chapter 2), and is valid for the time-shift operator if the lag times are not too dissimilar. For equal source strengths $s_A(t) = s_B(t) = s(t)$, the equations can be combined to yield a pseudo-multi-component data matrix for the shear waves

$$\mathbf{D}^{(9)}(t)=\left(\mathbf{d}_A(t)\quad \mathbf{d}_B(t)\quad 0\right)=T_{PS}(\theta_S)\mathbf{C}_Y^T(\theta_S)\mathbf{C}_{Z'}(\phi_N)\Lambda_S^{(9)}(t)\mathbf{C}_{Z'}^T(\phi_N)\mathbf{C}_Y(\theta_S)\begin{pmatrix}1&0&0\\0&1&0\\0&0&0\end{pmatrix}. \quad (8.26)$$

where the source signature has now been absorbed into the time-shift operator for convenience. This collapses to the four-component matrix

$$\mathbf{D}_S^{(4)}(t)=T_{PS}(\theta_S)\begin{pmatrix}\cos\phi_\parallel & \cos\phi_\perp\\ \sin\phi_\parallel & \sin\phi_\perp\end{pmatrix}\begin{pmatrix}\lambda_1(t) & 0\\ 0 & \lambda_2(t)\end{pmatrix}\begin{pmatrix}\cos\phi_\parallel & \sin\phi_\parallel\\ \cos\phi_\perp & \sin\phi_\perp\end{pmatrix}, \quad (8.27)$$

which is the same as the offset multi-component case of (8.14).

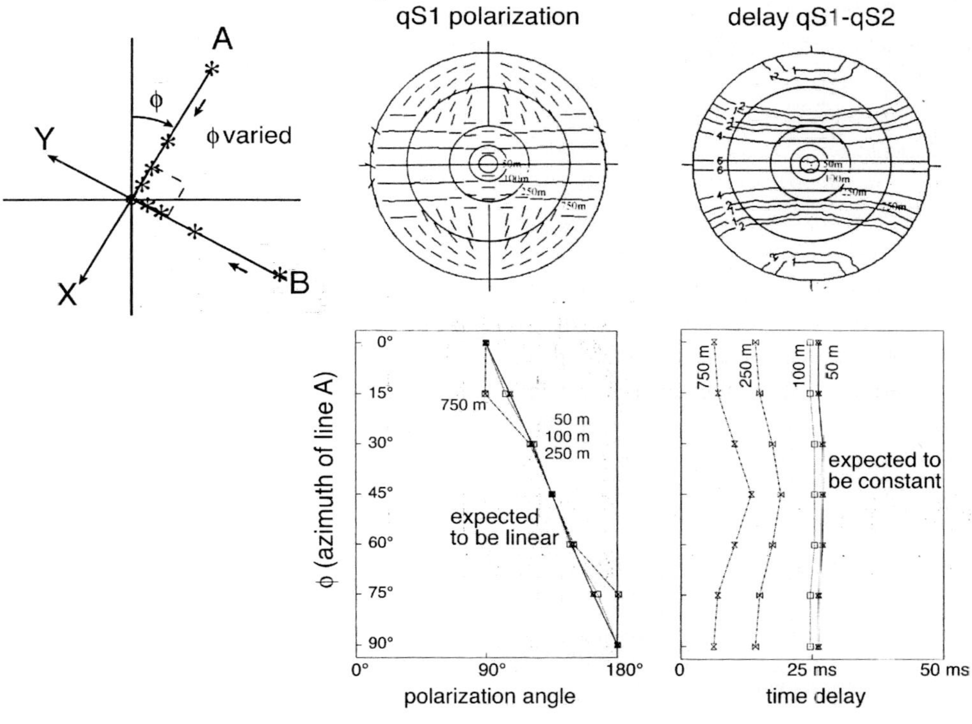

Figure 8.13. Acquisition geometry together with plots of shear-wave behavior to illustrate the limitations of pseudo-rotation analysis. Orthogonal line azimuths are aligned along the X and Y receiver axes. Ideally, anisotropy estimates for each axis orientation ϕ should be consistent if the technique is to be of value. Shots are at 50 m, 100 m, 250 m, and 750 m offsets. Only the first two give satisfactory agreement for the shear wave lag times. The method provides adequate estimates of fracture strike at all orientations (MacBeth 1996).

A similar equation (without the offset dependence) was proposed by Mallick and Frazer (1991) for OSS data based on the formulation of Schoenberg et al. (1988). In application, each common-offset trace set is corrected for incidence angle, prior to applying a multi-component rotation. This procedure is identical to that used for the offset VSP analysis in equation (8.6), and the interpretation of the shear wave polarizations is also the same. The offset limitations of the technique are shown in Figure 8.13 and 8.14, with the results of application to a dataset in Figures 8.15. The technique is suited to several fixed-offset shot positions but also to walkaway VSP lines. At small offsets it reduces to the familiar form of the Alford rotation used in Chapter 5. Equation (8.27) could also be adapted to treat shot lines at oblique angles, provided the offset azimuths are known accurately.

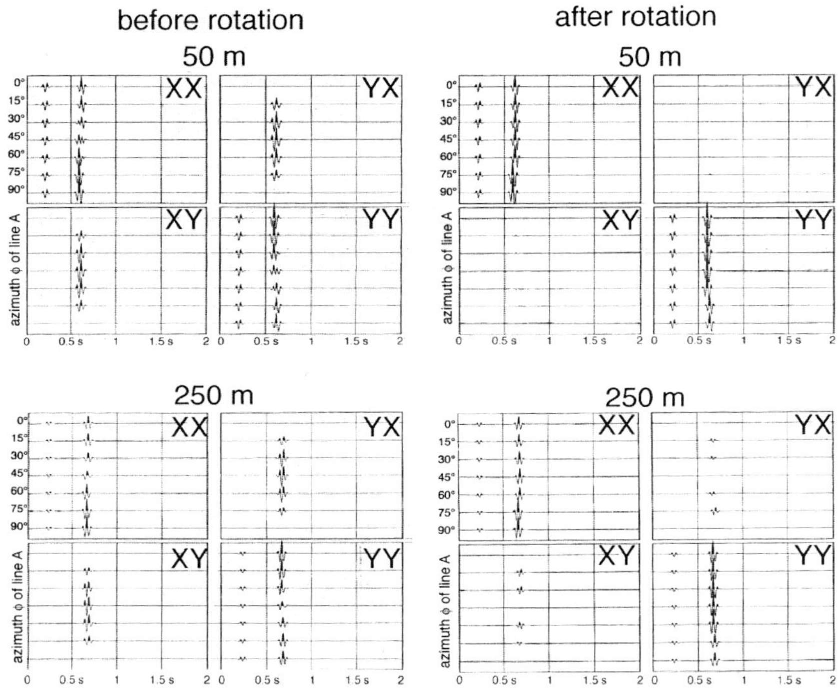

Figure 8.14. Pseudo-rotation of two fixed, orthogonal marine offset VSPs. The data matrix is formed from the two horizontal components (X and Y) and the two orthogonal source offsets ($X = A$ and $Y = B$) in Figure 8.13. Alford rotation is adapted for oblique incidence using equation (8.26). The larger offset data leave a residual due to inconsistency in the birefringence at the two offsets (MacBeth 1996b).

8.3.2 Variations with line azimuth

An alternative approach to the above for determination of fracture strike and birefringence is to monitor the transverse receiver component as a function of source position along a specific line. Ideally, this method takes the form of a fixed-offset azimuthal VSP; however, the technique can be adapted to deal with a walk-away VSP for which offset variations must also be assessed. The technique is only appropriate if the walk-away line does not pass through the receiver position, so that a range of azimuths is covered (Figure 8.16). For each source location, the receiver is re-oriented so that the X-axis points towards the source, i.e., along the radial direction (Figure 8.16). The angle between this radial direction – given by an azimuth ϕ_j and the fracture strike ψ – changes with each source location, which in turn changes the relative contribution of the two shear waves and the magnitude of the transverse component $d_T(t)$. The amplitude of this component can be written

$$d_T\left(t,\phi_j\right) = \frac{T_{PS}}{2}\left(s(t+\Delta\tau) - s(t)\right)\Omega\left(\Delta\phi_j,\theta\right) \ , \tag{8.28}$$

where

$$\Omega\left(\Delta\phi_j,\theta\right) = \frac{\sin\Delta\phi_j \cos\Delta\phi_j \cos\theta}{1-\sin^2\Delta\phi_j \sin^2\theta} \tag{8.29}$$

and $\Delta\,\phi_j = \psi - \phi_j$.

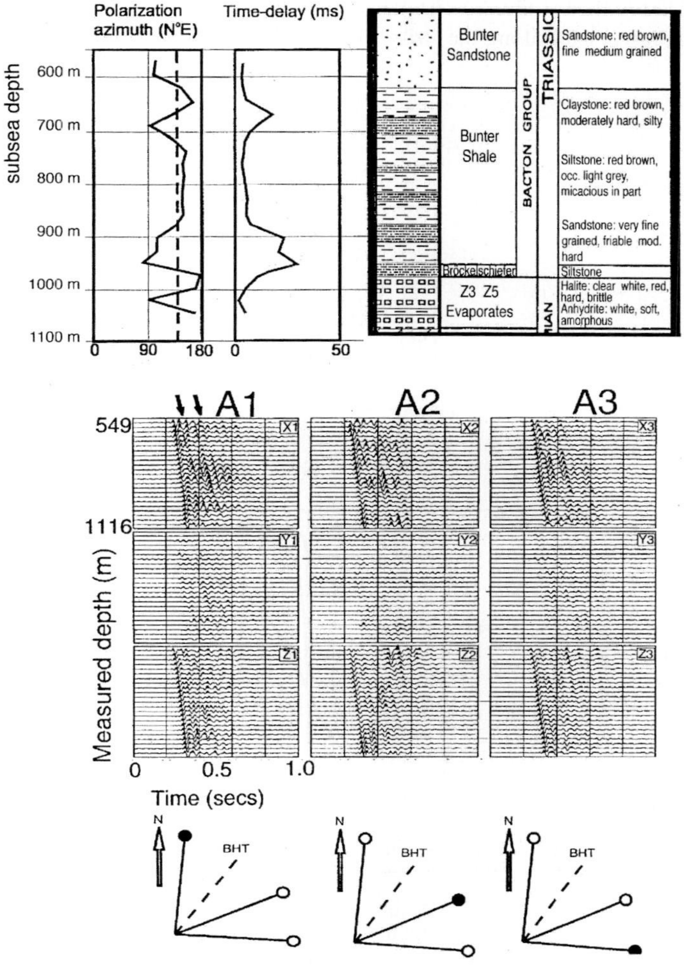

Figure 8.15. Three-component recordings from three fixed-offset (700 m) marine VSPs, together with their pseudo-rotation results. In the display, each offset has been corrected to let the X-axis point in the radial direction. Converted shear waves are marked by arrows. They originate from the subsurface and close to the seabed (MacBeth 1996b).

The equation shows that an adequate shear-wave conversion is required before this effect becomes significant, although once established the variation with offset and azimuth is controlled by the strength and orientation of the anisotropy. There are therefore two groups of directional variations. The first function is the difference between the waveforms, which is zero when crossing the line singularity (Figure 8.17) and then reverses polarity. Elsewhere, it depends upon the shape and bandwidth of the wavelet. The second function, $\Omega(\Delta\,\phi_j,\,\theta)$, varies predominantly as $\sin 2(\psi - \phi_j)$ for small θ. It is zero at the fracture strike or normal to it and reverses polarity at these directions (Figure 8.17). Fracture strike determination must therefore rely on correctly picking the location of the polarity change in the data (Figure 8.18). Once determined, the two shear waves can be separated by aligning the receiver axes along the strike directions, followed by cross-correlation analysis (see Appendix A).

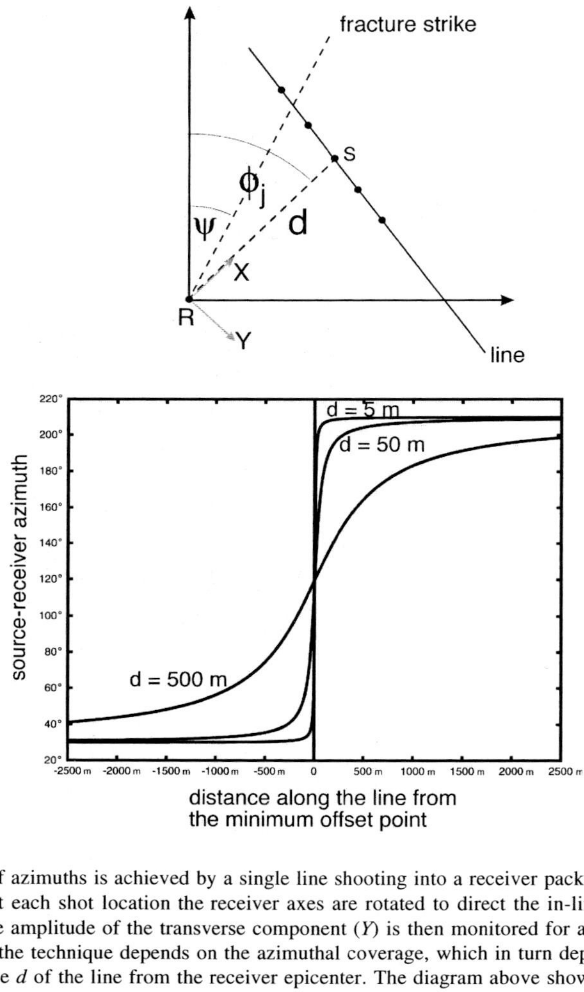

Figure 8.16. A spread of azimuths is achieved by a single line shooting into a receiver package lying to one side of the line. At each shot location the receiver axes are rotated to direct the in-line (X) axis towards the source. The amplitude of the transverse component (Y) is then monitored for anisotropy analysis. The success of the technique depends on the azimuthal coverage, which in turn depends on the line length and distance d of the line from the receiver epicenter. The diagram above shows how the spread is affected by the choice of d for a split spread of 2500 m length.

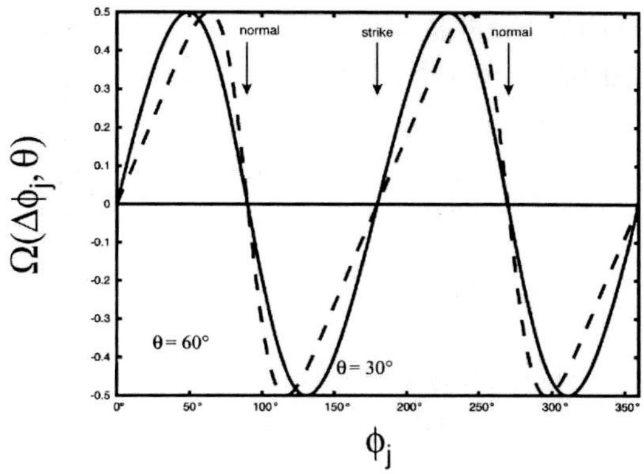

Figure 8.17. The amplitude of the transverse component is modulated by a function of source-receiver azimuth ϕ_j. It is zero at the fracture strike, regardless of the incidence angle θ.

The technique is simple in that only one component need be monitored. Polarity change is certainly a more robust indication than a reduction in amplitude, since it is not complicated by the effects of absorption and geometric spreading with offset.

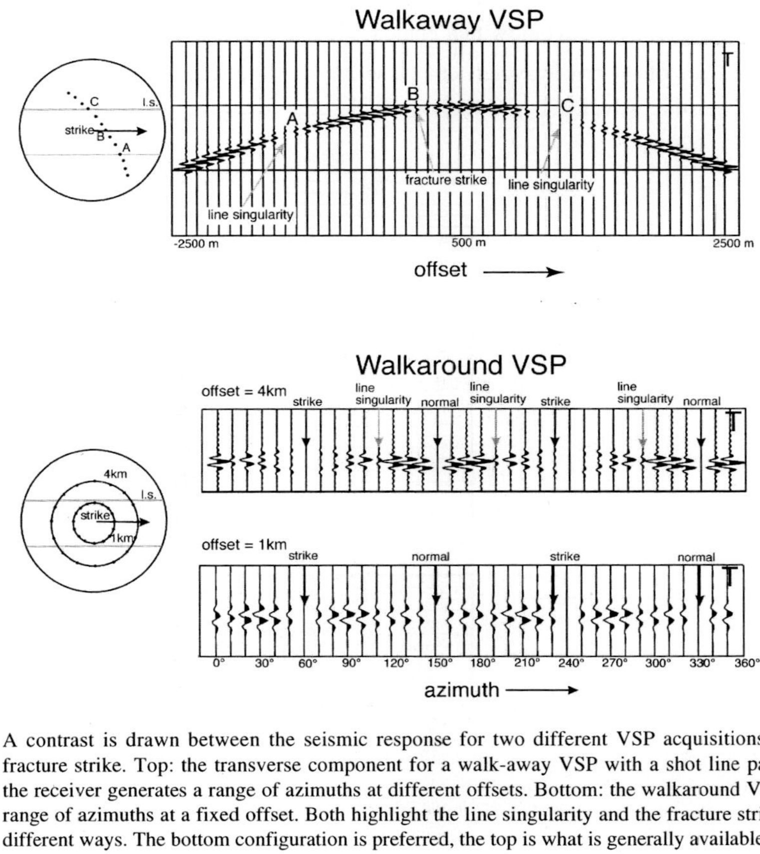

Figure 8.18. A contrast is drawn between the seismic response for two different VSP acquisitions for detecting fracture strike. Top: the transverse component for a walk-away VSP with a shot line passing close to the receiver generates a range of azimuths at different offsets. Bottom: the walkaround VSP generates a range of azimuths at a fixed offset. Both highlight the line singularity and the fracture strike (normal) in different ways. The bottom configuration is preferred, the top is what is generally available.

8.3.3 Local shear-wave conversion

This technique is very specific to low/high contrasts and mainly applicable to fractured carbonates (MacBeth et al. 1998), though it has also been considered for monitoring gas bearing zones (Gal'Perin 1985). It is most suited to walk-away VSP, for which it has an overall lateral range of a few hundred meters beyond the borehole, providing an average of the rock properties 30 m from the receiver. The focus is a P-S conversion that occurs locally at the target zone and is unaffected by the overburden or multiples. The converted S wave travels a short path to the receiver, and arrives just behind the P wave. The S wave can be detected by its distinctive horizontal motion, and in particular the transverse component energy (Figure 8.19). The TIH anisotropy in the reservoir zone controls the amplitude of this component. Interpretation of the effect requires careful quality control to ensure other possible causes such as tool coupling or lateral refraction are eliminated (MacBeth et al. 1998).

As in the previous section, the method monitors the amplitude of the transverse component after rotation of the receiver axes to point towards the source location. Dependence on the fractured layer of interest is controlled by the difference of the two shear waves, which is very sensitive to anisotropy, incidence angle and line azimuth. The distinction from Section 8.3.2 is that the shear wave lag is small. This permits a practical measure of the amplitude to be made by calculating the average amplitude over a time window that encloses the arrival: $\langle d_T(R)\rangle = \left(\sum_i d_T^2(R,t_i)\right)^{1/2}$. The average of the difference of the two wavelets can be written as a function of the time delay. For example, assuming a seismic wavelet shaped like one cycle of a sine wave, (8.14) gives

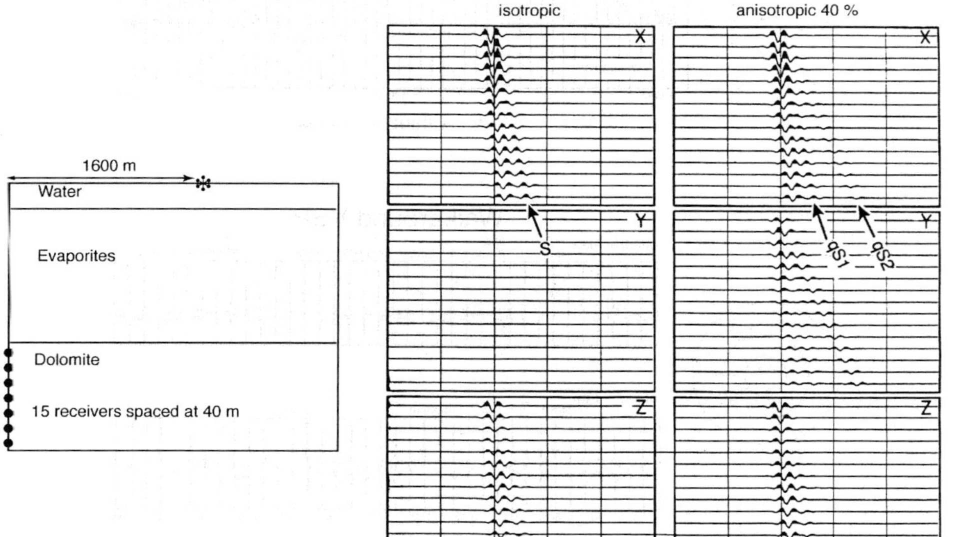

Figure 8.19. Full-wave synthetic seismograms for an isotropic and anisotropic dolomite half-space, highlighting the development of the local shear-wave conversion with penetration into the high-velocity layer. Note the large difference in shear wave moveouts for the case with anisotropy (MacBeth et al. 1998).

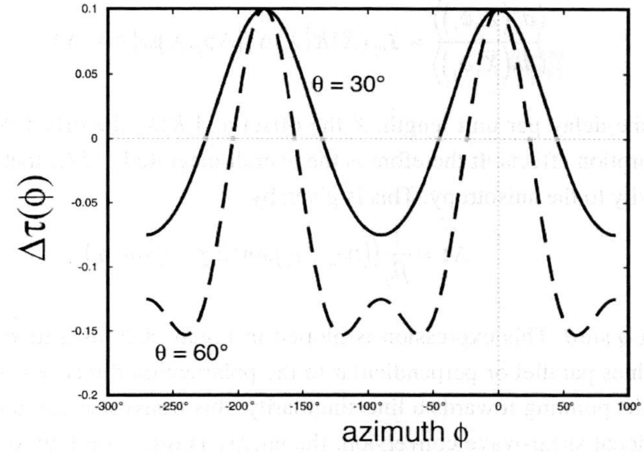

Figure 8.20. For the local shear wave conversion, the amplitude of the transverse component is proportional to the product of the lag time $\Delta t(\Delta\phi_j, \theta)$ and the function $\Omega(\Delta\phi_j, \theta)$ in Figure 8.17. This lag also varies with azimuth, producing a number of distinct zeros (at the line singularity), and changes of polarity. This behavior also varies with the incidence angle θ.

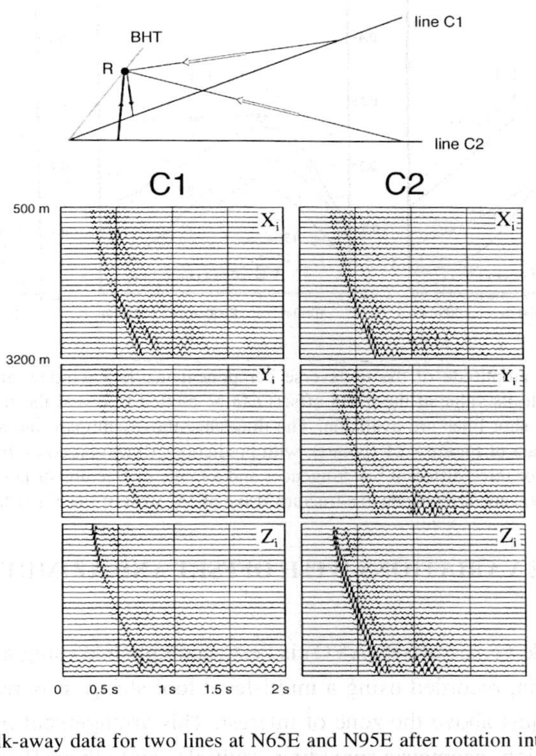

Figure 8.21. Observed walk-away data for two lines at N65E and N95E after rotation into the radial and transverse directions. Note the development of transverse energy, and its different evolution on each line. In this case the receiver package is placed inside a short horizontal borehole deviating at N35E into a gas bearing carbonate (MacBeth et al. 1998).

$$\frac{\langle d_T(X,\phi_j)\rangle}{\langle d_T(\bar{X},\phi_j)\rangle} \propto T_{PS}(X)K(X)\Delta\tau(\Delta\phi_j,X)\Omega(\Delta\phi_j,X) \ , \tag{8.30}$$

where $\Delta\tau$ is the time-delay per unit length, X the offset and $K(X)$ the offset-dependent geometric spreading and absorption effects. It therefore is the overall magnitude of $\Delta\tau$ that directly determines the general sensitivity to the anisotropy. This is given by

$$\Delta\tau \propto \frac{1}{\beta_0}\left((\eta e_n - e_t)\sin^2 2\chi - e_t \sin^2\chi\right) \ , \tag{8.31}$$

where $\cos\chi = \cos\Delta\,\phi_j\,\sin\theta$. This expression is plotted in Figure 8.20 as a function of azimuth. As before, for offset lines parallel or perpendicular to the polarization directions, or for incidence angles giving ray paths pointing towards a line singularity, this transverse component becomes zero. In the case of the local shear-wave conversion, the energy is often too low to adequately detect a polarity change (Figure 8.21). Instead, it is best to use the relative amplitudes between common offsets from two or more lines to judge the orientation (Figure 8.22).

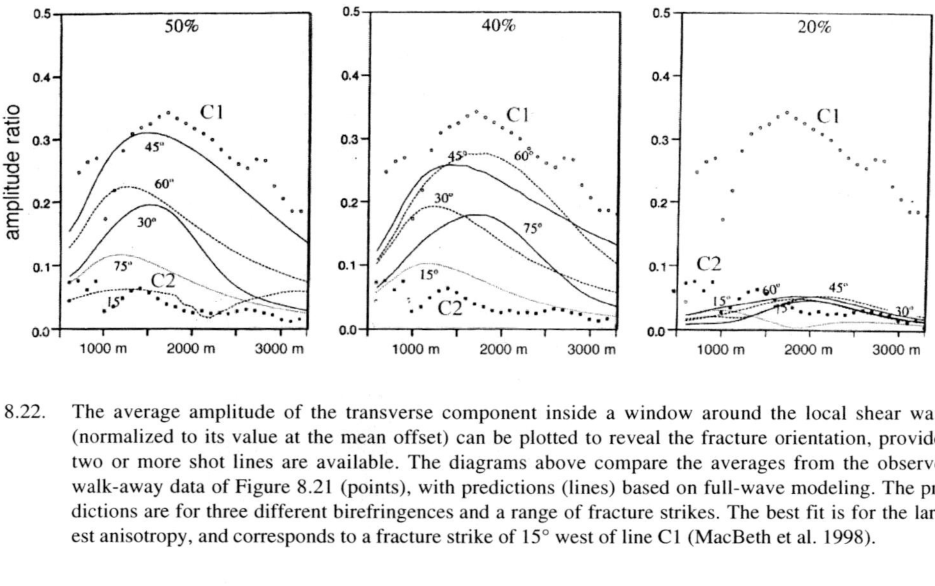

Figure 8.22. The average amplitude of the transverse component inside a window around the local shear wave (normalized to its value at the mean offset) can be plotted to reveal the fracture orientation, provided two or more shot lines are available. The diagrams above compare the averages from the observed walk-away data of Figure 8.21 (points), with predictions (lines) based on full-wave modeling. The predictions are for three different birefringences and a range of fracture strikes. The best fit is for the largest anisotropy, and corresponds to a fracture strike of 15° west of line C1 (MacBeth et al. 1998).

8.4 AMPLITUDE VARIATIONS WITH OFFSET AND AZIMUTH FOR TIH AND TIV

Analysis of amplitude versus offset (AVO) is best accomplished using a multiple offset or walk-away VSP configuration, recorded using a multi-level tool string. It is recommended that the receivers are positioned just above the zone of interest. This arrangement means that the source radiation pattern, absorption, geometric spreading, multiples and other offset-dependent effects, that usually overwhelm the response, can be corrected using the downgoing-wavefield deconvolution as described in Section 8.2.2. Furthermore, recordings across the receiver array also allow wavefield separation based on the techniques in Section 7.1, and open up the possibility of using both P-P and the P-S waves (Figure 8.23). By ensuring that the receivers are placed a small distance above the target, lateral effects are also minimized. The local phase slowness method and interval moveout analysis provide supplementary anisotropy estimates above and at the target level. Output from the procedure is several high quality receiver gathers for subsequent AVO interpretation, representing conditions close to the borehole and thus suited to modeling calibration using the logs. If fracture analysis is a primary objective, several walk-away lines with different azimuths are necessary. On land with multi-component sources, the S-S normal incidence reflectivity is sufficiently strong for fractures to obviate the need for AVO. The techniques above are suited to both offshore and onshore environments.

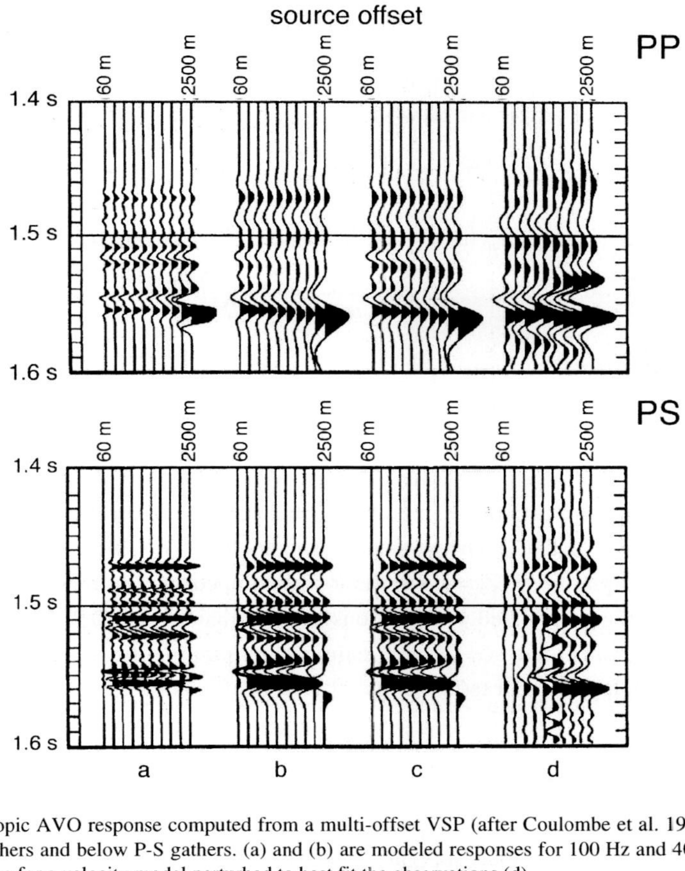

Figure 8.23. Isotropic AVO response computed from a multi-offset VSP (after Coulombe et al. 1992). Above are P-P gathers and below P-S gathers. (a) and (b) are modeled responses for 100 Hz and 40 Hz, (c) is the response for a velocity model perturbed to best fit the observations (d).

8.4.1 Isotropic response

The isotropic form of the AVO technique stems from the study of Ostrander (1984), who showed that P-wave data could be used to detect changes in the velocity ratio α/β at an interface. This provides an additional interpretative dimension to normal incidence reflectivity, and thus acts as a more direct hydrocarbon indicator. It has the potential to distinguish between bright spots caused by gas sands and other non-productive rock changes. Another paper central to this work is Shuey (1985), who used a weak-contrast approximation to evaluate the dependence of the isotropic AVO response on the jumps in the P and S wave velocities and density. His relationship is the basis of lithological interpretation in most routine AVO analyses (Castagna 1993). The extension to anisotropy builds on this existing framework, following a similar approach to evaluate expressions for both weak anisotropy and weak contrasts. They can thus be regarded as an extension of Shuey's formulae. As a consequence, the impact of anisotropy cannot be fully understood without first understanding the isotropic response, and multi-component analysis is not complete without considering both the P- and S wave reflectivity.

8.4.1.1 P-P reflectivity

The classical AVO equation for two isotropic media with elastic properties α_1, β_1, ρ_1, and α_2, β_2, ρ_2, and moderate incidence angles can be written

$$R_P = R_{0P} + G_P \sin^2\theta_P + H_P \tan^2\theta_P \sin^2\theta_P \ , \tag{8.32}$$

where θ_P is the incidence angle, R_{0P} the normal-incidence reflection coefficient (AVO intercept)

$$R_{0P} = \frac{\rho_2\alpha_2 - \rho_1\alpha_1}{\rho_2\alpha_2 + \rho_1\alpha_1} \approx \frac{1}{2}\frac{\Delta Z_P}{\hat{Z}_P} \ , \tag{8.33}$$

and G_P the short-offset gradient (justified for incidence angles less than 20°), i.e.,

$$G_P = \frac{1}{2}\frac{\Delta\alpha}{\hat{\alpha}} - 2\tilde{\eta}\left(\frac{\Delta\rho}{\hat{\rho}} + 2\frac{\Delta\beta}{\hat{\beta}}\right) . \tag{8.34}$$

The prefix Δ denotes a difference in properties between the lower and upper medium and the average is denoted by a "hat" (^). The parameter $\tilde{\eta}$ refers to $\hat{\beta}^2/\hat{\alpha}^2$, which is – to the level of approximation used – equal to $\hat{\eta}$. $Z_P = \rho\alpha$ is the P-wave impedance. In these expressions, a weak contrast means that the normalized perturbations are less than about 10 %. In most AVO studies only the first-order gradient term G_P is used, estimated by plotting the reflected amplitude against $\sin^2\theta_P$. The additional higher order term specified by the coefficient H_P

$$H_P = \frac{1}{2}\frac{\Delta\alpha}{\hat{\alpha}} \tag{8.35}$$

is significant for offsets resulting in incidence angles beyond 30°. The magnitude of both trigonometric terms for G_P and H_P is shown in Figure 8.24. The crossplot of intercept R_{0P} against the gradient G_P is fundamental to AVO interpretation (Castagna 1993). Within this domain, Rutherford and Williams (1989) define a three-class system for shale/gas-sands (Figure 8.25). Class I anomalies have a large positive R_{0P}; the reflection coefficient reduces to a smaller positive value with offset. Class II anomalies have a small positive or negative R_{0P} that gets more negative at larger offsets with the possibility of a sign reversal. Class III anomalies have a negative R_{0P} with a reflection coefficient that increases with offset (this is the original bright spot). Castagna et al. (1998) propose a class IV for high impedance gas-sands, characterized by a negative R_{0P} and a reflection coefficient that decreases in magnitude with offset. Procedures for robust attribute analysis are based on this fundamental classification. In particular, class III responses can be highlighted by the product of R_{0P} and G_P, which is positive for both the top and base of a shale-encased gas sand. Other weighted combinations of these dimensions provide further discriminants (Castagna and Smith 1994).

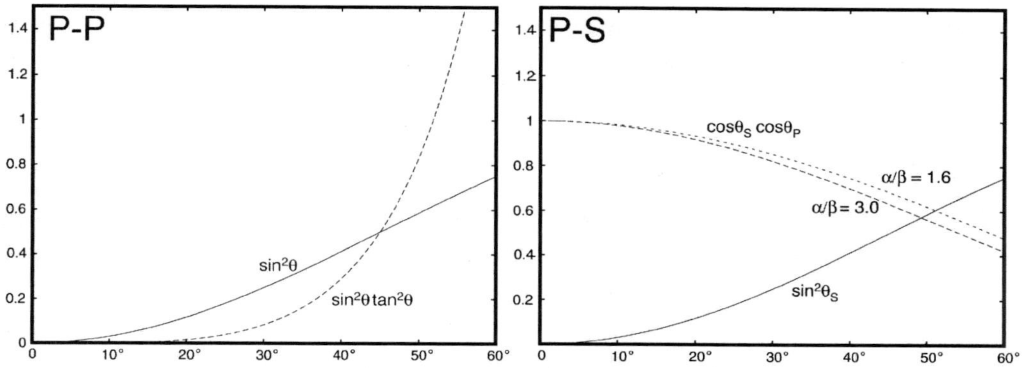

Figure 8.24. Relative magnitudes of the trigonometric terms common to the isotropic and anisotropic P-P and P-S equations, demonstrating the limits of application for the first and second order approximations.

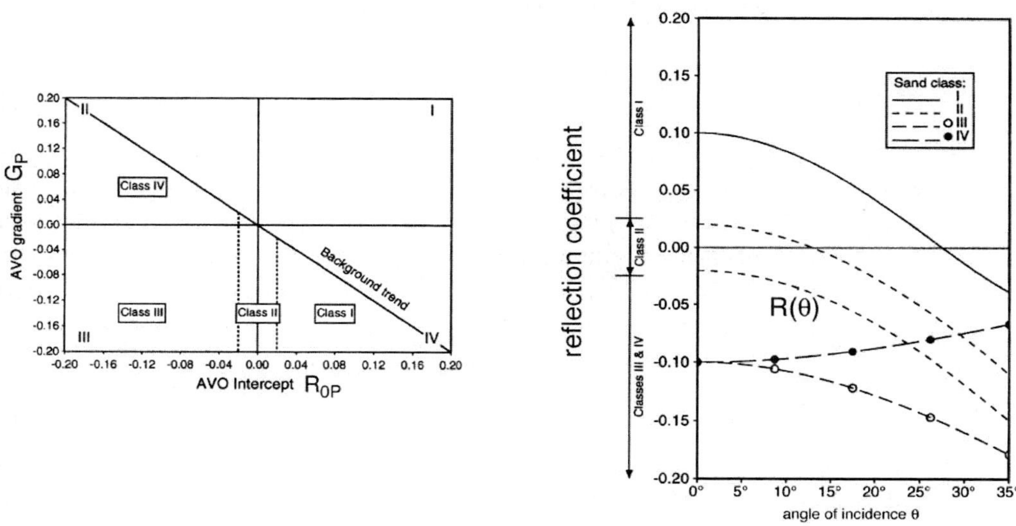

Figure 8.25. Left: isotropic P-P AVO classification based on a cross plot of intercept R_{0P} and gradient G_P – after Castagna et al. 1998.
Right: plane wave reflection coefficients for each Rutherford-and-Williams (1989) gas-sand classification. Class IV sands are a new addition by Castagna and Swan (1998).

8.4.1.2 S-S reflectivity

A distinction must be made between incident SH waves that generate reflected and transmitted SH waves only, and incident SV waves that generate both reflected and transmitted P and SV waves. For the former, Aki and Richards (1981) give

$$R_{SH} = R_{0S} + G_{SH}\sin^2\theta_S + H_{SH}\tan^2\theta_S\sin^2\theta_S \ , \tag{8.36}$$

where

$$R_{0S} = \frac{\rho_1\beta_1 - \rho_2\beta_2}{\rho_1\beta_1 + \rho_2\beta_2} \approx -\frac{1}{2}\frac{\Delta Z_S}{\hat{Z}_S} \ , \tag{8.37}$$

and

$$G_{SH} = H_{SH} = \frac{1}{2}\frac{\Delta\beta}{\hat{\beta}} \ . \tag{8.38}$$

For SV waves,

$$R_{SV} = R_{0S} + G_{SV}\sin^2\theta_S + H_{SH}\tan^2\theta_S\sin^2\theta_S \ , \tag{8.39}$$

where

$$G_{SV} = 2\frac{\Delta\rho}{\hat{\rho}} + \frac{7}{2}\frac{\Delta\beta}{\hat{\beta}} \ . \tag{8.40}$$

and $H_{SV} = -G_{SH}$. Pure mode SV and SH waves have the same normal-incidence reflectivity.

8.4.1.3 P-S reflectivity

Incident P-waves convert only to SV waves, leading to (Aki and Richards 1981)

$$R_{PSV} = -\frac{\sin\theta_P}{\cos\theta_S}\left(G_0 + G_1\cos\theta_P\cos\theta_S + G_2\sin^2\theta_P\right) \ , \tag{8.41a}$$

or

$$R_{PSV} = -\frac{\sin\theta_P}{\cos\theta_S}\left(G_0 + G_1\cos(\theta_P + \theta_S)\right) \ , \tag{8.41b}$$

where

$$R_{0PS} = \frac{1}{2}\frac{\Delta\rho}{\hat{\rho}} \ , \tag{8.42}$$

$$G_1 = \sqrt{\hat{\eta}}\left(\frac{\Delta\rho}{\hat{\rho}} + 2\frac{\Delta\beta}{\hat{\beta}}\right) , \tag{8.43}$$

and

$$G_2 = -\hat{\eta}\left(\frac{\Delta\rho}{\hat{\rho}} + 2\frac{\Delta\beta}{\hat{\beta}}\right) , \tag{8.44}$$

In these expressions, the emergence angle θ_S for the shear-waves must be calculated as a function of the P-wave incidence angle θ_P using

$$\cos\theta_S = \sqrt{1 - \eta\sin^2\theta_p} \ . \tag{8.45}$$

The individual trigonometric terms for this conversion coefficient are plotted in Figure 8.23. Evidently, the term $\cos\theta_S \cos\theta_P$ is dominant. The conversion coefficient has been further approximated for near-offsets by Zaengle and Frasier (1993)

$$R_{PSV} \approx 4\sqrt{\hat{\eta}}\sin\theta_S\left(R_{0S} - \frac{1-2\sqrt{\hat{\eta}}}{8\sqrt{\hat{\eta}}}\frac{\Delta\rho}{\hat{\rho}}\right) . \tag{8.46}$$

When $\hat{\eta}$ = 1/4 (α/β = 2), the equation is a scaled version of the normal incidence shear wave reflectivity in (8.12). An alternative is offered by Donati and Martin (1998), who suggest the linearized approximation valid for an incidence range from 0°–90°

$$R_{PSV} = \sin\theta_P\left(R_0 + R_1\cos\theta_P\right) , \tag{8.47}$$

where

$$R_0 = \hat{\eta}\frac{\Delta\mu}{\hat{\mu}} - \frac{\Delta\rho}{\hat{\rho}} , \tag{8.48}$$

and

$$R_1 = -\sqrt{\hat{\eta}}\left(1 - \frac{\hat{\eta}}{2}\right)\frac{\Delta\mu}{\hat{\mu}} . \tag{8.49}$$

The formulation permits the construction of an η-section (α/β-section) and a shear-rigidity section (μ-section) if the connection between shear modulus and density is known.

8.4.2 P-P response in the presence of anisotropy

Wright (1987), Banik (1987) and Thomsen (1988) were among the first to demonstrate the strong influence of TIV anisotropy on AVO. Indeed, it has been shown that the AVO gradients above can be reversed by the anisotropy, and hence significantly change the classification of the AVO anomaly. As shales are highly TIV anisotropic, this effect is of major importance and cannot be ignored. If the overburden consists predominantly of shales or clay, then TIV anisotropy also means that the phase angle (necessary for calculating the reflectivity) and group angle (for the geometric ray path) are no longer identical. Correction for this difference uses equation (3.28), δ from moveout analysis, and a suitable guess at the anellipticity using the data from Chapter 3. There is also a considerable body of work to show that the azimuthal variation of the AVO response can be used to detect fractures. Examples of support comes from the numerical study of Mallick and Frazer (1991), modeling by Chang and Gardner (1997), onshore surveys by Lynn et al. (1995) and Perez and Gibson (1996), and finally the offshore work by MacBeth et al. (1999) and Harwood et al. (1998).

8.4.2.1 Shale/layer-related TIV anisotropy

Rüger (1997) gives

$$R_P^{TIV} = R_{0P} + \left(G_P + \frac{1}{2}\Delta\delta\right)\sin^2\theta_P + \left(H_P + \frac{1}{2}\Delta\varepsilon\right)\tan^2\theta_P \sin^2\theta_P \quad ,> \tag{8.50}$$

which incorporates a correction given by Thomsen (1993). The normal-incidence coefficient uses vertical velocities. This equation predicts that to first order the intercept remains unchanged, and that the effect of the anisotropy is concentrated only in the AVO gradient. The parameter δ controls the AVO behavior at small to moderate angles. It has been shown in Chapter 2 that for shales, δ can be expected to be either positive or negative. Thus an anisotropic shale over an isotropic sandstone will produce a "false negative" if δ is positive ($\Delta\delta = -\delta_{\text{shale}}$), or a "false positive" if it is negative ($\Delta\delta = +\delta_{\text{shale}}$). There is therefore a chance that the shale will down-weight a bright gas sand response. Thomsen (1993) draws a qualitative conclusion from this, observing that among bright reflections there is more confidence in the isotropic interpretation than among less bright reflections. A VSP example for shales is given in Figure 8.26 and 8.27.

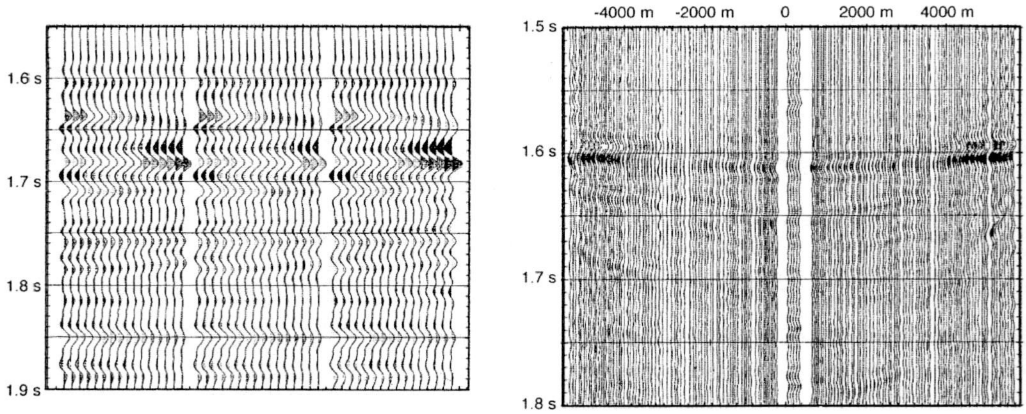

Figure 8.26. Left hand diagram: ray CMP synthetics for isotropy (left) and for two flavors of shale anisotropy with different anellipticity.
Right hand diagram: AVO walkaway data after processing. These upgoing reflections are shown after flattening on the direct P arrival time and shifting to vertical two-way time.
All after Leaney (1994).

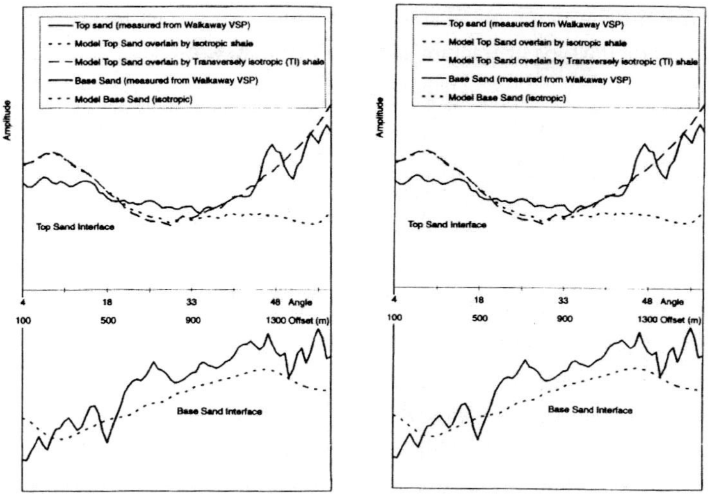

Figure 8.27. Further results from Leaney (1994), showing predictions for isotropic and anisotropic shales, and their affect on the top and base sand AVO response. Comparison against the observed results from a walk-away VSP, indicate that the base-of-sand reflector is in good agreement with the isotropic model, while the top-of-sand response agrees well with the anisotropic shale model.

8.4.2.2 Crack/fracture-related TIH anisotropy

For TIH anisotropy the reflectivity depends on azimuth. Rüger (1998) and Zillmer et al. (1997, 1998) give

$$R_P^{TIH} = R_{0P} + \left(G_P + G_P^h(\phi)\right)\sin^2\theta_P + \left(H_P + H_P^h(\phi)\right)\tan^2\theta_P\sin^2\theta_P \quad . \tag{8.51}$$

For an isotropic overburden

$$G_P^h(\phi) = -\eta_f\left(\zeta_f e_n - \left(\frac{2\tilde{\eta}}{\eta_f} - 1\right)e_t\right)\sin^2\phi \tag{8.52}$$

and

$$H_P^h(\phi) = -\eta_f\left(\left(1-\eta_f\right)e_n\sin^4\phi - \left(\zeta_f e_n + e_t\right)\sin^2\phi\cos^2\phi\right) \quad . \tag{8.53}$$

The parameters η_f and ζ_f are evaluated only for the fractured formation, and ϕ is the angle between the line and the fracture strike. $\tilde{\eta}$ is $\hat{\beta}^2/\hat{\alpha}^2$, where ^ is an average of the overburden- and target values. The effect of the fracturing is to enhance both the gradient and higher order terms by an amount dependent upon the fracture indices e_n and e_t (Figure 8.28). The combination of fracture indices means that an estimate of the AVO gradient is not sufficient for independent estimation of e_n and e_t. Indeed, an estimate of e_t can only be obtained if it is assumed that the anisotropy arises from thin brine- or oil-filled cracks (for which e_n is much smaller than e_t). However, the combination of fracture indices in the $\sin^2\theta$ gradient term can be either positive or negative, leading to a brightening or a dimming perpendicular to the fracture strike. The reflectivity response is thus highly dependent upon the type of fracture and its fill (Figure 8.29).

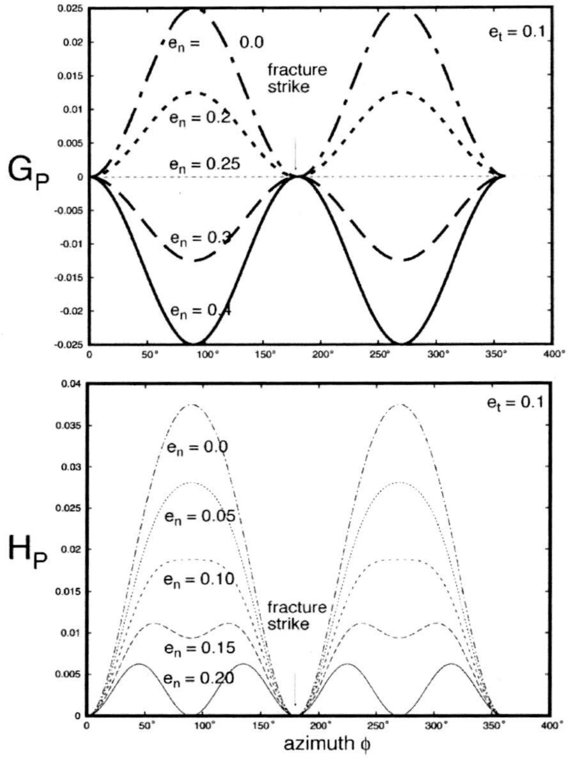

Figure 8.28. Azimuthal variation of the gradient and second-order AVO terms for an isotropic/TIH interface, as a function of the fracture response defined by the indices e_n and e_t. The gradient is isotropic parallel to the fracture strike and at a maximum perpendicular to the fracture strike (either positive or negative for the case of the gradient). Interestingly, the second-order term changes its periodicity at small values of e_n (i.e., for thin, brine filled cracks).

Calculation shows that *total* gas saturation always leads to a dimming perpendicular to the fracture strike, while liquid fill can produce either a dimming or a brightening. Although the overall response from P-P AVO is not expected to be strong, – since the coefficients consist of the difference of the fracture indices (Figure 8.30) – it is sensitive to crack/fracture fill with discrimination between oil and brine being possible for certain types of cracks/fracturing. For high-impedance targets such as chalk, the depth of variation for the base of the layer is more pronounced due to ray bending. This produces a clearer indication of the AVO response of the target. An analysis of gas sands is given by Sayers and Rickett (1997).

For near-offset AVO analysis only the first-order term is required. Indeed, this choice is also through necessity, as the larger offsets possess rapid azimuthal variations and can only be detected using acquisitions with a finer directional sampling. The $\sin^2\phi$ variation (hence an $A+B\cos 2\phi$ variation) of the first-order gradient term means that an analysis along only a few different line azimuths is adequate to determine the combination of fracture indices $\eta_f\left(\zeta_f e_n - \left(2\hat{\eta}/\eta_f - 1\right)e_t\right)$ causing the variation. Although two orthogonal walk-away lines are sufficient to determine whether an effect is present, three lines are the minimum mathematical requirement for individual resolution of the fracture response. Rüger (1998) suggested fitting an ellipse to the measured azimuthal variation, in a strategy analogous to the NMO analysis of Chapter 7. In practice, a robust measure can also be obtained by a partial average of the ellipse parameters fitted at each offset, or fitting of an ellipse to the weighted partial stack.

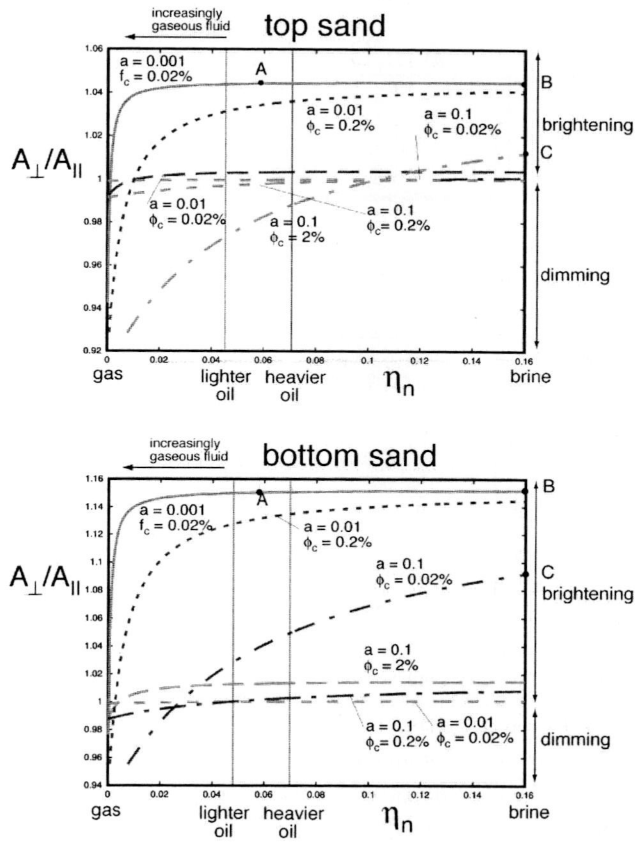

Figure 8.29. Ratio of reflected P-P amplitudes perpendicular and parallel to the crack strike for continuous variation of the fill ($\eta_n = \rho_{liq}V^2_{liq}/\rho V^2_P$), and different aspect ratios (a) and porosities (ϕ). These predictions are made using the first-order gradient terms. Solid circles correspond to synthetic seismogram examples in Figure 8.30. Adapted from MacBeth (1999b).

A further possibility is given by Mallick et al. (1996), who considered three arbitrary lines with reflected amplitudes A_1, A_2 and A_3 at a common offset. This produces an estimate of the fracture azimuth

$$\hat{\phi} = \frac{1}{2}\tan^{-1}\left(\frac{(A_1 - A_3)\sin^2\phi_{12} - (A_1 - A_2)\sin^2\phi_{23}}{(A_1 - A_2)\sin\phi_{23}\cos\phi_{23} - (A_1 - A_3)\sin\phi_{12}\cos\phi_{12}}\right) \pm n\pi \ , \tag{8.54}$$

where ϕ_{ij} is the angle between lines i and j. The formula is most accurate when the lines are 45° apart, for which

$$\hat{\phi} = \frac{1}{2}\tan^{-1}\left(\frac{A_1 + A_3 - 2A_2}{A_1 - A_3}\right) \pm n\pi \ . \tag{8.55}$$

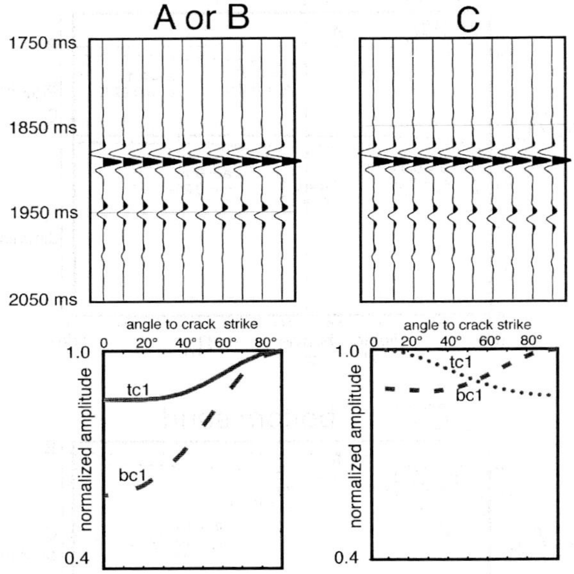

Figure 8.30. Simulated azimuthal gathers for a 150 m thick cracked chalk reservoir. Averaged amplitudes for top (tc) and base (bc) events are shown. The cracks are brine- or oil filled (A or B in Figure 8.29) with an aspect ratio of 0.001, or brine filled with an aspect ratio of 0.1 (C in Figure 8.29). Seismograms are normalized with respect to the largest amplitude for the top event (MacBeth 1999b).

If the numerator and denominator are both zero, then there is isotropy. If more than three shot lines are present or a walkaround VSP is used, equation (8.51) can be fitted directly to any offset. Alternatively, arbitrary triplets of azimuths can be selected, from which the fracture strike can be determined. By selecting all independent n–1 triplet combinations, a matrix equation can be set up from which a least-squares solution can be determined.

8.4.2.3 Fractured TIV – orthorhombic anisotropy

For TIV overburden and a fractured TIV reservoir, Teng and Mavko (1996), Zillmer et al. (1997, 1998) and Rüger (1997) have given expressions. Rüger's generalization of Corrigan (1990) gives

$$R_P^{ORTH} = R_{0P} + \left(G_P + \frac{1}{2}\delta + G_P^O(\phi)\right)\sin^2\theta_P + \left(H_P + \frac{1}{2}\Delta\varepsilon + H_P^O(\phi)\right) , \tag{8.56}$$

where

$$G_P^O(\phi) = G_P^h(\phi) - \frac{1}{2}\Delta\delta \sin^2\phi \ . \tag{8.57}$$

The azimuthal variation is given by similar terms as in the TIH case and is thus amenable to the same techniques. A shale caprock does not obscure the azimuthal variations of a fractured reservoir.

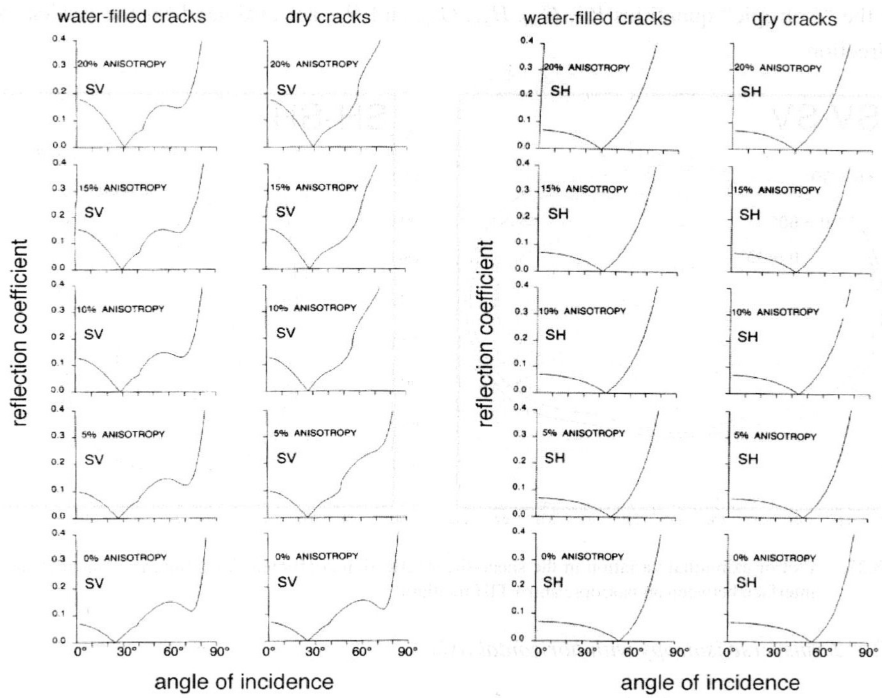

Figure 8.31. (a) SV-SV reflection coefficients for an isotropic/TIH interface. Wave propagation is in the plane normal to the fracturing.
(b) as (a), but for SH-SH reflections.
The most significant effect is in the SV-SV response, which correlates with the anisotropy. Fill appears to be of secondary importance. The general conclusion is that S-S AVO is not a sensitive indicator of the anisotropy, and only normal incidence reflectivity need be used.
Extracted from Yardley et al. (1991).

8.4.3 S-S response in the presence of anisotropy

8.4.3.1 Transverse isotropy with vertical axis

The S-wave AVO response was given by Thomsen (1993) and then later by Rüger (1996). The results are for a shear-wave polarized perpendicular to the vertical (symmetry) plane (the $S_\perp$ mode defined in Chapter 3) or parallel (the $S_\parallel$ mode). As in the isotropic case, the $S_\perp$ mode gives rise to only transmitted and reflected $S_\perp$ waves, and the $S_\parallel$ mode to additional qP contributions. Thus,

$$R_{\parallel}^{TIV} = R_{0S} + \left(G_{SH} + \frac{\Delta\gamma}{2} \right) \sin^2\theta_S + \left(H_{SH} + \frac{\Delta\gamma}{2} \right) \tan^2\theta_S \sin^2\theta_S \ , \tag{8.59}$$

and

$$R_{\perp}^{TIV} = R_{0S} + \left(G_{SH} + \frac{(\Delta\varepsilon - \Delta\delta)}{2\hat{\eta}} \right) \sin^2\theta_S + H_{SV} \tan^2\theta_S \sin^2\theta_S \ , \tag{8.60}$$

where the "isotropic" quantities R_{0S}, G_{SH}, H_{SH}, G_{SV}, and H_{SH} are evaluated for the vertical propagation direction.

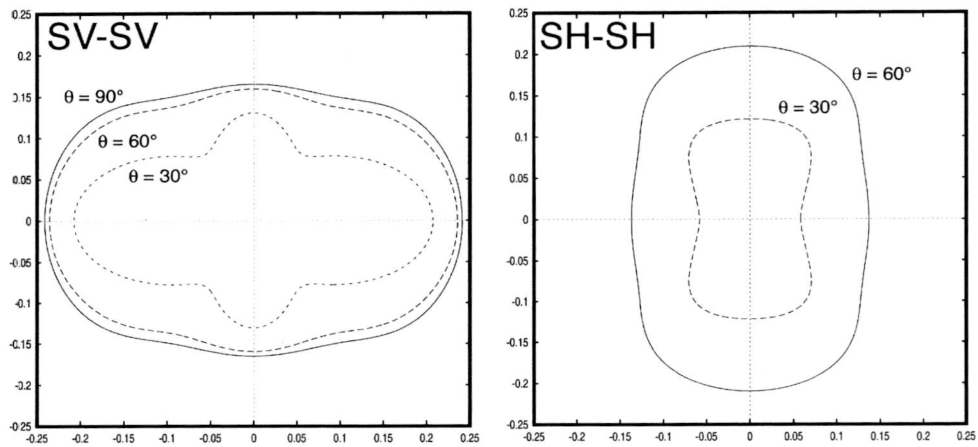

Figure 8.32. Plot of azimuthal variation in the shear-shear reflection coefficient for different incidence angles at the interface between an isotropic and a TIH medium.

8.4.3.2 *Transverse isotropy with horizontal axis*

Rüger (1996) discusses for four cases, viz., the two shear-wave polarizations parallel and perpendicular to the fracture strike. For a line along the strike direction, propagation is in a plane of isotropy, so that the relations (8.36) and (8.39) can be used to give

$$R_{\parallel}^{TIH}(\text{strike}) = R_{0S} + G_{SV} \sin^2\theta_S + H_{SV} \tan^2\theta_S \sin^2\theta_S \ , \tag{8.61}$$

and for waves polarized perpendicular to the propagation plane

$$R_{\perp}^{TIH}(\text{strike}) = R_{0S} + G_{SH} \sin^2\theta_S + H_{SH} \tan^2\theta_S \sin^2\theta_S \ , \tag{8.62}$$

where again G_{SH} and H_{SH} are evaluated with vertical velocities.

For a line perpendicular to the strike direction, propagation is in a vertical symmetry plane, and waves polarized perpendicular to this plane yield

$$R_{\parallel}^{TIH}(\text{normal}) = R_{0S} + \left(G_{SH} - \frac{e_t}{4}\right)\sin^2\theta_S + \left(G_{SH} - \frac{e_t}{4}\right)\tan^2\theta_S \sin^2\theta_S \ , \tag{8.63}$$

and for waves polarized parallel to this plane

$$R_{\perp}^{TIH}(\text{normal}) = \left(R_{0S} + \frac{e_t}{4}\right) + \left(G_{SV} + \eta_f \hat{\eta}\left((2-3\eta_f)e_n + e_t\right) - \frac{7}{4}e_t\right)\sin^2\theta_S + \left(H_{SV} - \frac{e_t}{4}\right)\tan^2\theta_S \sin^2\theta_S . \tag{8.64}$$

Numerical computations of these coefficients are displayed in Figure 8.31, and their azimuthal behavior is illustrated in Figure 8.32. The magnitude of the anisotropy is a primary controlling factor for these coefficients, and importantly there are major azimuthal variations also at normal incidence, for which the effects described in Chapter 5 are relevant. Thus, comparison of normal incidence reflectivity for SV and SH waves is sufficient to determine the fracture index e_t of the target zone. Indead, the most significant AVO response is mainly confined to the post-critical regime.

8.4.4 P-S response in the presence of anisotropy

8.4.4.1 *Transverse isotropy with vertical axis*

Conversion can take place only between P- and SV waves, as every vertical plane is a symmetry plane. Li, Kühnel and MacBeth (1996) give

$$R_{PSV} = -\frac{\sin\theta_P}{\cos\theta_S}\left(\left(R_{0PS} + R_0^V\right) + \left(G_1 + G_1^V\right)\cos\theta_P\cos\theta_S + \left(G_2 + G_2^V\right)\sin^2\theta_P\right) , \tag{8.65}$$

where

$$R_0^V = \frac{\Delta\delta}{2(1-\hat{\eta})} , \tag{8.66}$$

$$G_1^V = \frac{-\sqrt{\hat{\eta}}}{2(1-\hat{\eta})}\Delta\delta - 2\hat{\eta}\Delta\gamma , \tag{8.67}$$

and

$$G_2^V = \frac{\Delta\delta}{2} - \frac{1}{2(1-\hat{\eta})}(3\Delta\delta - 2\Delta\varepsilon) + 2\hat{\eta}\Delta\gamma . \tag{8.68}$$

8.4.4.2 *Transverse isotropy with horizontal axis*

For propagation in the vertical plane along the fracture strike, the results are the same as in the isotropic case. For propagation in a vertical plane perpendicular to the fracture strike

$$R_{PSV} = -\frac{\sin\theta_P}{\cos\theta_S}\left(\left(R_{0PS} + R_0^H\right) + \left(G_1 + G_1^H\right)\cos\theta_P\cos\theta_S + \left(G_2 + G_2^H\right)\sin^2\theta_P\right) , \tag{8.69}$$

where

$$R_0^H = \frac{-2\eta_f(\zeta_f e_n + e_t)}{1-\hat{\eta}} , \tag{8.70}$$

$$G_1^H = 2\hat{\eta}e_t + \frac{\sqrt{\hat{\eta}}}{1-\hat{\eta}}\eta_f(\zeta_f e_n + e_t) , \tag{8.71}$$

and

$$G_2^H = -2\hat{\eta}e_t - \eta_f(\zeta_f e_n - e_t) + \frac{\eta_f}{1-\eta_f}(4(2-3\eta_f)e_n + 3e_t) . \tag{8.72}$$

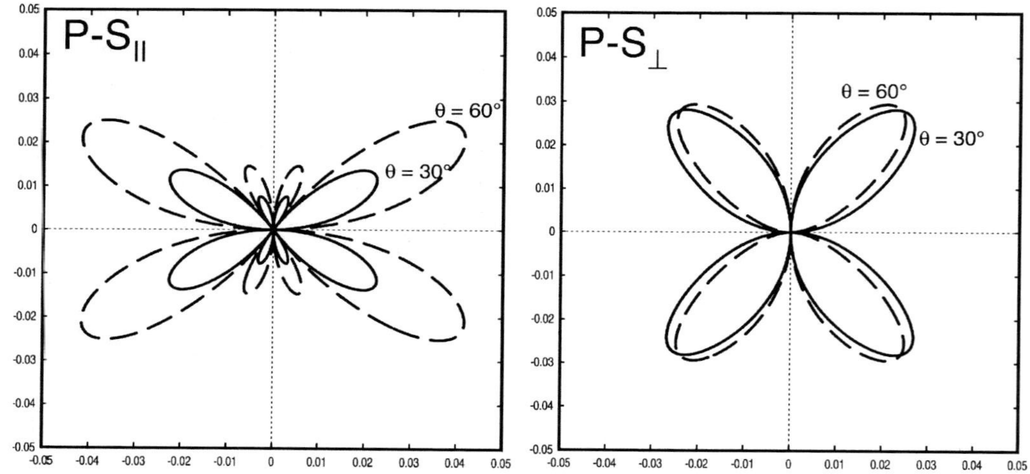

Figure 8.33. Converted-wave reflection coefficients for an isotropic/TIH interface and different incidence angles. There is a strong and distinctive azimuthal variation.

Higher order terms have been excluded from these expressions. The azimuthal variation of the P-S$_{\perp}$ and P-S$_{\parallel}$ coefficients are shown in Figure 8.33. They show four distinctive main lobes but also much minor complexity in the P-S$_{\parallel}$ case. The behavior is variable and has the potential for providing fracture details. The response is generally brighter than the P-P AVO as the fracture indices appear as a sum in the coefficients. At present interpretation of these patterns remains an active area of research.

CHAPTER 9 – THE ROAD AHEAD

Central to this book is the description of small-scale lithological constituents by an anisotropic replacement medium, with a primary focus on those that create transverse isotropy – the only type of anisotropy required in most practical applications. Transverse isotropy exists in two different flavors that depend upon the cause of the anisotropy, each with its own distinctive set of descriptive parameters. Shales and sedimentary sequences give a TIV anisotropy, distinguished by azimuthally independent but offset dependent seismic signatures. Vertical cracks and fractures give TIH anisotropy, with both azimuthally and offset dependent seismic signatures. The difference in these signatures has shaped the development of the two separate streams of VSP method described in this book. Further sub-division has led to a healthy selection of possible approaches, with the choice largely depending upon what acquisition design has been afforded. Indications are that financial restrictions could change, with research and marketing literature drawing attention to a new wave of developments in VSP acquisition capability. In the near future this will lead to a dramatic enhancement in what is currently possible with seismic anisotropy analysis.

In the context of these encouraging advances, there are two research areas anticipated to impact directly on the VSP methods of the future: the first relates to improved definition of the anisotropy parameters and their ability to predict lithology more precisely, the second recognizes the advantages of using seismic anisotropy to monitor temporal changes of a reservoir.

In the hydrocarbon industry, VSP is rarely acquired with anisotropy-related lithology description as a primary focus. Indeed, it is common for anisotropy to be quoted as a secondary, but more usually a tertiary, objective in survey specifications – supplementing existing arguments for improving the reservoir image. Thus, walk-away or rig/normal incidence VSPs are shot regularly to establish a velocity model. The notable exception in this practice is the multi-component near-offset VSP described in Chapters 4, 5, and 6, which can only be shot for fracture analysis. As a consequence, most seismic anisotropy analyses are performed using acquisition geometries designed for a different purpose, with a non-optimal azimuth and offset distribution. Within these current restrictions, the techniques described in this book provide options to cover almost every possibility in both offshore and onshore surveys. Figure 9.1 helps select the best method(s) to use, given a particular acquisition and objective for the seismic anisotropy analysis.

9.1 FIXED NEAR-OFFSET VSP

Onshore, multi-component VSP is the prime choice, offering highest fidelity and definition for crack/fracture analysis. The technique works by detecting the presence of fractures through the descriptive fracture parameter e_t. This parameter is the product of the fracture porosity and slip compliance. With this approach, not only can a carbonate reservoir be recognized as fractured, but the fracture intensity evaluated and orientation determined. Measurements are possible over a large depth range to provide calibration and control. The survey is particularly attractive as it provides a direct and straightforward interpretation of the fracturing.

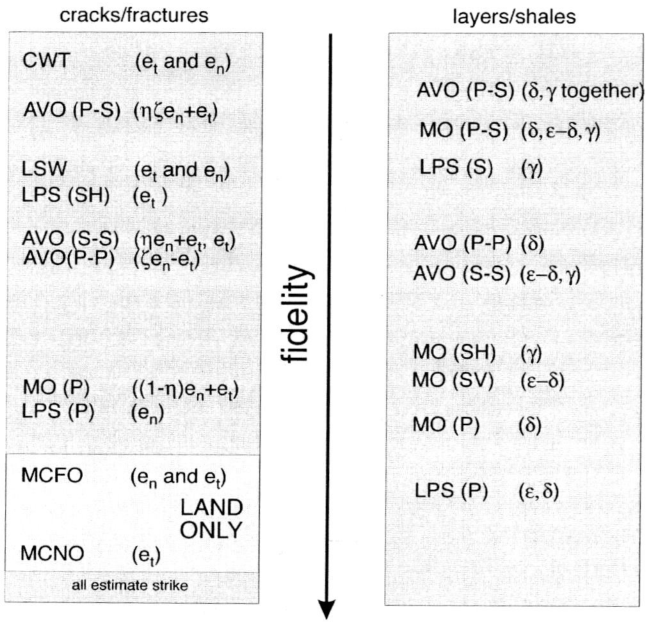

Figure 9.1. Qualitative assessment of the techniques described in this book, ranked according to the anticipated clarity or fidelity of the results using current technology. The anisotropy parameters they estimate are shown in brackets.
MCNO – multi-component (both source and receiver) near-offset;
MCFO – multi-component far offset;
LPS – local phase-slowness method;
AVO – amplitude-versus-offset method;
MO – moveout;
CWT – converted-wave techniques for overburden evaluation;
LSW – local shear wave investigation for target anisotropy.
LPS, AVO and MO use several lines of walk-away VSP, MO can also use the combination of rig- and normal incidence VSP. Only a single line of walk-away VSP is necessary for TIV estimation. Changes in technology adding additional depth aperture and shot coverage can be expected to substantially change this profile.

The VSP is cost-effective when shot as part of a larger surface seismic survey with horizontal vibrators (ARIS or OMNIPULSE sources are less common). The VSP survey normally takes between 40 and 60 hours to complete, which can cause difficulties with rig time. This can limit the VSP survey to six components or a narrow depth range (but see below). To combat this, an interesting solution was offered by Edelmann (1992) using the CIPHER technique. This involves two horizontal vibrators with orthogonal orientations activated simultaneously, but with a phase shift of 90°. This creates a circularly polarized shear wave that rotates and samples all azimuths equally (Figure 9.2). An alternative, but in VSP still untried, method is the slip-sweep method (Rozemond 1996) where the vibrator sweeps are overlapped, one starting before the other is finished.

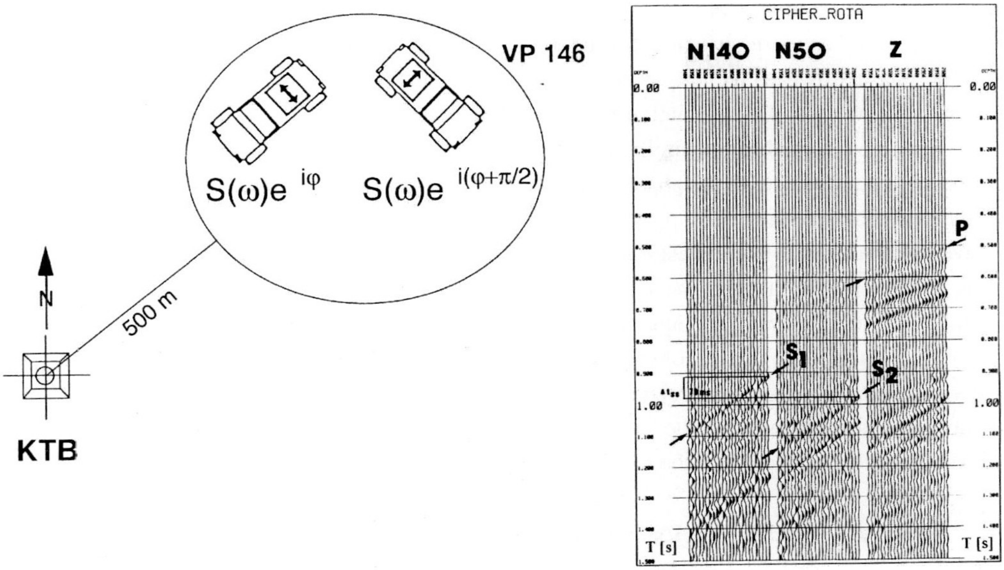

Figure 9.2. An interesting method for saving rig time is to use a two-vibrator arrangement, in which both are controlled by the same sweep but with a phase difference of 90°. This creates a circularly polarized shear wave that rotates through all polarizations – after Gut et al. (1994). The three-component trace set shows the results after receiver rotation to separate the two shear waves.

9.2 FIXED FAR-OFFSET VSP

Multi-component sources at near and far offset can resolve the additional fracture parameter e_n, providing a link to fracture fill and/or geometry. Several fixed far-offsets can also be used if non-vertical fractures are suspected: a natural extension of this design is a full walkaround – fixed offset, varying azimuth. If multi-component sources are not available or are impractical, then converted waves using pseudo-rotation analysis with orthogonal fixed source offsets provides a good alternative.

9.3 WALK-AWAY VSP

Walk-away VSP is the most flexible acquisition configuration for anisotropy analysis, and is common to a large proportion of the methods in this book. In practice, it has been most commonly used to evaluate anisotropy in shales/sedimentary layers for a specific target interval pinpointed by the receiver array. Offshore or onshore this is the natural choice, since one survey line is sufficient to determine the Thomsen parameters ε and δ from the direct P-waves using the local phase-slowness method. At sea, this acquisition need take only a few hours of boat time for 300 shot points. The P-wave moveout can supply supplementary information on δ, with non-hyperbolic moveout then yielding the anellipticity ε–δ, while P-P AVO gives overlapping information. In principle, the remaining Thomsen parameter γ can be obtained from direct or converted shear waves, using the theory described in this book.

The phase slowness technique is best applied to data from an array positioned in a vertical portion of the well, and the deviation part should be avoided unless location is known very accurately. For fracture analysis, if multi-component sources are not available or cannot be used, the next best option is to extend the phase slowness approach. Using direct P-waves, e_n can be obtained together with the fracture strike, provided several shot lines at different azimuths are acquired. The P-wave moveout and P-P AVO estimate the magnitude of different combinations of e_n and e_t. Processing is aided by the fact that the azimuthal behavior of both attributes can be fitted by an ellipse. The local phase slowness technique also offers the theoretical possibility of determining e_t from converted S waves. The use of converted S waves or S reflections places added emphasis on wavefield separation, which in turn relates to the depth aperture. As an alternative, evaluation of the fracture strike using converted shear waves can be obtained from a single shot line provided it does not pass through the location of the receiver array, giving a range of offset and azimuth.

9.4 RIG- AND NORMAL INCIDENCE VSP

The relatively inexpensive combination of rig and normal incidence boat source is often shot in a deviated well. This provides an ideal scenario to obtain δ using the P-wave moveout.

9.5 THE FUTURE IMPACT OF RECENT ADVANCES IN BOREHOLE SENSORS

A critical element in this equation is the rig time and hence loss of production involved with the VSP. Application of borehole seismic techniques has been restricted to the geometries shown above due to the relatively high costs of data acquisition. Quite recently, this situation has changed by an explosion of new downhole receiver technology. This brings access to larger-scale geometries for the first time. These technology advances have culminated in a dramatic increase in the number of 3C receiver levels that can be suspended on a tool string. Indeed, receiver array size has improved from the five-level array used in the 1990s, to six levels, eight levels (1995), twelve levels (1998, and currently in widespread operational use), 16 levels, 32 levels (active) to multi-level (passive) arrays with 80 or even 200 levels of clamped receivers operating at 3 to 15 m intervals (Figure 9.3). The latest can be deployed on tubing and in either vertical or horizontal wells. There is even the possibility of using a well tractor to help pull or push the tools in high angle wells, where current operations dictate costly drill pipe or coiled tubing. The local phase-slowness technique benefits enormously from this advance, although the impact on the multi-component near-offset VSP is reduced by the cost of the source. Overall, these provide a major enhancement for anisotropy analysis, heralding a new era of enhanced quality.

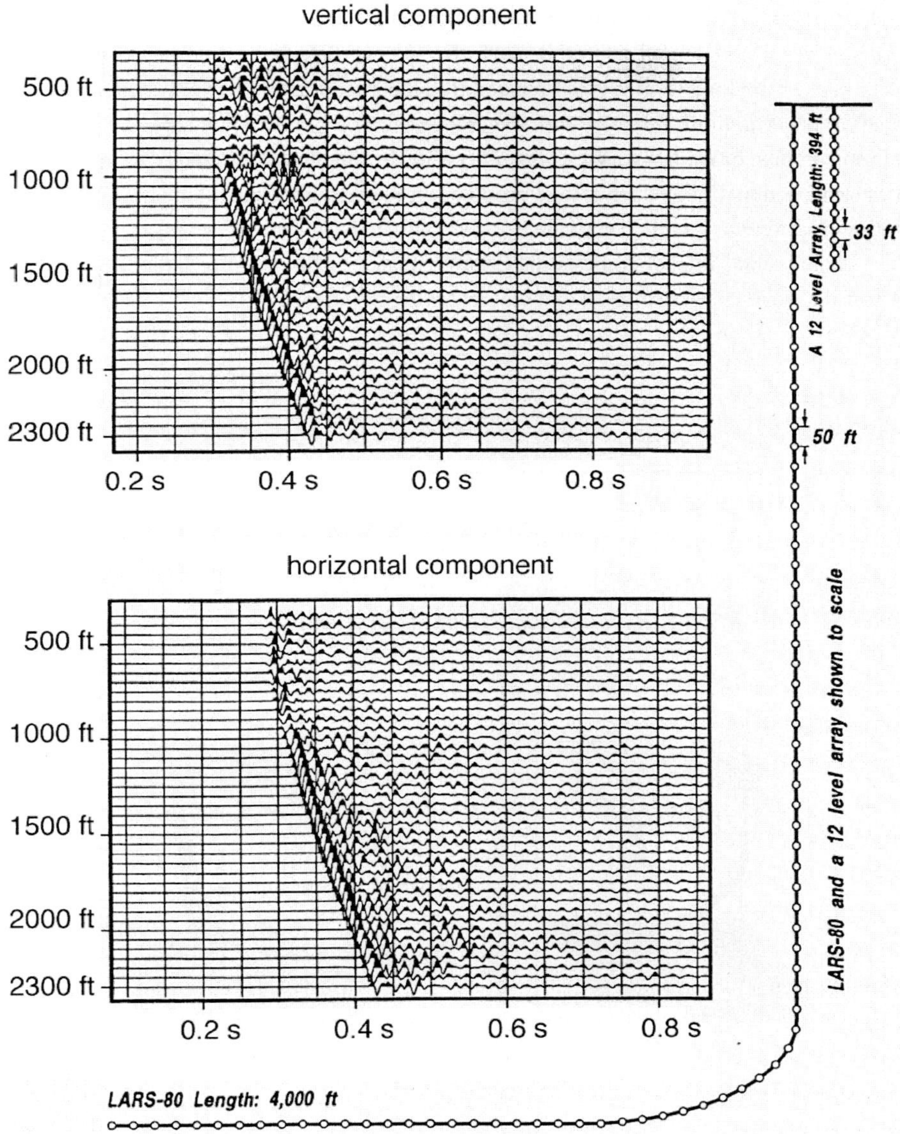

Figure 9.3. Downhole technology has now developed to the state that it can place 80–200 three-component receiver levels downhole. The schematic highlights this dramatic change in perspective. The particular tool has 80 levels and is from P/GSI (1998). The example trace set is taken from a smaller 40-level sub-array employed on tubing.

For studies of seismic anisotropy the advantages are

(a) large aperture and good spatial sampling means a more effective wavefield separation (Omnes and Clough 1998), thus opening the door to more efficient processing of converted shear waves and accurate utilization of the converted wave techniques than is currently possible;

(b) a depth range of anisotropy values using the local phase-slowness method;

(c) fewer source repeatability problems – for both walk-away and near-offset data. For example, when a multi-component source is used, the near-surface correction procedure will be more accurate. In the long term it is anticipated that this could help the multi-component near-offset VSP to become more flexible, with the source being replaced by cheaper alternatives such as airguns in pits, or explosive located near to walls or trenches, or a shock gun.

(d) in marine acquisition, a drop in cost provides a greater source spread in a full 3D-VSP configuration (Figure 9.4), and hence a rich azimuth coverage for fracture analysis.

(e) The fundamental impact consists in better estimates of the fracture indices e_n and e_t, and the Thomsen parameters ε, δ and γ over a greater depth range.

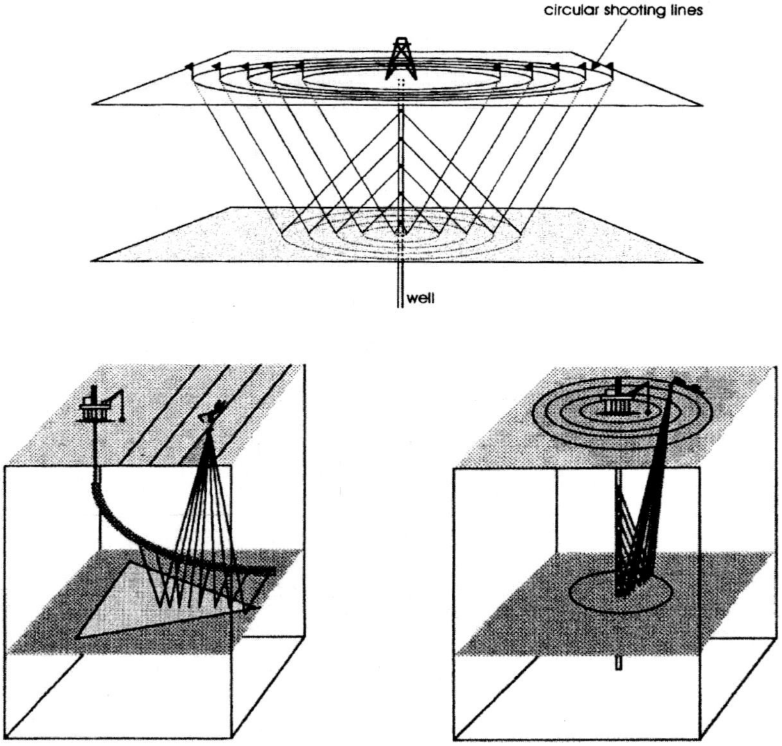

Figure 9.4. It is now possible to shoot 3D-VSP over a wide depth and shot aperture, providing excellent azimuth and offset coverage for anisotropy analysis. For a vertical well, a circular shoot is usually preferred because it reduces acquisition time as the boat does not have to turn. The radius of the circle decreases while the tool is maintained in a fixed position (lower right). An interesting survey was recently acquired in which the tool depth was decreased as the radius was increased, thus ensuring a constant incidence angle and no AVO effect (upper diagram) (Omnes and Clough 1998). For a deviated well, usually a set of parallel lines are shot. The tool string need not be moved with the current generation of downhole tools, thus decreasing the previous survey times substantially. It can take 50 to 80 hours to shoot 6000 to 8000 shot points (Reid Well services news brochure, 1995). Routine use of 3D VSP will considerably enhance seismic anisotropy analysis.

9.6 INTERPRETATION OF THE ANISOTROPY PARAMETERS

More robust and reliable anisotropy measurements contribute significantly to building a finer sub-division of lithology classification on the basis of the fracture indices e_n and e_t and the Thomsen parameters ε, δ and γ. For this to be effective, fundamental research must play a role in ensuring that the link with the underlying lithological constituents is correctly established. For example, the relative magnitudes of e_n and e_t are known to vary with the content of the cracks/fractures, depending upon whether they are thin or fat cracks, rough fracture surfaces, partially welded or smooth fracture planes. Only by specifying the particular model can an accurate statement about the contents be made from seismic data alone. No satisfactory scheme has yet emerged to resolve this ambiguity. Research must help to calibrate the relationship between e_n and e_t, perhaps through more extensive laboratory experiments such as those of Hsu and Schoenberg (1993) and Pyrak-Nolte et al. (1990). Moreover, recent research has shown that the predictive capability of current models can be improved by including the fluid flow process. Interconnections between fractures, cracks and pores cause a flow of fluids that tends to weaken the material, and hence increase its compliance in a specific direction. These new advances are from two basic directions:

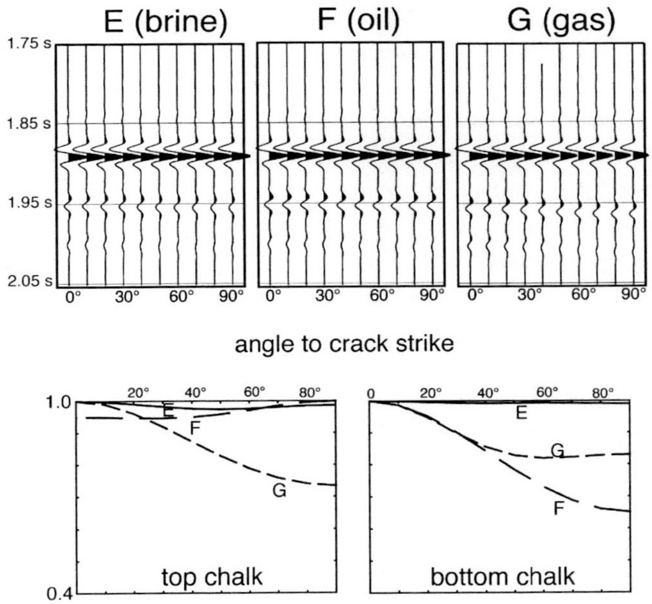

Figure 9.5. Recent theory predicts that attenuation caused by fluid flow in cracks or fractures can lead to different fluid fills being distinguished. Here, the anisotropic reflection response for the base reservoir event is highlighted using synthetic azimuthal gathers. The reservoir is chalk, with the cracks with an aspect ratio of 0.01, and filled with 100% brine (E), oil (F) or gas (G). Averaged amplitudes are shown for each event. There is little discrimination for the top chalk reflector (MacBeth 1999b).

(a) Fluid flow affects e_n but not e_t, because shear stress does not provide a change in volume, while normal stress does. Fluid flow processes such as squirt flow (Mavko and Nur, 1979), flow between cracks, differently oriented crack sets (Hudson et al. 1996a) or cracks and pores

(Thomsen 1995), cause e_n to become complex and lead to flow-related attenuation (MacBeth 1999b). This significantly changes the P-P AVO response, thus there is the possibility that oil and brine fill can actually be distinguished (Figure 9.5). This encouraging result contradicts the early Hudson theory described in this book and requires further research.

(b) Another area open to scrutiny is Hudson's prediction that e_t remains constant with fluid type. Indeed, Potters et al. (1999) has presented field evidence showing correlation of anisotropy highs with a known gas cap. Van der Kolk (1999) has introduced a new model to explain this particular observation as being due to the viscous element in the squirt flow between crack/fracture sets. Interestingly, this also causes the fractured medium to become dispersive and attenuating.

For shales, a finer classification on the basis of the relative magnitudes in the three-parameter Thomsen system is expected to provide the ability to distinguish conditions of maturity, overpressure, compaction and contributions from bedding-parallel cracks (Vernik and Liu 1996). For this, more empirical observation and more integrated studies are needed. Steps have been made towards this goal already. It has been established that ε and γ are usually positive, while δ can be positive, negative, or very small. Indeed, for a layered sequence, δ is close to zero (and exactly zero when α/β is constant throughout the stack). Most shales are positively anelliptic, meaning that ε–δ is positive. Bedding-parallel fractures can contribute to a negative value of anellipticity. Areas for further development are:

(a) Sedimentary sequences are still difficult to assess, and much research remains before this category of anisotropy is properly understood. A key factor in common with shales appears to be the degree of α/β fluctuation. For layered sequences, the impact of amplitude loss at seismic frequencies due to frequency-selective multiple scattering must be considered. This has been shown to be significant by O'Doherty and Anstey (1971), and progress is being made towards understanding the resultant frequency dependent anisotropy (Werner and Shapiro 1996). Understanding the interplay between attenuation, dispersion, and anisotropy remains a subject of ongoing research.

(b) If the long-wavelength anisotropy is known to arise from sedimentary sequences consisting of two interbedded facies, the individual constituents can be resolved by inverting Backus theory (Schoenberg 1994). For layers that are not exactly the same, but statistically distributed about some mean, future research must determine how well these individual facies elements can be resolved given their spread in elastic properties (Schoenberg, personal communication).

9.7 TIME-LAPSE VSP

One major future direction for the application of seismic anisotropy is time-lapse surveying. Many time-lapse studies currently concentrate on interpreting reflectivity, assuming the reservoir changes are manifested as changes of impedance. Such effects are confined to rocks with susceptible behavior. They generally work well for sandstones, but less well for carbonates. Seismic anisotropy is sensitive to cracks, fractures, thin pores and defects in the rock that possess only a small porosity but contribute to anisotropy significantly because they are aligned. These elements are highly compliant and thus likely to register a major change due to an applied stress (Nur and Simmons 1969). If these cracks or fractures change their geometry, orientation distribution, or contents (thus affecting fracture indices e_n and e_t), it is expected that this change of response can be detected and monitored using VSP techniques sensitive to the anisotropic signature. To this end, Zatsepin and Crampin (1997) have extended crack models to include evolution under tri-axial stress combined with fluid communication between the cracks and pores (Figure 9.6). This technique is particularly relevant to the stiff carbonate rocks, where the Gassmann equations predict small changes of P-wave impedance.

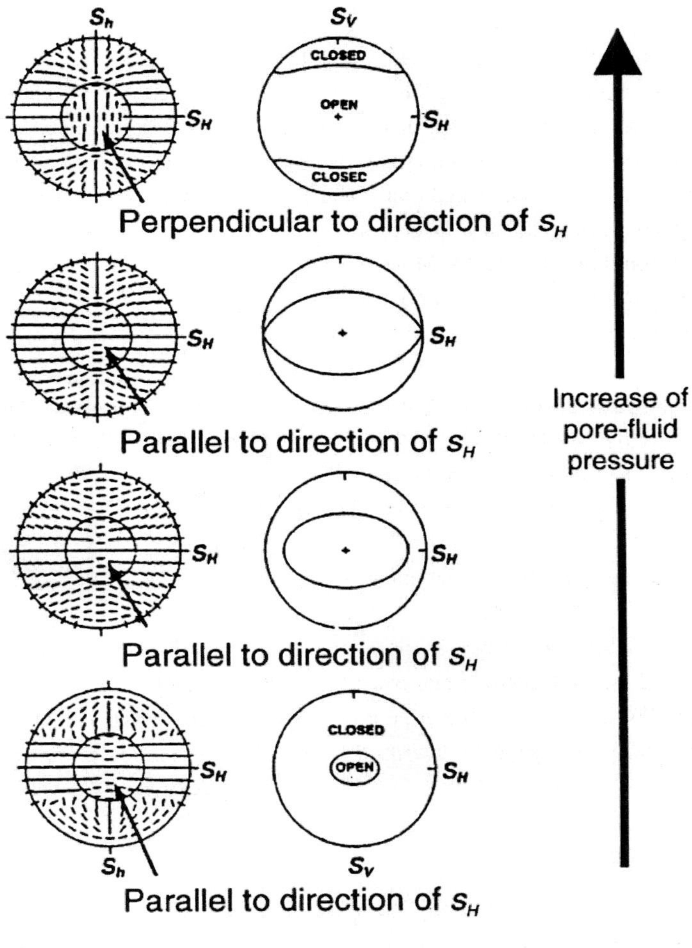

Figure 9.6. The theory of Zatsepin and Crampin (1997) predicts changes in the shear wave behavior for cracked media subjected to changing pore-fluid pressure. In this example, the particular stress conditions ensure that the fastest shear wave polarization in these stereographic projections points towards the maximum horizontal stress (SH) at low pressures. When the pore pressure is equal to the maximum horizontal stress, the "fast" polarization changes its orientation by 90°. The diagrams on the right hand column illustrate how the orientation distribution of open and closed cracks changes under these conditions.

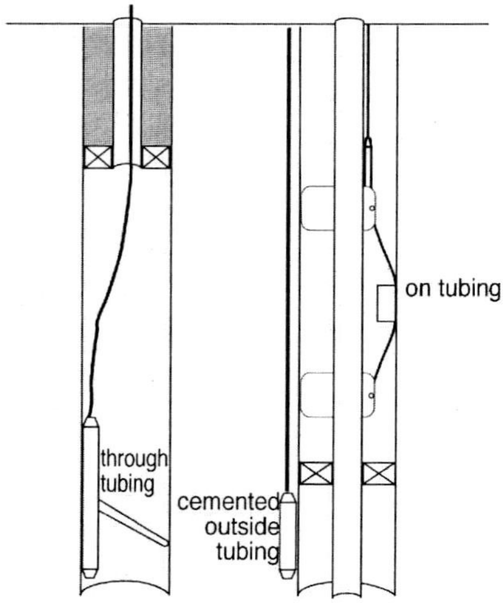

Figure 9.7. Advances in downhole technology have produced a range of instruments that can be used to record either as permanent installations or on wireline, with the production tubing still in place. This opens the doors to time-lapse VSP. A slimline tool can be guided through the tubing, flushed down by the mud; an annular shaped sonde fits over the tubing; a geophone can be bolted or welded on the outside of the casing for permanent installation (Bathellier and Czernichow 1997).

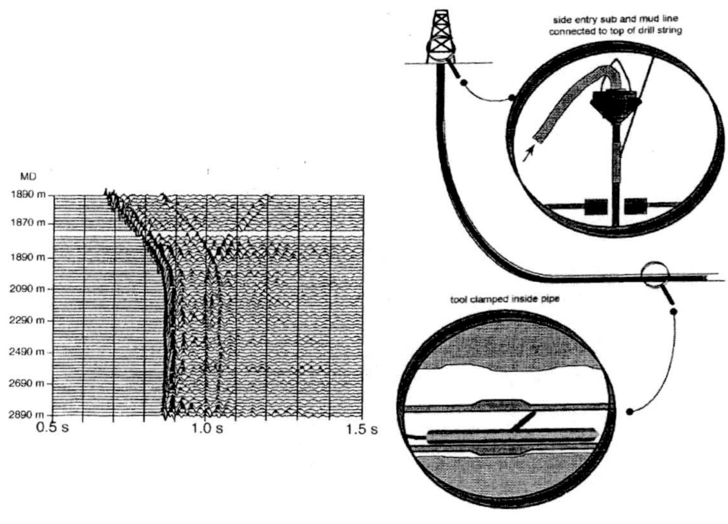

Figure 9.8. A slimline tool can be used in highly deviated or horizontal wells without the need for being pipe conveyed or using well tractors. With the drill string already in the well, the VSP tool is inserted inside it. Example shown after Randall (1998).

In practice, the value of time-lapse VSP has only recently been recognized owing to the removal of restrictions that have prevented re-shooting in production wells. Technology advances have provided permanent sensors that can be placed outside the casing, between the casing and the tubing, on the tubing, and finally an ultra "slimline" version that can be fed inside the tubing and below the packer (Figures 9.7 and 9.8). Indeed, sondes designed for permanent or wireline use can be installed at each step of the well completion process. Permanent sensors have the additional benefit of eliminating position-related variability between surveys. Examples of time-lapse VSPs to date include monitoring of a gas storage reservoir in the Paris Basin (Deflandre et al. 1995); a nine-component surface spread shot in the Cymric field, California, to monitor heavy oil steam injection (Figure 9.9) (Winterstein et al. 1998), and another nine-component survey with a 3D source spread to monitor CO_2 injection in the Vacuum field, New Mexico. In the latter area, time-lapse changes have already been observed in the surface data (Roche et al. 1997).

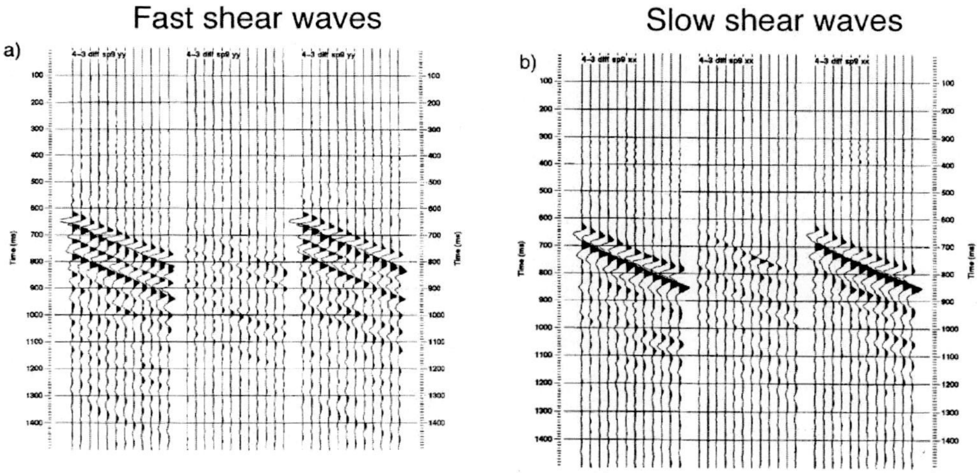

Figure 9.9. An example of using a 12-level array in a multi-component experiment involving an ARIS impact source. This was part of an acquisition of 31 shot points around three wells which lasted 60 hours. The three-component traces are shown before (left) and after (right) steam injection in this time-lapse experiment, with the difference before and after in the middle. Time-delay differences of up to 6 ms were observed, together with a small amplitude change.
(a) slow shear waves
(b) fast shear waves.
After Winterstein et al. (1998).

In conclusion, the many major technological and theoretical advances made over the last few years indicate an accelerated growth in the research areas above, with good case studies becoming available over the next few years. This enhanced definition ensures a long and prosperous future for the VSP method as a tool for lithology description in bridging the logging scale to surface seismic scale.

APPENDIX – SHEAR-WAVE BIREFRINGENCE ANALYSIS

Ideally, birefringence analysis should always be performed using multi-component data, that is three- component receivers in combination with orthogonal source motions. The processing steps described in Chapters 4, 5 and 6 then provide the recipe for extracting a high-resolution birefringence profile, with potential for lithological correlation and interpretation. However, in the early days of anisotropy analysis, multi-component sources were not available, and much of the analysis effort concentrated on the three-component receivers. With the greater uncertainty involved, solutions were more case-specific with a wide variety of algorithms being developed. Today, despite many multi-component advances, three-component receiver analysis remains an active research topic due mainly to the greater availability of such datasets, both in land and offshore seismics, and the current trend of using converted shear waves. Analyses of such data focus on interpreting the birefringence signature formed from the projection of the two shear wave arrivals onto the individual receiver components.

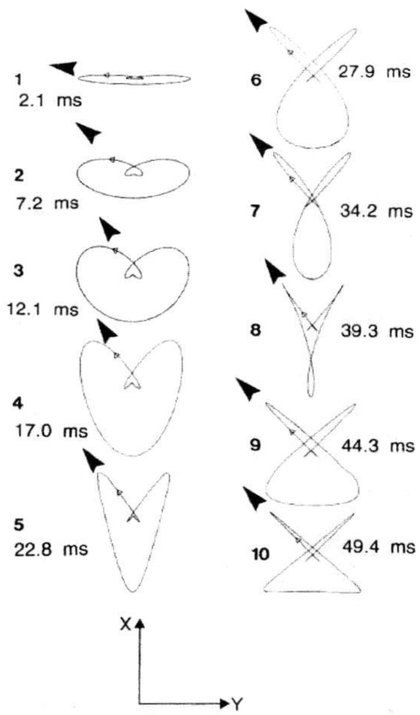

Figure A.1. A sequence of progressively split shear waves for a 15 Hz minimum phase wavelet. The directions of initial onset are marked by the large arrowheads, and the sense of motion is indicated by the open triangles. The lag time for each is shown (MacBeth and Crampin 1991a).

A.1 POLARIZATION DIAGRAMS

Polarization diagrams were popularized by Stuart Crampin and co-workers in the early 1980s, primarily for use on earthquake data. Analysis is performed by direct visual inspection of the particle motion from each set of horizontal recordings. The polarization of the faster shear wave is estimated by assessing the direction of the initial linear onset (Figure A.1). The lag time is determined by identifying the slow shear-wave onset signaled by an abrupt change in the particle motion after the first break. Rotation of the receiver to align with the fast shear wave polarization helps with this assessment (Figure A.2). The minimum lag that can be estimated is between 0.25 *T* (Naville, 1986) and 0.7 *T* (Schulte and Edelmann 1988), where *T* is the dominant signal period.

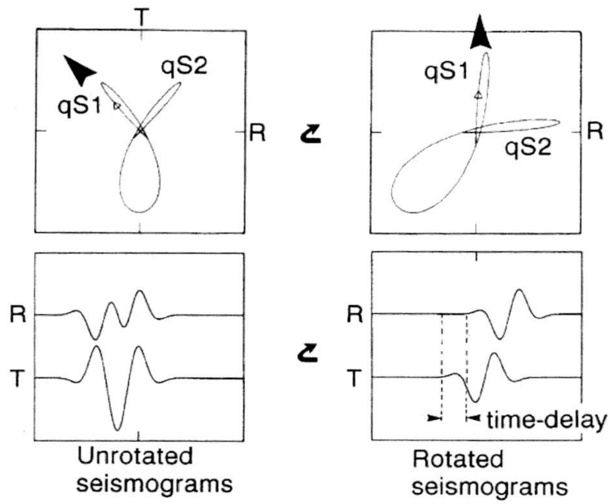

Figure A.2. Polarizations and lag times can be evaluated by visual inspection of polarization diagrams, preferably plotted in the radial and transverse coordinates. The initial onset (large arrowhead) is marked. Rotation into this direction provides an accurate measure of the lag time (MacBeth and Crampin 1991a).

Polarization diagrams are relatively straightforward to interpret in a medium with strong anisotropy, giving rise to a familiar cruciform pattern. For impulsive sources this is a satisfactory approach, but in the presence of interfering effects, changing source wavelet (Figure A.3) or small lags that give elliptical or circular polarization diagrams, there is more emphasis on the subjective side of the approach. It can become difficult to disentangle the effects of birefringence from the other phenomena, although the procedure can be made more objective by assigning a reliability factor to each estimate (Chen, Booth and Crampin 1987). Experience of recognizing shear wave birefringence in a wide range of settings is essential. For commercial VSP applications the subjective and time-consuming nature of the approach has prevented widespread application. However use is still advocated as a quality control check for multi-component processing.

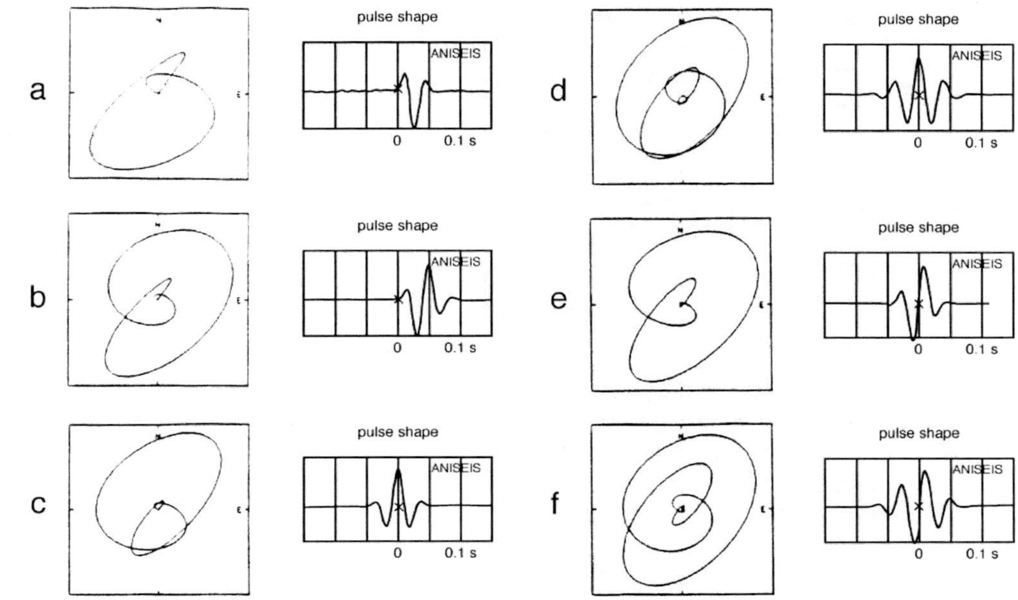

Figure A.3. Polarization diagrams are very sensitive to changes in the source wavelet. This makes consistent interpretation of the birefringence signature difficult. The above example displays sets of horizontal recordings simulated for an identical polarization direction (Campden 1990).

A.2 NEURAL NETWORKS

Dai and MacBeth (1998) applied a neural network to the task of visual interpretation of the polarization diagrams, in an attempt to make the method more objective. The feed-forward network is organized into four layers of nodes. Input nodes are assigned to each sample in the time series within the window of interest, with the output being the polarization angle and lag time. The optimal network structure has 100 nodes in the input layer, followed by layers with 25, 10, and 2 nodes. This network architecture is determined mainly through trial and error, but guided to some degree by the knowledge that generalization is increased and memory reduced by limiting the number of hidden nodes.

An insufficient number of hidden nodes leads to a long learning process or to a lack of convergence in the training stage. Too many hidden nodes introduce noise in the output. Weights are initially adjusted in a process involving training, which presents the network with a set of test input and outputs determined by visual analysis (as in Section A.1) or by calculations of synthetic seismograms. Learning occurs when the network weights are forced to assume certain values by repeated exposure to the same stimulus. Once trained, the network can be applied very quickly. Figures A.4 and A.5 show examples of the results on field data, together with a weight map showing the various interconnections between the layers.

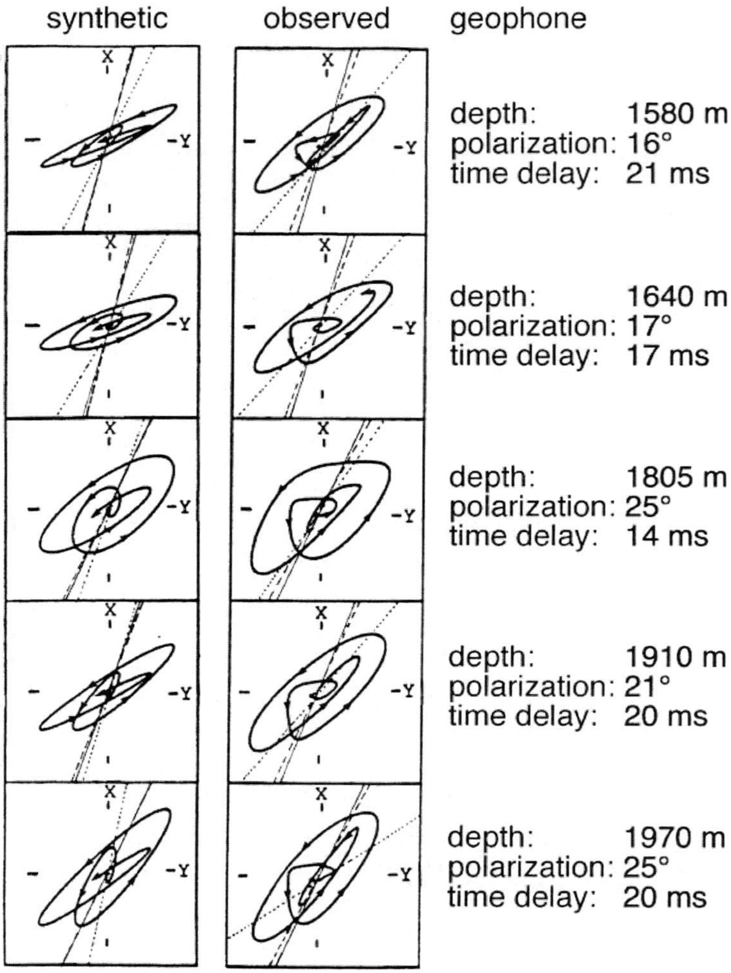

Figure A.4. Example of neural-network output for field data. A selection of five horizontal recordings is shown together with the synthetic data chosen by the network as having the closest fit. Lines on each plot mark the polarization azimuth chosen by the network (dashed line), polarization azimuth chosen by an independent technique (solid line on observed data), actual model polarization (solid line on the synthetic data), and polarization direction chosen by visual examination (dotted line). From Campden (1990)

Although it is not possible to fully understand the logic underlying the network solution from these patterns, they do highlight those aspects of the seismic waveform that are relevant to the analysis. It appears that to a large extent the solution focuses on the interferences lobes. To make the neural network more robust, more training data is required, exploring different wavelets, coherent noise, source direction and offsets. The major drawback with further extensions using this type of network architecture is the increase in training time and a large network structure that is difficult to manage.

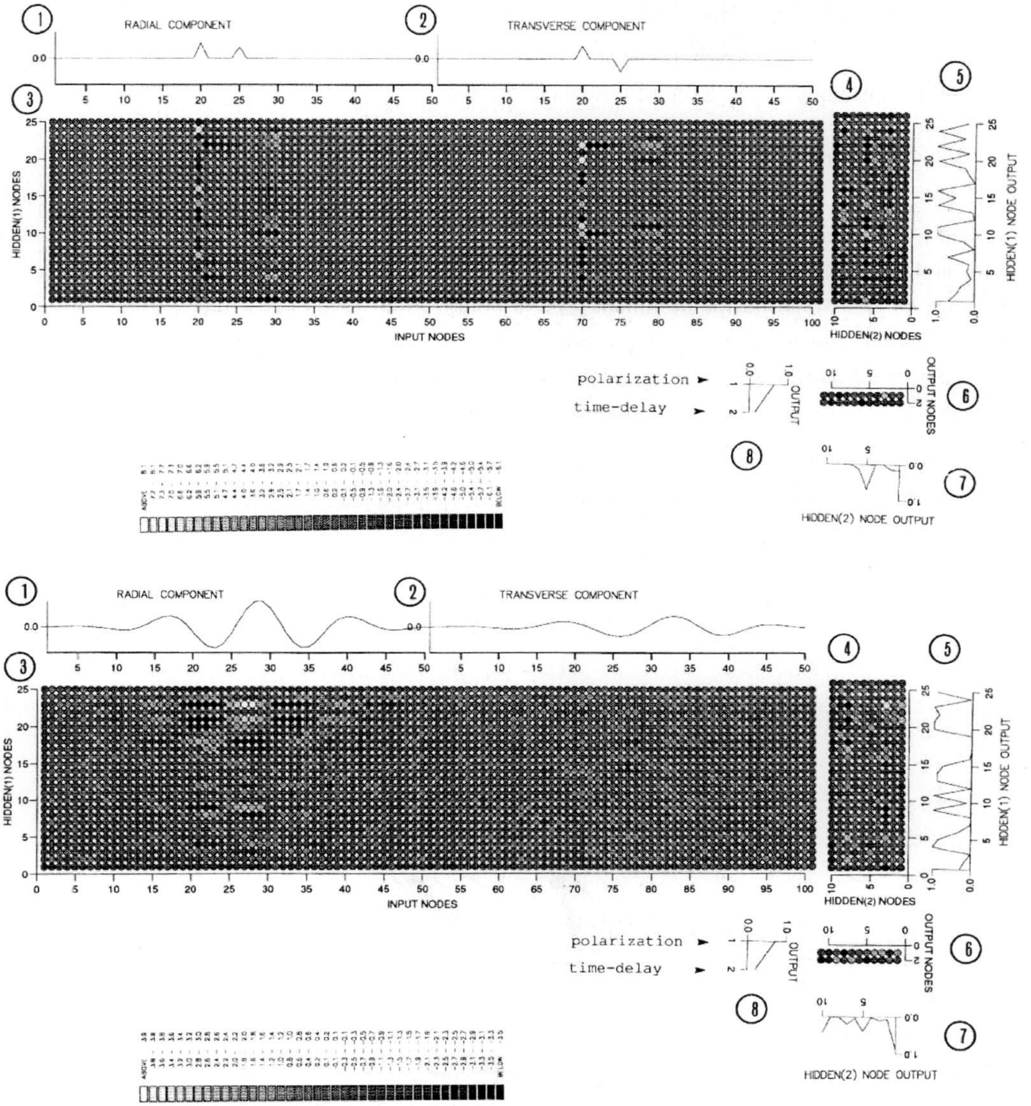

Figure A.5. Weight map for the trained neural network applied to a synthetic training set with an emergent (bottom) followed by an impulsive (top) source wavelet. (1) and (2) input traces, (3) mapping between input layer and first hidden layer; (4) mapping between first hidden layer and second hidden layer; (5) output from the first hidden layer; (6) mapping between second hidden layer and output layer; (7) output of second hidden layer; and (8) final output. In each weight map, the right hand row of circles represents the node threshold (Dai and MacBeth 1998).

A.3 COVARIANCE MATRIX

The evolution of the three-component particle motion with time can be tracked using techniques for polarization analysis based on the covariance matrix

$$\mathbf{P}_j = \sum_{i=j}^{N+j-1} \mathbf{d}(t_i)\mathbf{d}(t_i)^T \ , \tag{A.1}$$

defined for a sliding window of width N samples. A detailed expression for this matrix of individual auto- and cross-correlations is given in (4.38). Analysis of the instantaneous value of the matrix provides an assessment of the dominant type of polarization within the window (linear, elliptical but confined to a plane, or generally ellipsoidal), together with the orientation of this motion (Kanasewich 1981). This is achieved by the eigen-decomposition

$$\mathbf{P}_j = \mathbf{U}_j \Lambda_j \mathbf{U}_j^T \ , \tag{A.2}$$

with this real, symmetric, and positive definite matrix yielding real, non-negative eigenvalues

$$\lambda_1^j \geq \lambda_2^j \geq \lambda_3^j \ , \tag{A.3}$$

along the diagonal of Λ_j, and orthogonal eigenvectors grouped as columns in the matrix U_j

$$\mathbf{U}_j = \left(\mathbf{e}_1^j \quad \mathbf{e}_2^j \quad \mathbf{e}_3^j \right) \ . \tag{A.4}$$

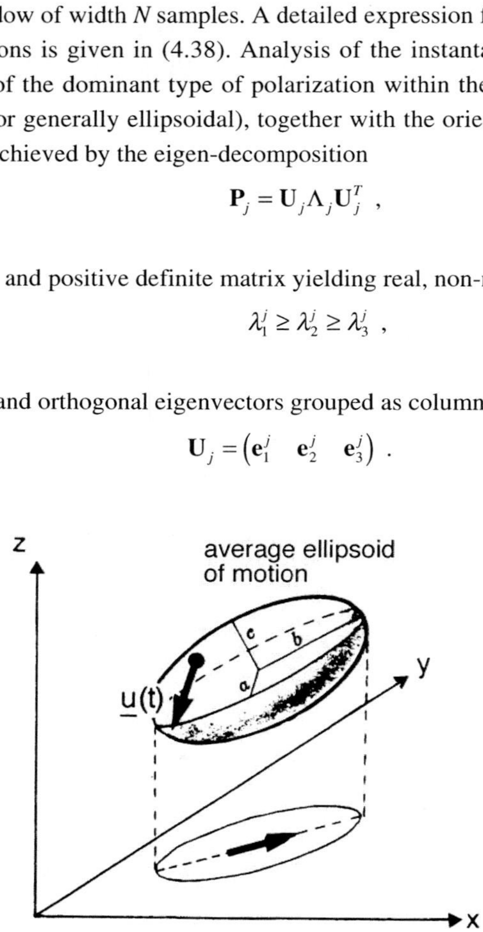

Figure A.6. The covariance matrix of auto- and cross-correlations between the receiver components can be analysed using eigenanalysis. This procedure estimates the length and orientation of the axes for the average particle-motion ellipsoid (MacBeth and Crampin 1991a).

Eigenanalysis can be achieved using standard numerical libraries such as NAG or MATHEMATICA. The relative magnitude of the eigenvalues reveals the nature of the average particle motion in the window of interest (Figure A.6). For example, if one eigenvalue is very much larger than the other two, the average motion within the window is predominantly linear. If one eigenvalue is very much smaller than the other two, the motion is confined predominantly to a plane. If all three eigenvalues are equal, the motion is spheroidal. Indeed, the average particle motion describes a spatial function defined by the expression

$$\begin{pmatrix} X & Y & Z \end{pmatrix} \begin{pmatrix} \lambda_1 & 0 & 0 \\ 0 & \lambda_2 & 0 \\ 0 & 0 & \lambda_3 \end{pmatrix} \begin{pmatrix} X \\ Y \\ Z \end{pmatrix} . \tag{A.5}$$

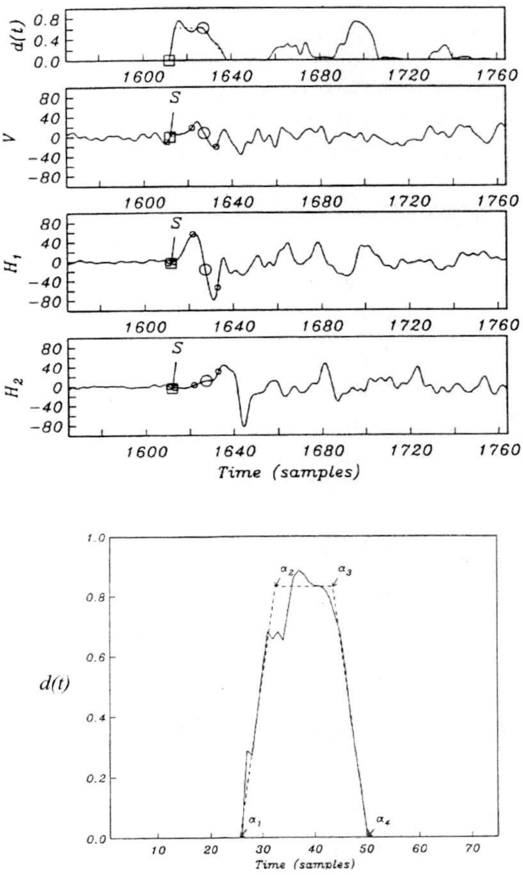

Figure A.7. Top: example of a linearity function $d(t)$ (constructed using a Watson axial distribution, and related to $F^L(t)$) for earthquake recordings. For the definition of F^L see the text. The lag time was large enough in this case for the procedure to discriminate between the two shear waves. Bottom: trapezoidal best-fit envelope used to define statistics for the lag time, defined as $\Delta\tau = \alpha_3 - \alpha_1$. After Aster, Shearer and Berger (1989).

This average particle motion ellipse has its semi-axes given by $\sqrt{\lambda_i}$, as each eigenvalue is proportional to the signal power in the direction of its eigenvector. Combinations of eigenvalues can thus be used to obtain a statistical measure of the polarization by calculating these properties for each position of the window and noting how they change in time. For evaluating the degree of linearity of the particle motion Aster, Shearer and Berger (1989) propose

$$F_j^L\left(\lambda_1^j,\lambda_2^j\right)=\frac{\lambda_1^j-\lambda_2^j}{\lambda_1^j+\lambda_2^j+\lambda_3^j}\ , \tag{A.6}$$

which lies between 0 and 1. To sharpen this function, they advocate combining this with the projection of the signal power along the major eigenvector given by a unit vector in the polarization direction $\hat{\mathbf{a}}$ defined by the window positioned over the initial linear onset. This yields the stronger discriminant

$$F_j^L(t)=F\left(\lambda_1^j,\lambda_2^j\right)\ \left(\hat{\mathbf{e}}_1^j\cdot\hat{\mathbf{a}}\right)^2\ . \tag{A.7}$$

The persistence of this linearity gives the shear wave lag time, with the onset of the slow shear wave prompting a sharp reduction in $F^L(t)$ (Figure A.7). A similar indicator can be provided by

$$S_j^L(t)=F\left(\lambda_1^j,\lambda_2^j\right)\ \left(\hat{\mathbf{e}}_1^j\cdot\hat{\mathbf{b}}\right)^2\ . \tag{A.8}$$

where $\hat{\mathbf{a}}$ is orthogonal to $\hat{\mathbf{b}}$ in the horizontal plane. Equation (A.7) can be adapted further by introducing a directionality function with $\hat{\mathbf{e}}_1^j\cdot\hat{\mathbf{b}}$ as the independent variable. The correct step size for each covariance matrix evaluation is critical: too smooth and the duration between the pulses is missed, too rough and spurious sub-pulses appear as the profile breaks up. The lag time is calculated by measuring the mean half-width of the linearity functions. Aster, Shearer and Berger (1989) constructed statistics for these profiles to aid calculation (Figure A.7).

The above technique can also be applied in the frequency domain (Samson and Olson 1980; Samson, 1983; Park, Vernon, and Lindberg 1987), which offers the choice of different frequency bands. The band with the largest signal-to-noise ratio is selected. Polarization analysis can also be extended to include arrays of receivers, either by simple averaging of the characteristics (Jurkevics 1989) or by array processing (see, e.g., Rutty and Greenhalgh 1986 for an application to dual receivers).

A.4 RECEIVER ROTATION DISCRIMINANTS

With receiver rotation being the only option to process the shear-wave birefringence, the idealization of (4.13) becomes

$$\mathbf{d}(t)=\begin{pmatrix}\cos\phi & \sin\phi\\ -\sin\phi & \cos\phi\end{pmatrix}\begin{pmatrix}s_1(t)\\ s_2(t)\end{pmatrix}, \tag{A.9}$$

where the fast and slow shear waves have been excited by the source in an unknown proportion. Implicit in this model is assumption is that the wavelets $s_1(t)$ and $s_2(t)$ are identical, so that they differ only by a scale factor and a lag (differential attenuation or velocity dispersion are not considered). Birefringence analysis consists of rotating the receiver about the Z-axis by an angle ϕ_R, giving the two-component recordings

$$\mathbf{d}'(t) = \begin{pmatrix} \cos(\phi - \phi_R) & \sin(\phi - \phi_R) \\ -\sin(\phi - \phi_R) & \cos(\phi - \phi_R) \end{pmatrix} \begin{pmatrix} s_1(t) \\ s_2(t) \end{pmatrix}, \tag{A.10}$$

until the recording resembles the expected form

$$\mathbf{d}'(t) = \begin{pmatrix} \kappa_1 s(t) \\ \kappa_2 s(t - \Delta\tau) \end{pmatrix}, \tag{A.11}$$

with the recovery of both polarization azimuth ϕ and the lag time $\Delta\tau$. Comparison between the expected formand observations for a range of trial values for f and Dt, followed by a search for the optimal choice, has been considered by Campden (1990). Alternatively, focusing solely on the rotation but not on the lag time, a range of discriminants has been proposed to detect this final form.

(a) One approach is to use only the fast shear wave to align the receiver axes. Thus, a small window must be chosen around the initial onset, and the P-wave techniques used in Section 4.3 can be employed. The size of the window ultimately limits the resolution for small lag times. The method by Shih, Schneider and Meyer (1986) is specifically tailored to shear waves. It measures the degree of rectilinear motion on the receiver axes. The criterion is the ratio of the sum of the projections of the differential motion, $\Delta\mathbf{d}_i = \mathbf{d}(t_{i+1}) - \mathbf{d}(t_i)$, onto arbitrarily oriented orthogonal x-, and y-axes rotated by an angle χ with respect to the receiver axes. This measure can be written

$$F(\chi) = \frac{\sum_{i=1}^{n-1} \left|\Delta d_i \cos(\chi - \theta_i)\right|}{\sum_{i=1}^{n-1} \left|\Delta d_i \sin(\chi - \theta_i)\right|}, \tag{A.12}$$

where θ_i is the angle which $\Delta\mathbf{d}$ makes with the Y-axis and Δd_i is its magnitude (Figure A.8). After the process has aligned the axes with the shear wave polarizations, a larger time window is chosen, and the slow wave is time-shifted towards the fast until this linearity function is maximized once more. The approach has been applied with some success to local earthquake data by Savage et al. (1987).

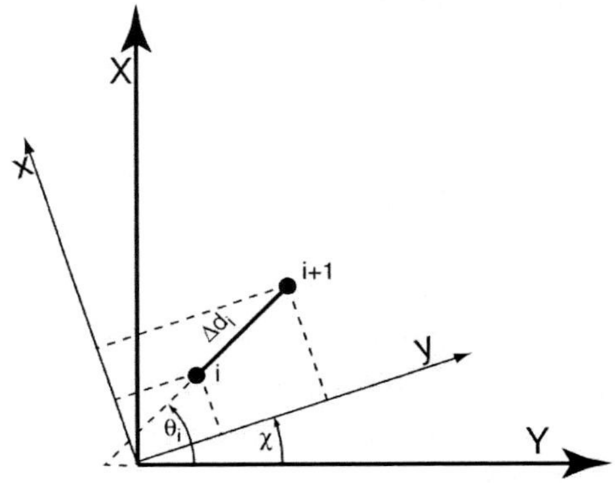

Figure A.8. The method of Shih, Schneider and Meyer (1986) uses as its basis a linearity function given by the ratio of the sums of the relative displacements Δd_i on the rotating (x-y) axes.

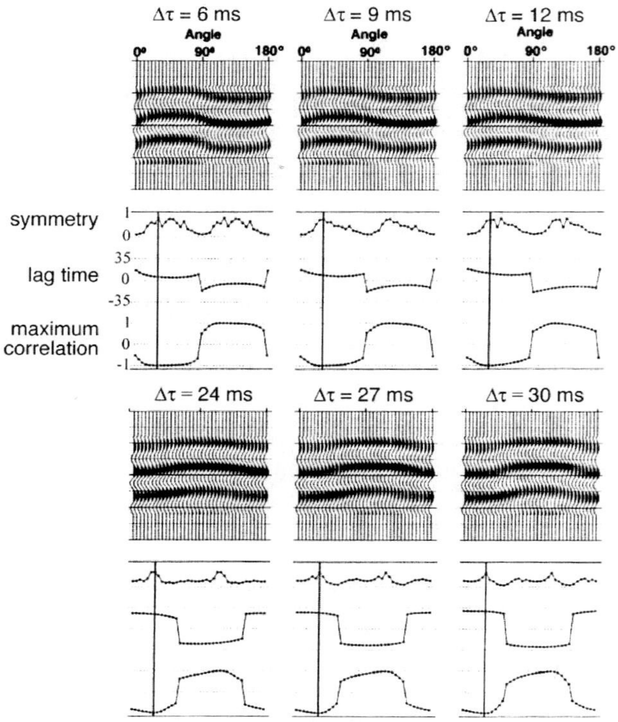

Figure A.9. Correlation traces and evolution of associated measures of symmetry, lag time of peak correlation, and the peak value, as a function of rotation angle. Six groups are plotted for different lag times in this synthetic example, with a common polarization azimuth of 30° (given by the solid vertical line). After Peron (1989).

(b) Discriminants originating from the cross-correlation function $A(\tau)$ are also possible (Naville, 1986; Garotta and Marechal 1987; Peron 1989). When the receiver axes are aligned with the principal axes of the anisotropy, the horizontal recordings have an almost identical wavelet. This point can be detected by monitoring the largest peak correlation (see 5.12), the maximum lag time, or maximal symmetry in the correlation function as a function of rotation angle. The symmetry indicator, in particular, is constructed by taking the average D of the squares of the difference between correlation values located each side of the peak:

$$D = \sum_{i=1}^{N} \left(A(\tau_{\max} - \tau_i) - A(\tau_{\max} + \tau_i) \right)^2 . \tag{A.13}$$

Peron (1989) suggests using a smoothed function $\overline{A}(\tau)$, formed by first fitting a cubic spline, then rescaling $\overline{A}$ to the amplitude of the peak correlation, and finally forming the symmetry indicator

$$S = \frac{1}{1 + 20D} , \tag{A.14}$$

with $0 < S < 1$. Tests on synthetic seismograms and field data by Peron (1989) favored the maximum correlation and time-delay functions as being smoother and more robust than symmetry (Figure A.9). The symmetry functions tend to exhibit spikes and reach maxima at incorrect angles, but are better behaved for larger lag times.

On a similar theme, Naville (1986) considered horizontal recordings at two separate depth levels

$$\mathbf{d}_i(t) = \begin{pmatrix} x_i(t) \\ y_i(t) \end{pmatrix} = \mathbf{C}(\phi)\begin{pmatrix} s_1^i(t) \\ s_2^i(t) \end{pmatrix}, \tag{A.15}$$

and

$$\mathbf{d}_{i+1}(t) = \begin{pmatrix} x_{i+1}(t) \\ y_{i+1}(t) \end{pmatrix} = \mathbf{C}(\phi)\begin{pmatrix} s_1^{i+1}(t) \\ s_2^{i+1}(t) \end{pmatrix}, \tag{A.16}$$

and the determination of the anisotropy properties between these two levels. He advocated a scan over rotation angle ϕ to determine the most symmetric cross-correlation between $x_i(t)$ and $x_{i+1}(t)$:

$$\sum_{i=1}^{N} x^i x^{i+1} = \sum_{i=1}^{N}\left(\cos^2\phi s_1^i s_1^{i+1} + \sin^2\phi s_2^i s_2^{i+1} + \sin\phi\cos\phi\left(s_1^i s_2^{i+1} + s_2^i s_1^{i+1}\right)\right) . \tag{A.17}$$

An additional estimate can also be obtained for the transverse component

$$\sum_{i=1}^{N} y^i y^{i+1} = \sum_{i=1}^{N}\left(\sin^2\phi s_1^i s_1^{i+1} + \cos^2\phi s_2^i s_2^{i+1} + \sin\phi\cos\phi\left(s_1^i s_2^{i+1} + s_2^i s_1^{i+1}\right)\right) . \tag{A.18}$$

The asymmetrical part of these equations comes from the cross product term that becomes zero when the rotation has aligned with the principal axes.

(c) A rotation attributed to Harrison (1993) is based on the energy-normalized cross-correlation (Neidell and Taner 1971)

$$\sigma(\phi_R, \Delta\tau) = \frac{\sum_j \sum_i A(\phi_j, \tau_i) G(\phi_j, \tau_i, \phi_R, \Delta\tau)}{\sqrt{\sum_j \sum_i A^2(\phi_j, \tau_i) \sum_j \sum_i G^2(\phi_j, \tau_i, \phi_R, \Delta\tau)}} , \tag{A.19}$$

where $A(\phi_j, \tau_i)$ is the actual cross-correlation between the radial and transverse components for a suite of possible rotations ϕ_j, and $G(\phi_j, \tau_i)$ is a function of the autocorrelation of the input signal. The analysis is performed by rotating (ϕ_R) and time-shifting ($\Delta\tau$) the recordings within a specific time window, evaluating the statistic, and then finding a maximum of this function in the two-dimensional space. This method was applied by Donati and Brown (1995) in the analysis of anisotropy in a 3C-2D dataset from a fractured carbonate reservoir in southwestern Venezuela (Figure A.10).

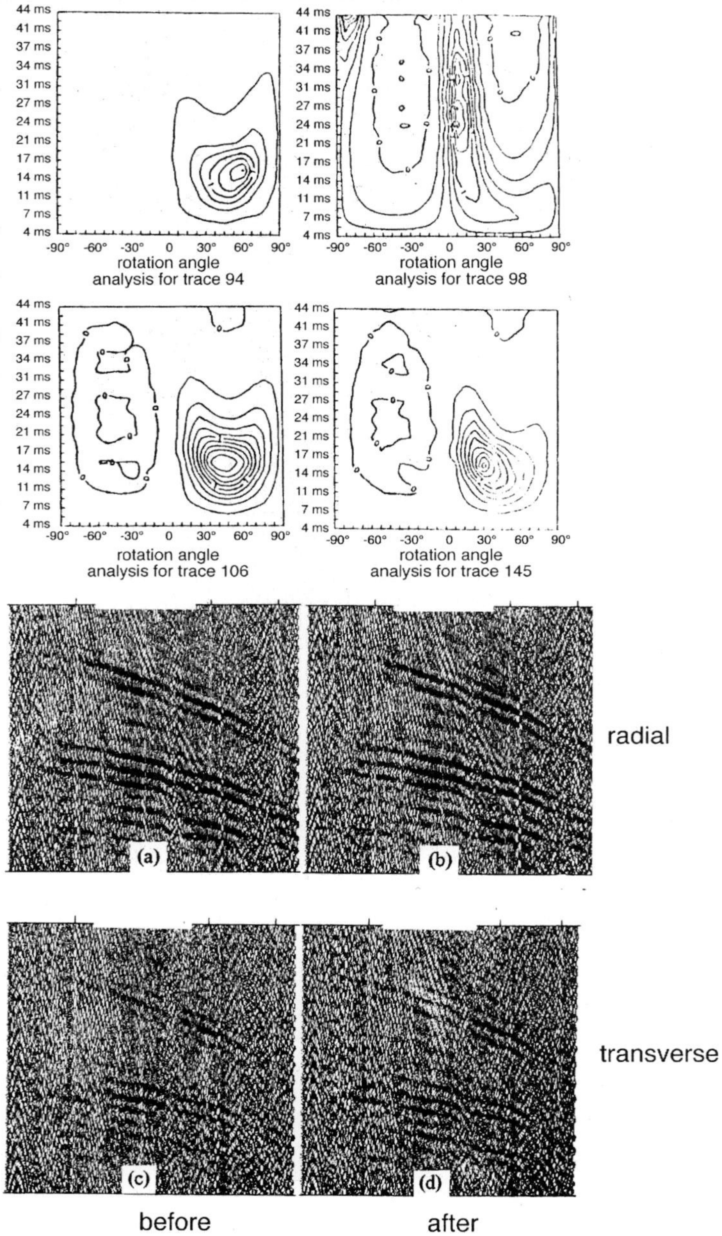

Figure A.10. Top: contours of rotation angle versus lag time for a converted wave dataset from the Barinas basin (Donati and Brown 1995). Below: the corresponding radial and transverse components for one of the shots before and after rotation into the best estimate of the shear wave directions.

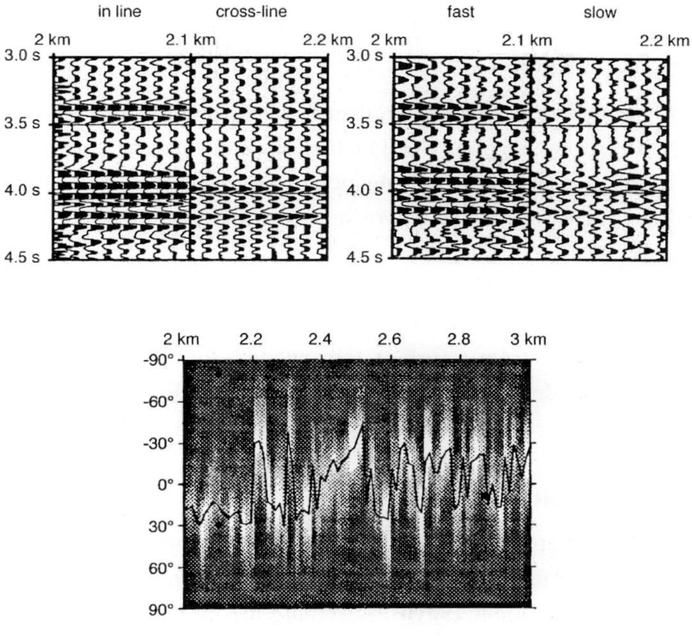

Figure A.11 Example of application to surface seismic data – after Michelena (1995). Stacked horizontal components before (left) and after (right) application of the energy-ratio method. Below: contoured energy ratio as a function of lateral distance and rotation angle, with a continuous line passing through the maxima (white).

(d) The converted-wave method of Michelena (1995) is based on the technique of Garotta and Granger (1988). Here, a rotation angle is chosen that gives the minimum (or maximum) energy ratio between the two receiver components. It assumes conversion from P-waves to an SV-wave with an initial in-line motion, before incidence on the anisotropic region. Thus in (A.11) the relative amplitudes of the two shear waves are determined by the fracture strike angle to the in-line direction $\kappa_1 = \sin\phi$ and $\kappa_2 = \cos\phi$. First rotate the recordings $\mathbf{d}(t)$ by an angle ϕ_R:

$$\mathbf{d}'(t) = \mathbf{C}(\phi_R)\mathbf{d}(t) = \begin{pmatrix} \cos\phi\cos(\phi-\phi_R) & \sin\phi\sin(\phi-\phi_R) \\ \cos\phi\sin(\phi-\phi_R) & -\sin\phi\cos(\phi-\phi_R) \end{pmatrix} \begin{pmatrix} s(t) \\ s(t-\Delta\tau) \end{pmatrix}, \tag{A.20}$$

and then evaluate the energy ratio of the two rotated components

$$E(\phi_R) = \frac{2E_S\cos^2\phi_R\cos^2\varepsilon + 2E_S\sin^2\phi_R\sin^2\varepsilon + E_C\sin2\phi_R\sin2\varepsilon}{2E_S\cos^2\phi_R\sin^2\varepsilon + 2E_S\sin^2\phi_R\cos^2\varepsilon - E_C\sin2\phi_R\sin2\varepsilon}, \tag{A.21}$$

where

$$E_S = \sum_{i=1}^{N} s^2(t_i) = \sum_{i=1}^{N} s^2(t_i - \Delta\tau), \tag{A.22}$$

$$E_C = \sum_{i=1}^{N} s(t_i)s(t_i - \Delta\tau) \ , \tag{A.23}$$

and $\varepsilon = \phi - \phi_R$. For an impulsive wavelet, then the cross-terms of the denominator and numerator are close to zero. In this case, this function has a distinct maximum and minimum in the range $-\pi/2$ to $\pi/2$ at ϕ and $\phi + \pi/2$. Evaluation of the rotated in-line and cross-line results for both angles determines which angle corresponds to the fracture strike. The procedure works particularly well when the source is nearly impulsive, with success and failure strongly depending upon the wavelet's peak frequency and bandwidth. Michelena (1995) shows that a minimum frequency of $\omega_0 = \frac{1}{2}\Delta\tau$ is necessary to resolve the lag time. He tested this technique in the field on a dataset from a surface-seismic study of a fractured carbonate reservoir in southwest Venezuela, determining the predominant direction picked up by the logs (Figure A.11).

A.5 SPECTRAL INTERFERENCE

Transformation of the birefringence signature into the frequency domain permits analysis of the spectral interference pattern formed by the interaction of the two closely spaced and similar shear waveforms (MacBeth and Crampin 1991b). Assuming the idealized model of (4.13), with a qS1 polarization azimuth ϕ and a source motion at an angle of ϕ_S measured against the in-line direction, the horizontal recordings are

$$d_X(t) = s(t)\cos\phi_S \cos\phi + s(t - \Delta\tau)\sin\phi_S \sin\phi \ , \tag{A.24}$$

and

$$d_Y(t) = -s(t)\cos\phi_S \sin\phi + s(t - \Delta\tau)\sin\phi_S \cos\phi \ , \tag{A.25}$$

where $s(t)$ is the source wavelet. After Fourier transformation this becomes

$$d_X(\omega) = \left(\cos\phi_S \cos\phi + \sin\phi_S \sin\phi e^{-i\omega\Delta\tau}\right)S(\omega) \ , \tag{A.26}$$

$$d_Y(\omega) = \left(-\cos\phi_S \sin\phi + \sin\phi_S \cos\phi e^{-i\omega\Delta\tau}\right)S(\omega) \ , \tag{A.27}$$

where $S(\omega)$ is the source spectrum. Both the amplitude and phase spectrum are modulated by terms involving ϕ_S, ϕ and $\Delta\tau$, the effect of which can be clearly detected in the power spectra of the X and Y components:

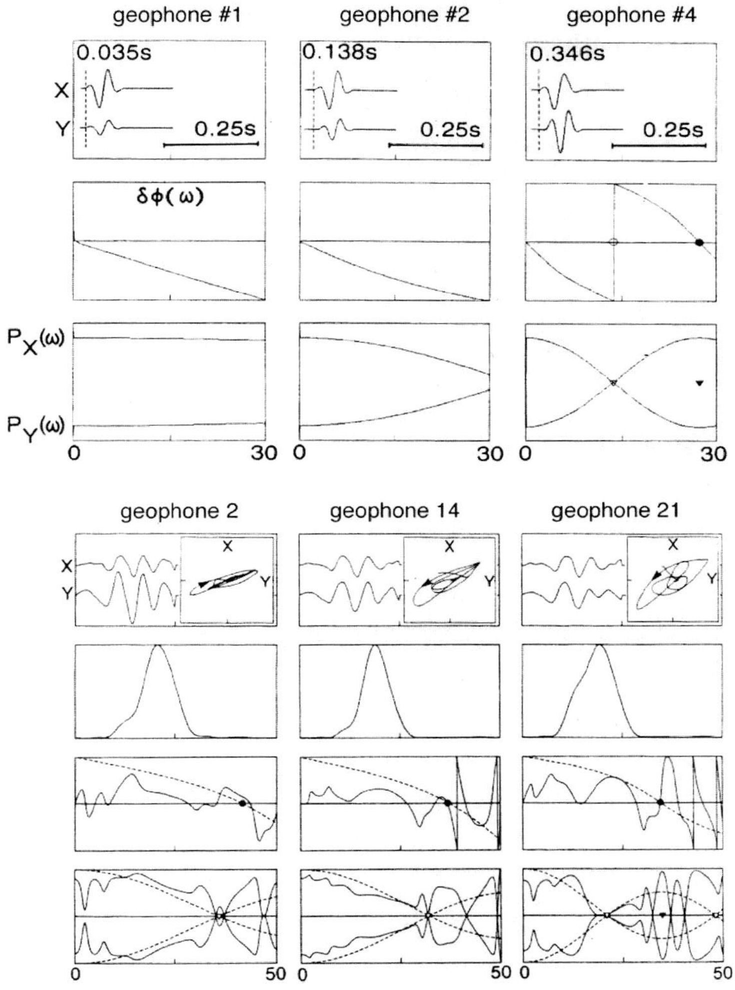

Figure A.12. Top: interference spectra for three depth levels in a synthetic zero-offset VSP with different lag times but the same polarization direction. Middle and bottom diagrams are for the phase and amplitude spectra, respectively. The discontinuities, intersections and turning points are regularly spaced and can be interpreted to give lag time. The depth of the oscillations in the amplitude spectrum gives the polarization. Lower diagram corresponds to field data, with estimated lag times of 8.5 ms, 9.5 ms and 14.5 ms, respectively (MacBeth and Crampin 1991b).

$$P_X(\omega) = \left(a^2 + b^2 + 2ab\cos\omega\Delta\tau\right)P_S(\omega) = I_X(\omega)P_S(\omega) \tag{A.28}$$

and

$$P_Y(\omega) = \left(1 - a^2 - b^2 - 2ab\cos\omega\Delta\tau\right)P_S(\omega) = I_Y(\omega)P_S(\omega) \ , \tag{A.29}$$

where $a = \cos\phi_S \cos\phi$ and $b = \sin\phi_S \sin\phi$. These power spectra are a combination of the source power spectrum $P_S(\omega)$ and the spectral interference functions $I_X(\omega)$ and $I_Y(\omega)$ which modulates the smooth source spectrum by a series of periodic oscillations. The depth of these oscillations, and hence the sensitivity to detection, depends on the factor ab (this product is maximal when $\phi_S = \phi = \pi/4$). The conservation of energy gives $P_S(\omega) = P_X(\omega) + P_Y(\omega)$, which helps to express the interference in terms of the observed quantities

$$I_X(\omega) = \frac{P_X(\omega)}{P_X(\omega) + P_Y(\omega)} , \tag{A.30}$$

and

$$I_Y(\omega) = \frac{P_Y(\omega)}{P_X(\omega) + P_Y(\omega)} . \tag{A.31}$$

The interference functions possess a sequence of turning points (Figure A.12); their positions f_k^T are related to the lag $\Delta\tau$ by

$$f_k^T = \frac{k}{2\Delta\tau} , \tag{A.32}$$

where $k = 0, 1, 2, 3, \ldots$. A minimum in $I_X(\omega)$ corresponds to a maximum in $I_Y(\omega)$, and vice-versa. Measurements of these common turning points give a direct estimate of the lag. The resolution limit for this technique is set by the position of f_1^T, which must be within the bandwidth of the signal. This gives a range of 5.6 ms $< \Delta\tau <$ 50 ms for a 50 Hz signal with a bandwidth of 80 Hz. Intersections occur before the turning points in the interference curves, and can be used to measure smaller lags. The spectral interference patterns can also be used to obtain estimates of a and b, and hence ϕ_S and ϕ. This requires measurement of the difference ΔI of $I_X(\omega)$ and $I_Y(\omega)$ at two adjacent turning points i and j. For odd-numbered turning points

$$\Delta^{odd} = \sqrt{\frac{\left(1 + \left|\Delta I^{odd}\right|\right)}{2}} , \tag{A.33}$$

and for even-numbered turning points

$$\Delta^{even} = \sqrt{\frac{\left(1 - \left|\Delta I^{even}\right|\right)}{2}} . \tag{A.34}$$

This yields

$$\phi_S = \frac{\cos^{-1}\left(\Delta^{odd}\right) + \cos^{-1}\left(\Delta^{even}\right)}{2} \tag{A.35}$$

and

$$\phi = \frac{\cos^{-1}(\Delta^{odd}) - \cos^{-1}(\Delta^{even})}{2} \; . \tag{A.36}$$

The requirement for measurements at two consecutive turning points implies that polarization is less well resolved than the time-delay. The phase spectra can be written

$$\Omega_X(\omega) = \arg(d_X(\omega)) = \Omega_S(\omega) + \Omega_k^X(\omega) \; , \tag{A.37}$$

and

$$\Omega_Y(\omega) = \arg(d_Y(\omega)) = \Omega_S(\omega) + \Omega_k^Y(\omega) \; , \tag{A.38}$$

where $\Omega_S(\omega)$ is the phase spectrum of the source, and

$$\Omega_k^X(\omega) = \tan^{-1}\left(\frac{-b\sin\omega\Delta\tau}{a + b\cos\omega\Delta\tau}\right) \tag{A.39}$$

and

$$\Omega_k^Y(\omega) = \tan^{-1}\left(\frac{d\sin\omega\Delta\tau}{c - d\cos\omega\Delta\tau}\right) , \tag{A.40}$$

with $c = \cos\phi_S \sin\phi$ and $d = \sin\phi_S \cos\phi$. Subtracting the source contribution yields

$$\delta\Omega(\omega) = \Omega_X(\omega) - \Omega_Y(\omega) = \Omega_k^X(\omega) - \Omega_k^Y(\omega) \; , \tag{A.41}$$

or

$$\delta\Omega(\omega) = \tan^{-1}\left(\frac{-\sin 2\phi_S \sin\omega\Delta t}{\sin 2\phi \cos 2\phi_S - \sin 2\phi_S \cos 2\phi \cos\omega\Delta\tau}\right) , \tag{A.42}$$

This relative phase spectrum has a number of zeros at the frequencies

$$f_k^Z = \frac{k}{2\Delta\tau} \; , \quad k = 0,\ 1,\ 2,\ 3,\ \ldots \; . \tag{A.43}$$

which coincide with the turning points of the power spectra. The phase spectra cross the horizontal axis twice every $\frac{1}{2\Delta\tau}$. The function provides the same time-delay resolution as the power spectra. The spectral interference method (Figure A.12) relies upon visual examination of patterns above, and thus encompasses some disadvantages of the polarization diagrams. A step towards automation is to search for the angles where minimal interference occurs (Campden 1990). Shear-wave lag is better resolved in the frequency domain (onset times converted to definite points), while polarization is better resolved in the time domain (first break is well defined).

A.6 SINGULAR VALUE DECOMPOSITION

Shieh (1997) re-arranged the three-component recording $\mathbf{d}(t)$ into a matrix $\mathbf{A}$ involving a window of N time samples

$$\mathbf{A} = \left(\mathbf{d}(t_1) \quad \mathbf{d}(t_2) \quad \mathbf{d}(t_3) \quad \ldots \quad \mathbf{d}(t_N)\right)^T . \tag{A.44}$$

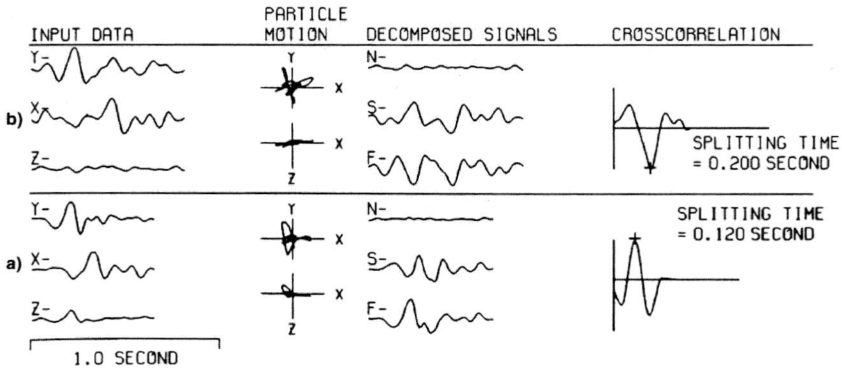

Figure A.13. The SVD technique decomposes the traces into a noise component and the two separate shear waves. The example above is from earthquake data by Shieh (1997).

The matrix contains three columns and N rows, and can be decomposed into contributions arising from the two shear waves and a noise element using the singular value decomposition (SVD)

$$\mathbf{A} = \mathbf{U}\Lambda\mathbf{V}^T , \tag{A.45}$$

where $\mathbf{U}$ is an $N\times N$ orthogonal matrix, Λ an $N\times 3$ diagonal matrix of eigenvalues, and $\mathbf{V}$ is a 3×3 orthogonal matrix. The smallest eigenvalue refers to the noise component $n(t)$, with the other two being the shear waves $s_1(t)$ and $s_2(t)$. An individual wave is found by combining the eigenvectors and eigenvalues

$$s_1(t) = \lambda_1\mathbf{U}(t,1) , \tag{A.46}$$

$$s_2(t) = \lambda_2\mathbf{U}(t,2) , \tag{A.47}$$

$$n(t) = \lambda_3\mathbf{U}(t,3) , \tag{A.48}$$

where $\mathbf{U}(t,1)$, $\mathbf{U}(t,2)$, $\mathbf{U}(t,3)$ are the first three columns of $\mathbf{U}$. Cross-correlation of $s_1(t)$ and $s_2(t)$ gives the requisite lag time. An example of this technique applied to earthquake data is given in Figure A.13.

A.7 WAVELET TRANSFORM

The wavelet transform is a time-frequency method that promises considerable increase in resolution. The method transforms the signal $x(t)$ according to

$$W(a,b) = \frac{1}{\sqrt{a}} \int_{-\infty}^{+\infty} x(\tau) h\left(\frac{\tau - b}{a}\right) d\tau \ , \tag{A.49}$$

where $h(a, b)$ is the basis function

$$h(a, t-b) = \mathrm{e}^{-\frac{1}{2}\left(\frac{t-b}{a}\right)^2} \mathrm{e}^{+j\omega\left(\frac{t-b}{a}\right)} , \tag{A.50}$$

with parameters a and b relating to scaling and time-shifting, respectively. In order to detect phase alignment, Niitsuma et al. (1993) introduce a phase-alignment index (PAI) defined as the average of the transformed phase along the parameter a. This function detects discontinuities of the signal caused by the onset of different arrivals. Niitsuma et al. (1993) analyse multiple shear wave birefringence (Section 6.2) by calculating the PAI for each angle of rotation, the results of which are shown in Figure A.14. They apply the technique to acoustic emission events in a geothermal reservoir. The polarization direction of the shear waves is evaluated by detecting the extinction point of each discontinuity component.

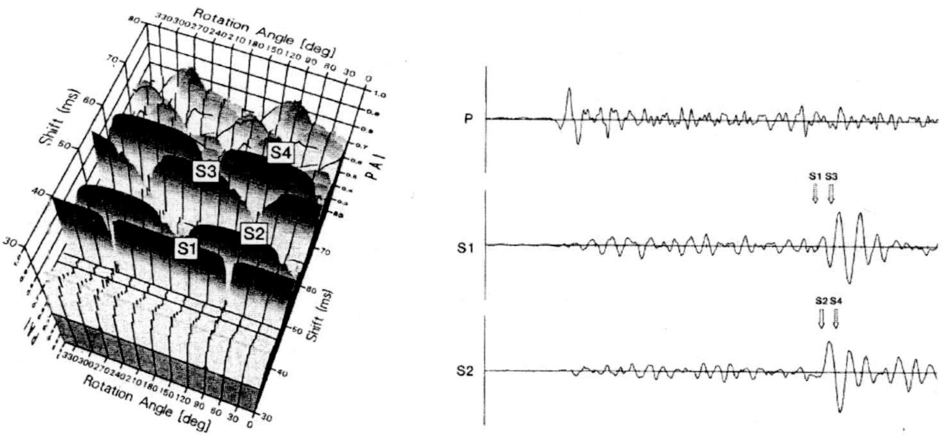

Figure A.14. Wavelet transform can provide a high-resolution image of the structure of multiple shear-wave splitting. This example is an acoustic emission event in a geothermal reservoir. The diagram at left corresponds to measures of the PAI (phase alignment index) taken as a function of receiver rotation angle and time.

REFERENCES

Aguilera, R., 1980. NATURALLY FRACTURED RESERVOIRS, Pennwell Publishing Company, Tulsa, Oklahoma.

Ahmed, H., Dillon, P.D., Johnstad, S.E. and Johnston, C.D., 1986. Northern Viking Graben multilevel three-component walkaway VSPs: a case history, First Break, 4, 9–27.

Aki, K., and Richards, P.G., 1980. QUANTITATIVE SEISMOLOGY, VOL.1, W.H. Freeman, San Francisco.

Al-Chalabi, M., 1974. An analysis of stacking, RMS, average, and interval velocities over a horizonatlly layered ground, Geophysical Prospecting, 22, 458–475.

Al-Dajani, A.A., Tsvankin, I., and Toksöz, M.N., 1998. Non-hyperbolic reflection moveout for azimuthally anisotropic media, SEG Expanded Abstracts, 1472–1482.

Alford, R.M., 1896. Shear data in the presence of azimuthal anisotropy: Dilley Texas, SEG Expanded Abstracts, 476–479.

Alkalifah. T., 1997. Velocity analysis using non-hyperbolic moveout in transversely isotropic media, 62, 1839–1854.

Alkalifah, T. and Larner, K., 1994. Migration error in transversely isotropic media: Geophysics, 59, 1405–1418.

Alkalifah, T., and Tsvankin, I., 1995. Velocity analysis for transversely isotropic media, Geophysics, 60, 1550–1566.

Alkalifah, T., Tsvankin, I., and Larner, K., 1996. Velocity analysis and imaging in transversely isotropic media: methodology and a case study, The Leading Edge of Exploration, May Issue, 371–378.

Altmann, S.L., 1986. ROTATIONS, QUATERNIONS, AND DOUBLE GROUPS, Clarendon Press, Oxford.

Ameen. M.S., 1995. Fractography: fracture topography as a tool in fracture mechanics and stress analysis, Geological Society Special Publication, No. 92.

Anderson, D.L., 1961. Elastic wave propagation in layered anisotropic media, J. of Geophysical Research, 66, 2953–2964.

Ansel, E.A., 1931. Das Impulsfeld der praktischen Seismik in graphischer Behandlung, Gerlands Beitr.angew. Geophys., 1, 117–136.

Armstrong, P.N., Chmela, W., and Leaney, W.S., 1995. AVO calibration using borehole data, First Break, 13, 319–328.

Assad, J.M., McDonal., J.A. and Tatham, R.H., 1990. Physical modelling of anisotropy caused by oriented inclusions, SEG Expanded Abstracts, 1413–1415.

Aster, R.C. Shearer, P.M., and Berger, J., 1990. Quantitative measurements of shear wave polarizations at the Anza seismic network, Southern California – implications for shear wave splitting and earthquake prediction, J. Geophysical Research, 95, 12449–12473.

REFERENCES

Backus, G.E., 1962. Long–wave anisotropy produced by horizontal layering, J. Geophysical Research, 66, 4427–4440.

Balch, A.H., and Lee, M.W., 1984. VERTICAL SEISMIC PROFILING, TECHNIQUE, APPLICATIONS AND CASE HISTORIES, Int.Hum.Resour.Dev.Corp., Boston.

Banik, N.C., 1984. Velocity anisotropy of shales and depth estimation in the North Sea Basin, Geophysics, 49, 1411–1419.

Banik, N.C., 1987. An effective anisotropy parameter in transversely isotropic media, Geophysics, 52, 1654–1664.

Baptie, B.J., 1996. Application of the cross-hole technique for analysis of shear waves in fractured limestone reservoirs, Ph.D. Thesis, University of Edinburgh.

Barton, C.C., 1995. Fractal analysis of scaling and spatial clustering of fractures, in FRACTALS IN THE EARTH SCIENCES, Eds. Barton C.C. and La Pointe, P.R., Plenum Press, New York.

Bathellier, E., and Czernichow, J., 1997. Downhole triaxial geophone sondes for reservoir monitoring, A Review, EAGE Expanded Abstracts, E049.

Batzle, M., and Wang, Z., 1992. Seismic properties of pore fluids, Geophysics, 57, 1396–1408.

Becker, D.F., Bishop, G.W., Peterson, D.N., Rendleman, C.A., and Shah, P.M., 1989. Implications of seismic anisotropy within the Pembrook Unit, Midland, Texas, SEG Research Workshop on recording and processing vector wavefield data, Snowbird, Utah, p. 60.

Ben-Menahem, A., Gibson, R.L., and Sena, A.G., 1991. Green's tensor and radiation patterns of point sources in general anisotropic inhomogeneous elastic media, Geophysical Journal International, 107, 297–308.

Berg, L.E., Magnus, I., and Vestby, J., 1995. Observation of seismic anisotropy from VSP measurements and compensation for the effect in migration of seismic data, EAGE Abstract, B023.

Berge, P.A., Mallick, S., Fryer, G.J., Barstow, N., Carter, J.A., Sutton, G.H., and Ewing, J.I., 1991. In situ measurement of transverse isotropy in shallow-water marine sediments, Geophysical Journal International, 104, 241–254.

Berryman, J.G., Grechka, V., and Berge, P.A., 1997. Analysis of Thomsen parameters for finely layered VTI media, SEG Expanded Abstracts, 941–944.

Booth, D.C., abd Crampin, S., 1985. Shear-wave polarizations on a curved wavefront at an isotropic free-surface, Geophys.J.R.astr.Soc., 83, 31–45.

Born, M., and Wolf, E., 1965. PRINCIPLES OF OPTICS, 3rd ed., Pergamon Press, New York.

Brekhovskikh, L.M., 1960. WAVES IN LAYERED MEDIA, Academic Press, San Diego, California.

Bruno, M.S., and Winterstein, D.F., 1994. Some influences of stratigraphy and structure on reservoir stress orientation, Geophysics, 59, 954–962.

Bubuska, V., and Cara, M., 1991. SEISMIC ANISOTROPY IN THE EARTH, Kluwer Academic Publishers, London.

Bush, I., 1990. Modelling of shear-wave anisotropy in the Paris Basin, Ph.D. Thesis, University of Edinburgh.

Byun, B.S., Corrigan, D., and Gaiser, J.E., 1989. Anisotropic velocity analysis for lithology discrimination, Geophysics, 54, 1564–1574.

References

Bush, I., Crampin, S., 1991. Paris Basin VSPs: case history establishing combinations of fine-layer (or lithlogic) anisotropy and crack anisotropy from modelling shear wavefields near point singularities, Geophys.J.int., 107, 433–447.

Campden, D.A., 1990. Analysis of multicomponent VSP data for shear-wave anisotropy, Ph.D. Thesis, University of Edinburgh.

Campden, D.A., Crampin, S., Majer, E.L., and McEvilly, T.V., 1990. Modelling the Geysers VSP: a progress report, The Leading Edge of Exploration, 36–39.

Castagna, J.P., 1993. AVO analysis – Tutorial and review, in Castagna, J.P. and Backus, M.M., Eds., Offset-dependent reflectivity – THEORY AND PRACTICE OF AVO ANALYSIS, SEG., 3–36.

Castagna, J.P., and Smith, S.W., 1994. Comparison of AVO indicators: a modelling study, Geophysics, 59, 1849–1855.

Castagna, J.P., Swan, H.W., and Foster, D.J., 1998. Framework for AVO gradient and intercept interpretation, Geophysics, 63, 948–956.

Cerveny, V., and Ravindra, R., 1971. THEORY OF SEISMIC HEAD WAVES, University of Toronto Press, Toronto.

Chang, C., and Gardner, G.H.F., 1997. Effects of vertically aligned subsurface fractures on seismic reflections: a physical model study, Geophysics, 62, 245–252.

Chen, T.C., Booth, D.C., Crampin, S., 1987. Shear wave polarizations near the North Anatolian fault, III Observations of temporal changes, Proceedings of the second international workshop on seismic anisotropy, Geophysical J. Royal Astronomical Society, 91, 287–311.

Chiburis, E.F., Kelamis, P.G., and Clement, W.G., 1988. Separation of P and S wavefields via the τ–p transform, SEG Expanded Abstracts, S1.3.

Cho, W.H., and Spencer, T.W., 1990. Separation of P and S events on multicomponent VSP data, SEG Expanded Abstracts, 124–128.

Cho, W.H., and Spencer, T.W. 1992. Estimation of polarization and slowness in mixed wavefields, Geophysics, 57, 805–814.

Choi, S., and Gangi, A.F., 1991. Effect of source imbalances on the determination of fracture orientations from VSP data, SEG Expanded Abstracts, 61–64.

Cliet, C., and Dubesset, M., 1987. Three-component recordings – interest for land seismic sources study, Geophysics, 52, 1048–1059.

Cliet, C., Brodov, L., Tikhonov, A., Marin, D., and Michon, D., 1991. Anisotropy survey for reservoir definition, Geophysical Journal International, 107, 417–427.

Corrigan, D., 1990,. The effect of azimuthal anisotropy on the variation of reflectivity with offset, 4th International Workshop on Seismic Ansiotropy, SEG, 1645.

Coulombe, C.A., Stewart, R.R., and Jones, M.J., 1992. Elastic wave AVO using borehole seismic data, SEG Expanded Abstracts, 864–866.

Coulombe, C.A., Stewart, R.R., and Jones, M.J., 1991. AVO analysis using the VSP, SEG Expanded Abstracts, 1080–1083.

Crampin, S., 1994. The fracture criticality of crustal rocks, Geophysical Journal International, 118, 428–438.

REFERENCES

Crampin, S., and Yedlin, M., 1981. Shear-wave singularities of wave propagation in anisotropic media, J.Geophys., 49, 43–46.

Crampin, S., Zatsepin, S.V., Slater, C., and Brodov, L.Y., 1996. Abnormal shear-wave polarizations as indicators of pressures and over pressures, EAGE Expanded Abstracts, X038.

Dai, H., and MacBeth, C., 1998. Split shear wave analysis using an artificial neural network, First Break 12, 605-613

Dankbaar, J.W.M., 1987. Vertical seismic profiling – separation of P and S waves, Geophysical Prospecting, 35, 803–814.

Davis, P.M., and Knopoff, L., 1995. The elastic modulus of media containing strongly interacting antiplane cracks, J. of Geophysical Research, 100 (B9), 18253–18258.

Deflandre, J.P., Laurent, J., Michon, D., and Blondin, E., 1995. Microseismic surveying and repeated VSPs for monitoring an underground gas storage reservoir using permanent geophones, First Break, 13, 129–138.

Dellinger, J. and Etgen, J., 1990. Wavefield separation in two-dimensional anisotropic media, Geophysics, 55, 914–919.

Dellinger, J., and Nolte, B., 1997. A crossed-dipole reciprocity ``paradox", The Leading Edge, 16, 1465–1466, 1468–1471.

Dellinger, J., Nolte, B.J., and Etgen, J.T., 1998. Symmetric Alford diagonalization, SEG Expanded Abstracts, 1673–1676.

de Parscau, J., 1991. Relationship between phase velocities and polarization in transversely isotropic media, Geophysics, 56, 1578–1583.

de Parscau, J., and Nicoletis, L., 1990. Transverse isotropy estimation from multi-offset VSPs, SEG Expanded Abstracts, 1439–1442.

Devaney, A.J., and Oristaglio, M.L., 1986. A plane-wave decomposition for elastic wave fields applied to the separation of P-waves and S-waves in vector seismic data, Geophysics, 51, 419–423.

Dillon, P.B., Ahmed, H., Roberts, T., 1988. Migration of mixed mode VSP wavefields, Geophysical Prospecting, 36, 825–846.

Di Siena, J.P., Gaiser, J.E., and Corrigan, D., 1984. Horizontal components and shear wave analysis of three-component VSP data, In: Toksöz, M.N., and Stewart, R.R., Eds., VERTICAL SEISMIC PROFILING, PART B., Advanced Concepts, Geophysical Press, 177–189.

Donati, M.S., and Brown, R.J., 1995. Birefringence study on 3C/2D: Barinas basin (Venezuela), SEG Expanded Abstracts, 723–726.

Donati, M.S. and Martin, N.W., 1998, Making AVO analysis for converted shear waves a practical issue, SEG Expanded Abstracts, 2060–2063.

Douma, J., and Helbig, K., 1987. What can the polarization of shear waves tell us?, First Break, 5, 95–105.

Douma, J., den Rooijen, H., and Schokking, F., 1990. Anisotropy detected in shallow clays using shear-wave splitting in a VSP survey, Geophysical Prospecting, 38, 983–998.

Duff, G.F.D., 1960. The Cauchy problem for elastic waves in an anisotropic medium, Philos. Trans.R.Soc.Ser.A:252, 249–273.

Ebrom, D., Tatham, R., Sekharan, K.K., and McDonald, J.A., and Gardner, G.H.F., 1990. Dispersion and anisotropy in laminated versus fractured media: an experimental comparison, SEG Expanded Abstracts, 1416–1419.

Edelmann, H.A.K., 1992. Circularly polarized shear waves used for VSP measurements, Geophysics, 57, 643–646.

Esmersoy, C., 1984. Polarization analysis, rotation and velocity estimation in three-component VSP, 236–255. In: VERTICAL SEISMIC PROFILING, PART B, ADVANCED CONCEPTS, Handbook of Geophysical Exploration, 14B, Toksöz, M.N. and Stewart, R.S., Eds.

Esmersoy, C., 1990a. Inversion of P and SV waves from multicomponent offset vertical seismic profiles, Geophysics, 55, 39–50.

Esmersoy, C., 1990b. Split-shear wave inversion for fracture evaluation, SEG Expanded Abstracts, 1400–1403.

Esmersoy, C., Hsu, K. and Schoenberg, M., 1989. Quantitiative analysis of fine layering effects. medium averaging versus synthetic seismograms, SEG Expanded Abstracts, 1083–1086.

Evans, R., 1984. Effects of the free-surface on shear wavetrains, Geophysical J. Royal Astronomical Society, 76, 165–172.

Fedorov, F.I., 1968. THEORY OF ELASTIC WAVES IN CRYSTALS, Plenum Press, New York.

Fehler, M., 1982. Interaction of seismic waves with a viscous liquid layer, Bulletin Seismological Society of America, 72, 55–72.

Flinn, E.A., 1965. Signal analysis using rectilinearity and direction of particle motion, Proc. IEEE, 53, 1874–1876.

Folstad, P.G. and Schoenberg, M., 1992. Low frequency propagation through fine layering, SEG Expanded Abstracts, 1278–1281.

Gaiser, J.E., 1990. Transversely isotropic phase velocity analysis from slowness estimates, J. of Geophysical Research, 95, 11,241–11,254.

Gajewski, D., 1993. Radiation from point sources in general anisotropic media, Geophysical Journal International, 113, 299–317.

Gal'Perin, 1985. VERTICAL SEISMIC PROFILING, SEG, Special Publication 12.

Gangi, A.F., 1978. variation of whole and fractured porous-rock permeability with confining pressure, Int.J.Rock.Mech.Min.Sci., and Geomech. Abstr., 15, 249–257.

Garbin, H.D., and Knopoff, L., 1975. The shear modulus of a material permeated by a random distribution of free circular cracks, Q.appl.Math., 33, 296–300.

Garotta, R., and Granger, P.Y., 1988. Acquisition and processing of 3C×3D data using converted waves, SEG Expanded Abstracts, S13.

Garotta, R.J., and Marechal, P., 1987. Shear wave polarization survey using converted waves, SEG Expanded Abstracts, S11.6.

Ghose, R., and Takahashi, 1992. Multiazimuthal seismic reception in a borehole: implications on field delineation of seismic anisotropy, SEG Expanded Abstracts, 34–37.

Gilbert, F., and Knopoff, L., 1961. The directivity problem for a buried line source, Geophysics, 26, 626–634.

Gillespie, P.A., Howard, C.B., Walsh, J.J., and Watterson, J., 1993. Measurement and characterization of spatial distributions of fractures, Tectonophysics, 226, 113–141.

Glangeaud, F., and Cliet, C., 1990. Surface wave identification and filtering in multicomponent seismics, SEG Expanded Abstracts, 1557–1560.

Glangeaud, F. and Mari, J.L. 1994. WAVE SEPARATION, Publication of the Institut Francais du Petrole.

Godin, L., 1998. Analysis of marine VSP, Edinburgh Anisotropy Project Annual Report, Volume 5.

Grechka, V.Y., and Obolentseva, I.R., 1993. Geometrical structure of shear wave surfaces near singularity directions in anisotropic media, 115, 609–616.

Grechka, V., and Tsvankin, I., 1998. 3D description of normal moveout in anisotropic inhomogeneous media, Geophysics, 63, 1079–1092.

Greenhalgh, S.A., and Mason, I.M., 1995. Orientation of a downhole triaxial geophone, Geophysics, 60, 1234–1237.

Gupta, I.N., 1973. Seismic velocities in rock subjected to axial loading up to shear fracture, J. Geophysical Research, 78, 6936–6942.

Gut, T., Sollner, W., Lüschen, E., and Edelmann, H.A.K., 1994. More reliable shear-wave data from VSPs using the CIPHER technique, First Break, 12, 123–129.

Hake, J.H., Gevers, E.C.A., van der Kolk, C.M., and Tichelaar, B.W., 1998. A shear experiment over the Natih field in Oman: Pilot seismic and borehole data, Geophysical Prospecting, 46, 617–646.

Hardage, B.A., 1991. VERTICAL SEISMIC PROFILING, PART A: PRINCIPLES, Handbook of Geophysical Exploration, Seismic Exploration. Volume 14A, 3rd edition, Pergamon Press, Oxford.

Harrison, M.P., 1993. Processing of P-SV surface-seismic data: anisotropy analysis, dip moveout and migration, PhD Thesis, University of Calgary.

Harwood, S., Rowbotham, P.S., Bush, I., Williamson, P.R., Houllevigue, H., and Kendall,, J.M., 1998. Uncertainties in the analysis of AVOA for determination of fracture orientation, Revue de l'Institut Francais du Petrole, 53, 621–627.

Haugen, G.U., and Schoenberg, M., 1995. The echo of a fracture, EAGE Expanded Abstracts, D020.

Hatchell, P.J., and Cowles, C., 1992. Flexural borehole modes and the measurement of azimuthal anisotropy, SEG Expanded Abstracts, 201–204.

Helbig, K., 1994. FOUNDATIONS OF ANISOTROPY FOR EXPLORATION SEISMICS, Volume 22, Handbook of Geophysical Exploration, Pergamon, Eds. K.Helbig and S.Treitel.

Helbig, K., 1958. Elastische Wellen in anisotropen Medien, Gerlands Beitrage zur Geophysik, 67, 177–211; 256–288.

Hermann, P., and Wapenaar, C.P.A., 1992. Wave-type separation in the space-frequency domain applied to VSP data, J. Seismic Exploration, 1, 161–172.

Hokstad, K., 1995. Iterative computation of VSP rotation angles, EAGE Expanded Abstracts, B020.

Hood, J., and Schoenberg, M., 1989. Estimation of fracturing from measure elastic moduli, J. Geophysical Research, 94, 15611–15618.

REFERENCES

Hornby, B.E., Schwartz, L.M. and Hudson, J.A., 1994. Anisotropic effective-medium modelling of the elastic properties of shales, Geophysics, 59, 1570–1583.

Horne, S.A., 1995. Applications of genetic to problems in seismic anisotropy, Ph.D. Thesis, University of Edinburgh.

Horne, S., and MacBeth, C., 1994. AVA observations in azimuthal VSPs, Expanded Abstracts, EAGE, Po51.

Horne, S., and MacBeth, C., 1997. Inversion of seismic anisotropy using genetic algorithms, Geophysical Prospecting, 42, 953–974.

Horne, S., MacBeth, C,. and Liu, E., 1997. "Unexpected'" shear-wave behaviour, Geophysics, 62, 1879–1883.

Horne, S., MacBeth, C,. Queen, J., Rizer, W., and Cox, V., 1997. Fracture characterization from near-offset VSP inversion, Geophysical Prospecting, 45, 141–164.

Horne, S.A., McGarrity, J.P., Sayers, C.M., Smith, R.L., and Wijnands, F., 1998. Fractured reservoir characterization using multi-azimuthal walkaway VSPs, SEG Expanded Abstracts, 1640–1643.

Hudson, J.A., 1994. Overall properties of a material with inclusions or cavities, Geophysical Journal International, 117, 555–561.

Hudson, J.A., and Liu, E., 1999. Effective elastic properties of heavily faulted structures, Geophysics, 64, 479–485.

Hsu, C-J., and Schoenberg, M., 1993. Elastic waves through a simulated fractured medium, Geophysics, 58, 885–915.

Hsu, K., Schoenberg, M., and Walsh, J., 1991. Anisotropy from polarization and moveout, SEG Expanded Abstracts, 1526–1529.

Hudson, J.A., Liu, E., and Crampin, S., 1996. The mechanical properties of materials with interconnected cracks and pores, Geophysical Journal International, 124, 105–112.

Horne, S., and MacBeth, C., 1997. AVA observations in azimuthal VSPs, Expanded Abstracts, 59th EAGE Meeting, P051.

Hudson, J.A., 1980. Overall properties of a cracked solid, Mathematical Proceedings, Cambeisge Philosophical Society, 88, 371–384.

Hudson, J.A., 1981. Wavespeeds and attenuation of elastic waves in material containing cracks, Geophysical J. Royal Astronomical Society, 64, 133–150.

Hudson, J.A., 1988. Seismic wave propagation through material containing partially saturated cracks, Geophysical J. Royal Astronomical Society, 92, 33–37.

Hudson, J.A., Liu, E., and Crampin, S., 1996a. The mechanical properties of materials with interconnected cracks and pores, Geophysical Journal International, 124, 105 – 112.

Hudson, J.A., Liu, E., and Crampin, S., 1996b. Transmission properties of a fault plane, Geophysical Journal International, 125, 559 – 566.

Hudson, J.A., Liu, E., and Crampin, S., 1997. The mean transmission properties of a fault with imperfect facial contact, Geophysical Journal International, 129, 720–726.

Imhof, M.G., Toksoz, M.N., Burch, C.I., and Queen, J.H., 1997,.Near-surface scattering from high velocity carbonates in West Texas, SEG Expanded Abstracts, P30.

Jones, L.E.A., and Wang, H.F., 1981. Ultrasonic velocities in Cretaceous shales from the Williston Basin, Geophysics, 46, 288–297.

Kaarsberg, E.A., 1959. Introductory studies of natural and artificial argillaceous aggregates by sound-propagation and X-ray diffraction methods, J. Geology, 67, 447–472.

Kaarsberg, E.A., 1968. Elasticity studies of isotropic rock samples, Trans.Soc.Mining.Eng., 241, 470–475.

Kachanov, M., 1992. Effective elastic properties of cracked solids: critical review of some basic concepts, Appl.Mech.Rev., 45, 304–335.

Kanasewich, E.R., 1981. TIME SEQUENCE ANALYSIS IN GEOPHYSICS, The University of Alberta Press.

Karrenbach, M., 1994. Multicomponent source equalization, SEG Expanded Abstracts, 1449–1452.

Kerner, S., Bouchon, M., and Coutant, O., 1999. Numerical simulation of the propagation of P waves in fractured media, Geophysical Journal International, 137, 197–206.

Kendall, R.R., and Kendall, J-M., 1996. Shear-wave amplitude anomalies in south-central Wyoming, The Leading Edge, August 1996, 913–920.

Kerner, C., Dyer, B. and Worthington, M., 1989. Wave propagation in a vertical transversely isotropic medium: field experiment and model study, Geophysical J. Royal Astronomical Society, 97, 295–309.

Kommedal, J.H., and Tjostheim, B.A., 1989. A study of different methods of wavefield separation for application to VSP data, Geophysical Prospecting, 37, 117–142.

Kramer, D., and Davis, T.L., 1992. Multicomponent VSPs for reservoir characterization, South Casper Creek field, Wyoming, The Leading Edge, September 1992, 31–35.

Kühnel, T., and MacBeth, C., 1999. Preliminary anisotropic analysis of a 3D-VSP, Edinburgh Anisotropy Project Annual Report, Volume 6, 453–484.

Landrum, R.A., Brook, R.A., and Sallas, J.J., 1994. Polarity convention for vibratory source/recording systems, Geophysics, 59, 315–321.

Lawson, C.L., and Hanson, R.J., 1974. SOLVING LEAST-SQUARES PROBLEMS, Prentice-Hall Inc.

Lawton, D.C., and Bertram, M.B., 1993. Field tests of 3-component geophones, Canadian J. Exploration Geophysics, 29, 125–131.

Leaney, W.S., 1990. Parametric wavefield decomposition and applications, SEG Expanded Abstracts, 1097–1100.

Leaney, W.S., 1994. Anisotropy and AVO from walkaways, SEG Expanded Abstracts, 105–109.

Leaney, S.W., and Esmersoy, C., 1989. Parametric decomposition of offset VSP wavefield, SEG Expanded Abstracts, 26–29.

Lee, M.W., 1987. Particle displacement on the wall of a borehole from incident plane waves, Geophysics, 52, 1290–1296.

Lefeuvre, F., Nicoletis, L., Ansel, V., and Cliet, C., 1992. Detection and measure of the shear wave birefringence from vertical seismic data: theory and applications, Geophysics, 57, 1463–1481.

Lefeuvre, F., and Queen, J.H., 1992. Propagator matrix processing single source 3-component VSPs, SEG Expanded Abstracts, 163–168.

REFERENCES

Lewis, C., Davis, T.L., and Vuillermoz, C., 1991. Three-dimensional multicomponent imaging of reservoir heterogeneity, Silo Field, Wyoming, Geophysics, 56, 2048–2056.

Li, X.-Y., 1992. Shear-wave splitting in reflection surveys: theory, method and case studies, PhD Thesis, University of Edinburgh.

Li, X-Y., 1999. A case study of azimuthal moveout analysis for fracture detection, Geophysics, 64, 1193–1210.

Li, X-Y., and Crampin, S., 1993. Linear-transform techniques for processing shear-wave anisotropy in four-component seismic data, Geophysics, 58, 240–256.

Li, X-Y., Kühnel, T., and MacBeth, C., 1996. Mixed mode AVO response in fractured media, SEG Expanded Abstracts, 1822–1825.

Li, X-Y., MacBeth, C., and Crampin, S., 1998. Interpreting non-orthogonal split shear waves for seismic anisotropy in multicomponent VSPs, Geophysical Prospecting, 46, 1–27.

Lin, S., and Mura, T., 1973. Elastic fields of inclusions in anisotropic media, Phys.Status Solidi, 15, 281–285.

Liu, E., and Crampin, S., 1990. Effects of the internal shear wave window: comparison with anisotropy induced splitting, J. Geophysical Research, 95, 11275–11282.

Liu, E., Crampin, S., and Queen, J.H., 1991. Fracture detection using crosshole surveys and reverse vertical seismic profiles at the Conoco Borehole test facility, Oklahoma, Geophysical Journal International, 107, 449–463.

Liu, E., Crampin, S., Queen, J.H., and Rizer, W.D., 1993a. Behaviour of shear waves in rocks with two sets of parallel cracks, Geophysical Journal International, 113, 509–517.

Liu, E., Crampin, S., Queen, J.H., and Rizer, W.D., 1993b. Velocity and attenuation anisotropy caused by microcracks and macrofractures in a multiazimuth reverse VSP, canadaian Jpuirna;l of Exploration Geophysics, 29, 177–188.

Lynn, H.B., 1991. Field measurements of azimuthal anisotropy: 60 meters, San Francisco Bay area, CA, and estimation of the horizontal stress' ratio from VS1/VS2, Geophysics, 56, 822–832.

Lynn, H.B., 1996. A geophysicist's view on seismic anisotropy, in Fjær, E., Holt, R., and Rathore, J.S., Eds., SEISMIC ANISOTROPY, SEG, 1–14.

Lynn, H.B., Bates, C.R., Simon, K.M., and van Dok, R., 1995. The effects of azimuthal anisotropy in P-wave 3D seismic, SEG Expanded Abstracts, 727–730.

Lynn, H.B., Simon, K.M., and Bates, C.R., 1996. Correlation between P-wave AVOA and S-wave traveltime anisotropy in a naturally fractured gas reservoir, The Leading Edge, 15, 931–935.

Lynn, H.B., Simon, K.M., Bates, C.R., Layman, M., Schneider, R., and Jones, M., 1995. Use of anisotropy in P-wave and S-wave data for fracture characterization in a naturally fractured gas reservoir, The Leading Edge, 14, 887–893.

Lynn, H.B., Simon, K.M., and Beckham, W., 1997. Fracture detection, mapping and analysis of naturally fractured gas reservoirs using P-wave reflection seismic, SEG Expanded Abstracts, 1210–1213.

Lynn, H.B., and Thomsen, L.A., 1990. Reflection shear-wave data collected near the principal axes of azimuthal anisotropy, Geophysics, 55, 147–156.

MacBeth, C., 1994. How can anisotropy be used for reservoir characterization?, First Break, 13, 31–37.

MacBeth, C., 1996a. Interpreting qS1 polarization due to intersecting fractures, SEG Expanded Abstracts, 758–761.

MacBeth, C., 1996b. Shear-wave analysis for azimuthal anisotropy using pseudo-rotation of marine VSP, 58th EAGE meeting, C022.

MacBeth, C., 1999a. A phase slowness approximation for vertical fracture determination using walkaway VSP, Expanded Abstracts, EAGE, P005.

MacBeth, C., 1999b. Azimuthal variation in P-wave signatures due to fluid flow, Geophysics, 64, 118–192.

MacBeth, C., and Arnold, G., 1998. The shock gun: VSP experiments with a new seismic source, The Leading Edge, 17, 183–188.

MacBeth, C., Boyd, M., Rizer, W., and Queen, J., 1998. Estimation of reservoir fracturing from marine VSP using local shear-wave conversion, Geophysical Prospecting, 46, 29–50.

MacBeth, C., and Crampin, S., 1991a. Comparison of signal processing techniques for estimating the effects of anisotropy, Geophysical Prospecting, 39, 357–386.

MacBeth, C., and Crampin, S., 1991b. Examination of a spectral method for measuring the effects of anisotropy, Geophysical Prospecting, 39, 667–690.

MacBeth, C., Jakubowicz, H., Kirk, W., Li, X-Y., and Ohlsen, F., 1999. Fracture-related amplitude variations with offset and azimuth in marine seismic data, First Break, 17, 13–26.

MacBeth, C., and Li, X-Y., 1996. Linear matrix operations for multicomponent seismic processing, Geophys. J. Int., 124, 189–208.

MacBeth, C., and Li, X-Y., 1999. AVD – an emerging new marine technology for reservoir chacaterization: acquisition and application, Geophysics, 64, 1153–1159.

MacBeth, C., Li, X-Y., and Horne, S., 1994. Limitations on near-surface correction for multicomponent offset VSP, SEG Expanded Abstracts, 10–104.

MacBeth, C., Li, X-Y., and Queen, J., 1993. Near-surface correction for nin-component VSP, 63rd SEG meeting, Expanded Abstracts, 125–128.

MacBeth, C., Li, X-Y., Zeng, X., Cox, D., and Queen, J., 1997. Processing of a nine-component near-offset VSP for seismic anisotropy, Geophysics, 62, 676–689.

MacBeth, C., and Yardley, G., 1992. Optimal estimation of crack-strike, Geophysical Prospecting, 40, 849–872.

MacBeth, C., Zeng, X., Li, X-Y., and Queen, J., 1995. Multi-component near-surface correction for land VSP data, Geophysical Journal International, 121, 301–315.

MacBeth, C., Zeng, X., Yardley, G., and Crampin, S., 1994. Interpreting data matrix asymmetry in near-offset, shear-wave VSP data, Geophysics, 59, 176–191.

Mallick, S., Craft, K.L., Meister, L.J., and Chambers, R.E., 1996. Computation of principal directions of azimuthal anisotropy from P-wave seismic data, SEG Expanded Abstracts, 1962–1865.

Mallick, S., and Frazer, L.N., 1991. Reflection/transmission coefficients and azimuthal anisotropy in marine seismic studies, Geophysical Journal International, 105, 241–252.

Marcus, M. and Minc, H., 1992. A SURVEY OF MATRIX THEORY AND INEQUALITIES, Dover Publications, New York.

REFERENCES

Marion, D., Mukerji, T., and Mavko, G., 1994. Scale effect on velocity dispersion: from ray to effective medium theories in stratified media, Geophysics, 59, 1613–1619.

Mars, J., Glangeaud, F., Lacoume, J.L., Fourmann, J.M., and Spitz, S., 1987. Separation of seismic waves, SEG Expanded Abstracts, 489–492.

Martin, M.A., and Davis, T.L., 1987. Shear-wave birefringence: a new tool for evaluating fractured reservoirs, The Leading Edge, 6, 22–25.

Mavko, G.M., and Nur, A., 1979.Wave attenuation in partially saturated rocks, Geophysics, 44, 161–178.

Meissner, R., 1965. P and SV waves from uphole shooting, Geophysical Prospecting, 13, 443-459.

Menke, W., 1984. GEOPHYSICAL DATA ANALYSIS: DISCRETE INVERSE THEORY, Academic Press, London.

Michelena, R.J., 1995. Quantifying errors in fracture orientation estimated from surface P-S converted waves, SEG Expanded Abstracts, 282–285.

Miller, D.E., Leaney, S., and Borland, W.H., 1994. An in situ estimation of anisotropic elastic moduli for a submarine shale, J. Geophysical Research, 99, 21,659–21,665.

Miller, D.E., and Spencer, C., 1994. An exact inversion for anisotropic moduli from phase slowness data, J. Geophysical Research, 99, 21,651–21,657.

Morland, L.W., 1974. Elastic response of regularly jointed media, Geophysical J. Royal Astronomical Society, 37, 435–446.

Mueller, M.C., 1991. Prediction of lateral variability in fracture intensity using multicomponent shear-wave surface seismic as a precursor to horizontal drilling in the Austin Chalk, Geophysical Journal International, 107, 409–415.

Muijres, G., 1998. Acoustic waves in cracked media, PhD. Thesis, University of Delft.

Murty, G.S., 1976. Reflection, transmission and attenuation of elastic waves at a loosely bonded interface of two half-spaces, Geophysical J. Royal Astronomical Society, 44, 389–404.

Musgrave, M.J.P., 1970. CRYSTAL ACOUSTICS, Holden-Day.

Nur, A., 1971. Effects of stress on velocity anisotropy in rocks with cracks, J. Geophysical Research, 76, 2022–203

Naville, C., 1986. Detection of anisotropy using shear-wave splitting in VSP surveys: requirements and applications, SEG Expanded Abstracts, 391–394.

Neidell, N.S., and Taner, M.T., 1971. Semblance and other coherency measures for multichannel data, Geophysics, 36, 482–497.

Nelson, R.A., 1995. GEOLOGIC ANALYSIS OF NATURALLY FRACTURED RESERVOIRS, Gulf Publishing Company, London.

Nicoletis, L., Cliet, C., and Lefeuvre, F., 1988. Shear-wave splitting measurements from multishot VSP data, SEG Expanded Abstracts, 527–530.

Niitsuma, H., Tsuyuki, K., and Asanuma, H., 1993. Discrimination of split shear waves by wavelet transform, Canadian J. Exploration Geophysics, 29, 106–113.

Nishizawa, O., 1982. Seismic velocity anisotropy in a medium containing oriented cracks – transversely isotropic case, J.Phys.Earth, 30, 331–347.

Noble, M.D., Lambert, R.A., Ahmed, H., and Lyons, J., 1987. The application of three-component VSP data on the interpretation of the Vulcan gas field and its impact on future field development (a case history), First Break, 6, 13–165.

Nur, A., and Simmons, G., 1969. Stress-induced velocity anisotropy in rock: an experimental study, J. Geophysical Research, 74, 6667–6674.

Nuttli, O., 1961. The effect of the earth's surface on the S wave particle motion, Bulletin Seismological Society of America, 51. 237–246.

O'Doherty, R.F. and Anstey, N.A., 1971. Reflections on amplitudes, Geophysical Prospecting, 9, 430-458.

Ohanian, V., and Beckham, W.E., 1992. Depth-variant fracture orientation from multicomponent VSP, SEG Expanded Abstracts, 177–180.

Ohlsen, F., and MacBeth, C., 1999. Elliptical anisotropy – regression or advance?, SEG Expanded Abstracts, 1600–1603

Ohlsen, F., MacBeth, C., Magnus, I., and Vestby, J., 1998. Anisotropy estimates from a North Sea VSP, SEG Expanded Abstracts, 280–283.

Omnes, G., and Clough, P., 1998. Data acquisition with a 12-level 48-channel borehole seismic system, The Leading Edge, July, 955–959.

Ostrander, W.J., 1984. Plane-wave reflection coefficients for gas sands at non normal angles of incidence, Geophysics, 49, 1637–1648.

Panziera, J.P., and Arnes, G., 1985. Three component vertical seismic profiles: polarized wave separation, SEG Expanded Abstracts, 58–60.

Park, J., Vernon, F.L., and Lindberg, C.R., 1987. Frequency dependent polarization analysis of high frequency seismograms, J. Geophysical Research, 92, 12664–12674.

Peacock, S., 1991. Seismic wave propagation in uncracked and cracked Carrara marble, EAGE Expanded Abstracts, 250–251.

Peacock, S., and Hudson, J.A., 1990. Seismic properties of rocks with distributions of small cracks, Geophysical Journal International, 102, 471–484.

Peng, C., Cheng, C.H., and Toksöz, M.N., 1993. Borehole effects on downhole seismic measurements, Geophysical Prospecting, 41, 883–912.

Perez, M.A., and Gibson, R.L., 1996. Detection of fracture orientation using azimuthal variation of P-wave AVO responses: Barinas field (Venezuela), SEG Expanded Abstracts, 1353–1356.

Perlberg, A.I. and Hornbostel, S.C. 1994. Applications of seismic polarization analysis, Geophysics, 59, 119–130.

Peron, J., 1990. Estimation of fracture direction from zero-offset VSP by two- and four-component rotation, 60th SEG meeting, Abstracts, 1443–1447.

Postma, G.W., 1955. Wave propagation in a stratified medium, Geophysics, 20, 780–806.

Potters, J.H.H.M., Groenendaal, H.J.J., Oates, S.J., and Hake, J.H., 1999. The 3D shear experiment over the Natih field in Oman: Reservoir Geology, data acquisition and anisotropy analysis, Geophysical Prospecting, , 47, 637-662.

Pratt, G., McGaughey, W.J., and Chapman, C.H., 1993. Anisotropic velocity tomography: a case study in a near-surface rock mass, Geophysics, 58, 1748–1763.

References

Pruett, R.C., 1989. Polarity documentation for multicomponent seismic sources and receivers: Presented at the Summer Research Workshop on Recording and Processing Vector Wavefield Data, SEG, Snowbird, Utah.

Puzirev, N.N., Trygubov, A.V., Brodov, L.Yu., Vedernikov, G.V., Lebedev, K.A., Obolentseva, I.R., Nefedkina, T.V., Hudobina, L.N., Sibiryakov, B.P., Kulichihina, T.N., Lebedeva, G.N., and Karzheva, L.V., 1985. SEISMIC PROSPECTING BY METHOD OF SHEAR AND CONVERTED WAVES, Nedra, Moscow.

Pickup, G.E., Ringrose, P.S., Jensen, J.L., and K.S. Sorbie, 1994. Permeability tensors for sedimentary structures, Mathematical Geology, P26, 227–250.

Pyrak-Nolte, L.J., Myer, L.R., Cook, N.G.W., 1990. Transmission of seismic waves across single natural fractures, J. Geophysical Research, 95, 8617–8638.

Queen, J.H., 1987. Orientation control in seismic anisotropy measurement, Workshop on Multicomponent Field Data Acquisition and Processing, Houston.

Queen, J.H., and Rizer, W.D., 1990. An integrated study of seismic anisotropy and the natural fracture system at the Conoco borehole test facility, Kay County, Oklahoma, J. Geophysical Research, 11255–11273.

Raikes, S.A., 1991. Shear wave characterization of the BP Devine test site, Texas, SEG Expanded Abstracts, 65–68.

Randall, N., 1998. A practical approach to VSP acquisition in highly deviated and horizontal wells, Expanded Abstracts, PETEX, Session 1, Paper 11.

Riznichenko, Y.V., 1949. On seismic quasi-anisotropy, Izv.Akad.Nauk.SSSR Ser.Geogr.Geofis., 13, 518–544.

Robertson, J.D., and Corrigan, D., 1983. Radiation patterns of a shear-wave vibrator in near-surface shale, Geophysics, 48, 19–26.

Roche, S.L., Davis, T.L., and Benson, R.D., 1997. 4D, 3C seismic study at Vacuum field, New Mexico, SEG Expanded Abstracts, 886–889.

Rozemond, H.J., 1996. Slip-sweep acquisition, SEG Expanded Abstracts, 64–67.

Rüger, A., 1996a. Variation of P-wave reflectivity with offset and azimuth in anisotropic media, SEG Expanded Abstracts, 1862–1865.

Rüger, A., 1996b. Analytical insight into shear-wve AVO for fractured reservoirs, SEG Expanded Abstracts, 1801–1804.

Rüger, A., 1997. Variation of P-wave reflectivity with offset and azimuth in anisotropic media, Geophysics, 935–947.

Rüger, A., 1997. P-wave reflection coefficients for transversely isotropic models with vertical and horizontal axis of symmetry,

Rüger, A., 1998. Variation of P-wave reflectivity with offset and azimuth in anisotropic media, Geophysics 63, 935–947.

Rutherford, S.R., and Williams, R.H., 1989. Amplitude-versus-offset variations in gas sands, Geophysics, 54, 680–688.

Rutty, M.J., and Greenhalgh, S.A., 1993. The correlation of seismic events on multicomponent data in the presence of coherent noise, Geophysical Journal International, 113, 343–358.

Ryan-Grigor, S., 1997. Empirical relationships between transverse isotropy parameters and VP/VS: Implications for AVO, Geophysics, 62, 1359–1364.

Ryan-Grigor, S., 1998. Empirical relationships between anellipticity and VP/VS in shales: potential applications to AVO studies and anisotropic seismic processing. SEG Expanded Abstracts, 208–567.

Sams, M.S., 1995. Attenuation and anisotropy: the effect of extra fine layering, Geophysics, 60, 1646 – 1655.

Sams, M.S. and Williamson, P.R., 1994. Backus averaging, scattering and drift, Geophysical Prospecting, 42, 541–564.

Samson, J.C., 1983. Pure states, polarized waves, and principal components in the spectral analysis of multiple, geophysical time-series, Geophysical J. Royal Astronomical Society, 72, 647–664.

Samson, J.C., and Olson, J.V., 1980. Some comments on the descriptions of the polarization states of waves, Geophysical J. Royal Astronomical Society, 61, 115–129.

Savage, M.K., Shih, X.R., Meyer, R.P., and Aster, R.C., 1987. Application of an automated, analytical method for the determination of she-wave anisotropy, Abstract at 19th General Assembly IUGG, Abstract vol. 1, 324.

Sayers, C.M., 1995. Anisotropic velocity analysis, Geophysical Prospecting, 43, 541–568.

Sayers, C.M., 1997. Determination of anisotropic velocity models from walkaway VSP data acquired in the presence of dip, Geophysics, 62, 723–729.

Sayers, C.M., 1998. Misalignment of the orientation of fractures and the principal axes for P and S waves in rocks containing multiple non-orthogonal fracture sets, Geophysical Journal International, 133, 459–466.

Sayers, C.M., and Ebrom, D.A., 1997. Seismic traveltime analysis for azimuthally anisotropic media: theory and experiment, Geophysics, 62, 1570–1582.

Sayers, C.M., and Rickett, J.E., 1997. Azimuthal variation in AVO response for fractured gas sands, 45, 165–182.

Sayers, C.M., van Munster, J.G., and King, M.S., 1990. Stress-induced ultrasonic anisotropy in Berea sandstone, Int.J.Rock Mech., 27, 429–436.

Schoenberg, M., 1980. Elastic wave behavior across linear slip interfaces, J. Acoustical Society of America., 68, 1516–1521.

Schoenberg, M., 1986. Fluid and solid motion in the neighborhood of a fluid-filled borehole due to the passage of a low-frequency elastic plane wave, Geophysics, 51, 1191–1205.

Schoenberg, M., 1994. Transversely isotropic media equivalent to thin isotropic layers, Geophysical Prospecting, 42, 885–915.

Schoenberg, M., and Douma, J., 1988. Elastic wave propagation in media with parallel fractures and aligned cracks, Geophysical Prospecting, 36, 571–590.

Schoenberg, M., Mallick, S., and Frazer, L.N., 1988. Synthetic shear VSPs in azimuthal anisotropic media, ASEG/SEG Int.Conf.on Exploration Geophysics, Expanded Abstracts, Expl.Geophys., 19, 151–157.

Schoenberg, M., Muir, F., and Sayers, C., 1996. Introducing Annie: A simple three-parameter anisotropy velocity model for shales, J. Seismic Exploration, 5, 35–49.

Schoenberg, M. and Protazio, J., 1992. 'Zoeppritz' rationalized and generalized to anisotropy, J. of Seismic Exploration, 1, 125–144.

Schoenberg, M. and Sayers, C.M., 1995. Seismic anisotropy of fractured rock, Geophysics, 60, 204 – 211.

Schulte, L., and Edelmann, H.A.K., 1988. Azimuthal anisotropy proven to be a useful approach for multicomponent shear wave data processing, SEG Expanded Abstracts, 1156–1158.

Seeman, B., and Horowicz, L., 1983. Vertical seismic profiling: separation of upgoing and downgoing acoustic waves in a stratified medium, Geophysics, 48, 555–568.

Sena, A.G., 1991. Seismic travel time equations for azimuthally anisotropic and isotropic media: estimation of interval elastic properties, Geophysics, 56, 2090–2101.

Shieh, C., 1997. Estimation of shear wave splitting time using orthogonal transformation, Geophysics, 62, 657-661.

Shih, X.R., Schneider, J.F., and Meyer, R.P., 1986. Shear-wave anisotropy observed from the Bucaramange Nest, Columbia, EOS, 67, 1117.

Shuck, E.L., Davis, T.L., and Benson, R.D., 1996. Multicomponent 3D characterization of a coalbed methane reservoir, Geophysics, 61, 315–330.

Shuey, R.T., 1985. A simplification of the Zoeppritz equations, Geophysics, 50, 609–614.

Simaan, M., 1984. ADVANCES IN GEOPHYSICAL DATA PROCESSING, VERTICAL SEISMIC PROFILES, Jai Press Inc., London.

Slater, C., 1997. Estimation and modelling of anisotropy in vertical and walkaway seismic profiles at two North Caucasus Oil fields, Ph.D. Thesis, University of Edinburgh.

Smidt, J.M., 1989. VSP processing with full downgoing wavefield deconvolution applied to the total wavefield, First Break, 7, 247–257.

Sondergeld, C.H., and Rai, C.S., 1992. Laboratory observations of shear-wave propagation in anisotropic media, The Leading Edge, 11, 38–43.

Spencer, T.W., and Chi, H.C., 1991. Thin-layer fracture density, Geophysics, 56, 833–843.

Taner, M.T., and Koehler, F., 1969. Velocity spectra – digital computer derivation of velocity functions, Geophysics, 34, 859–881.

Tatham, R.H., and McCormack, M.D., 1991. MULTICOMPONENT SEISMOLOGY IN PETROLEUM EXPLORATION, SEG.

Taylor, D.B., 1987. Double contour integration for transmission from point sources through anisotropic layers as used in ANISEIS software, Geophysical J. Royal Astronomical Society, 91, 373–381.

Teng, L., and Mavko, G., 1996. Fracture signatures on P-wave AVOZ, SEG Expanded Abstracts, 1818–1821.

Thomsen, L., 1986. Weak elastic anisotropy, Geophysics, 51, 1954–1966.

Thomsen, L., 1988. Reflection seismology over azimuthally anisotropic media, Geophysics, 53, 304–313.

Thomsen, L., 1993. Weak anisotropic reflections, in Backus, M., Ed., OFFSET-DEPENDENT REFLECTIVITY: THEORY AND PRACTICE OF AVO ANALYSIS, SEG, 103–111.

Thomsen, L., 1995. Elastic anisotropy due to aligned cracks in porous rock, Geophysical Prospecting, 43, 805–829.

Thomsen, L., 1998. Converted-wave reflection seismology over anisotropic, inhomogeneous media: SEG Expanded Abstracts, 2048–2051.

Thompson, J.M., 1950. Transmission of elastic waves through a stratified solid medium, J.Appl.Phys., 21, 89–93.

Thornburgh, H.R., 1930. Wavefront diagrams in seismic interpretation, American Association of Petroleum Geologists Bulletin, 14, 185–200.

Tsvankin, I., 1995. Inversion of moveout velocities for horizontal transverse isotropy, SEG Expanded Abstracts, 735–738.

Tsvankin, I., 1996. Effective parameters and P-wave velocity for azimuthally anisotropic media, SEG Expanded Abstracts, 1850–1853.

Tsvankin, I., 1997. Reflection moveout and parameter estimation for horizontal transverse isotropy, Geophysics, 62, 614–629.

Tsvankin, I. 2001, SEISMIC SIGNATURES AND ANALYSIS OF REFLECTION DATA IN ANISOTROPIC MEDIA, Volume 29, Handbook of Geophysical Exploration, Pergamon, Eds. K.Helbig and S.Treitel

Tsvankin, I.D., and Chesnokov, E.M. 1990. Synthesis of body wave seismograms from point sources in anisotropic media, J. of Geophysical Research, 95, 11317–11331.

Tsvankin, I., and Thomsen, L., 1994. Non-hyperbolic reflection moveout in anisotropic media, Geophysics, 59, 1290–1304.

Van der Kolk, C.M., Guest, W.S., and Potters, J.H.H.M., 1999. The 3D shear experiment over the Natih field in Oman. V The effect of fracture-filling fluids on shear propagation, Geophysical Prospecting, In Press.

Vernik, L., 1993. Microcrack-induced versus intrinsic elastic anisotropy in mature HC-source shales (short note), Geophysics, 58, 1703–1706.

Vernik, L., and Liu, X., 1997. Velocity anisotropy in shales: a petrophysical study, Geophysics, 62, 521–532.

Vernik, L., and Nur, A., 1992. Ultrasonic velocity and anisotropy of hydrocarbon source rocks, 57, 727–735.

Voigt, W., 1928. LEHRBUCH DER KIRSTALLPHYSIK, TEUBNER, Leipzig.

Werner, U., and Shapiro, S.A., 1996. Frequency dependent anisotropy of seismic waves in isotropic mulilayering, EAGE Expanded Abstracts 61st meeting, C011.

White, J.E., 1983. UNDERGROUND SOUND: APPLICATION OF SEISMIC WAVES, Elsevier, Oxford.

White, J.E., Martineau-Nicoletis, L., Monash, C., 1983. Measured anisotropy in Pierre Shale, Geophysics, 31, 709–725.

Wild, P., and Crampin, S., 1991. The range of effects of azimuthal isotropy and EDA anisotropy in sedimentary basins, Geophysical Journal International, 107, 513–529.

Willis, H.A., Rethford, G.L., and Bielanski, E., 1986. Azimuthal anisotropy: Occurrence and effect on shear-wave data quality, SEG Expanded Abstracts, Session S9.7.

Winterstein, D.F., 1986. Anisotropy effects in P-wave and SH-wave stacking velocities contain information on lithology, Geophysics, 51, 661–672.

Winterstein, D.F., 1990. Velocity anisotropy terminology for geophysicists, Geophysics, 55, 1070–1088.

Winterstein, D.F., De, G.S., and Meadows, M.A., 2001. Twelve years of birefringence in nine-component VSP data, Geophysics, 66, 582–597.

Winterstein, D.F. and Meadows, M.A., 1991a. Shear wave polarizations and subsurface stress directions at Lost Hills field, Geophysics, 56, 1331–1348.

Winterstein, D.F., and Meadows, M.A., 1991b. Changes in shear-wave polarization azimuth with depth in Cymric and Railroad Gap oil fields, Geophysics, 56, 1349–1364.

Winterstein, D.F., Meadows, M.A., and De, G.,S. 1996. An overview of vertical birefringence in sedimentary rock, Keynote address, Seventh International Workshop on Seismic Anisotropy, Miami.

Winterstein, D.F., and Paulsson, B.N.P., 1990. Velocity anisotropy in shale determined from cross-hole seismic and vertical seismic profile data, Geophysics, 55, 470–479.

Winterstein, D., Stefani, J., Johnston, S., and Julander, D., 1998. 9C time-lapse VSP monitoring of steam injection at Cymric Oil field, The Leading Edge, 17, 527–530, 533–537.

Wolff, C.J.M., and de Wardt, J.P., 1981. Borehole position uncertainty – analysis of measuring methods and derivation of systematic error model, SPE 9223, 2339–2349.

Wong, T., Fredrich, J.T., and Gwanmesia, G.D., 1989. Crack aperture statistics and pore space fractal geometry of Westernly granite and Rutland Quartzite: Implications for an elastic contact model of rock compressibility, J. Geophysical Research, 94, 10267–10278.

Worthington, M.H., Jituboh, C.K., Chang, C., and Chang, S.K., 1992. Borehole coupling effects on clamped geophone responses: Experiments and theory, EAGE Extended Abstracts, 192–193.

Wright, J., 1987. The effects of transverse isotropy on reflection amplitude versus offset, Geophysics, 52, 564–567.

Yardley, G.S., 1993. Processing and modelling of shear-wave VSPs in anisotropic structures: case studies, Ph.D. Thesis, University of Edinburgh.

Yardley, G.S., Graham, G., and Crampin, S., 1991. Viability of shear-wave amplitude versus offset studies in anisotropic media, Geophysical Journal International, 107, 493–503.

Zaengle, J.F., and Frasier, C.W., 1993. Correlation and interpretation of P-P and P-SV data, Zamora gas field, Yolo county, California, in Backus, M.M., Ed., OFFSET-DEPENDENT REFLECTIVITY – THEORY AND PRACTICE OF AVO ANALYSIS, Society of Exploration Geophysicists, 333-345.

Zatsepin, S.V., and Crampin, S., 1997. Modelling the compliance of crustal rock – I. Response of shear-wave splitting to differential stress, Geophysical Journal International, 129, 477–494.

Zeng, X., 1994. Shear-wave data processing for anisotropy, Ph.D. Thesis, University of Edinburgh.

Zeng, X., and MacBeth, C., 1992. Shear wave anisotropy estimations in the Romashkino field, Russia, SEG Expanded Abstracts, 295–298.

Zeng, X., and MacBeth, C., 1993a. Accuracy of shear–wave splitting estimates from near–offset VSP data, Proceedings of the 5IWSA, Canadian Journal of Exploration Geophysics., 29, 246–265.

Zeng, X., and MacBeth, C., 1993b. Algebraic processing techniques for estimating shear–wave splitting in zero-offset VSPs – theory, Geophysical Prospecting, 41, 1033–1066.

Zeng, X., and MacBeth, C., 1996. A case example of near-surface correction for multicomponent VSPs, Geophysical Prospecting, 44, 889–910.

Zhang, Y., 1995. Short window minimax norm optimization for multisource multicomponent S-wave data in presence of azimuthal anisotropy, SEG Expanded Abstracts, 360–363.

Zhang, Y., 1996. Nonhyperbolic converted wave velocity analysis and normal moveout, SEG Expanded Abstracts, 1555-1558.

Zillmer, M., Gajewski, D., Kashtan, B.M., 1997. Reflection coefficients for weak anisotropic media, Geophysical Journal International, 129, 389–398.

Zillmer, M., Gajewski, D., and Kashtan, B.M., 1998. Anisotropic reflection coefficients for a weak-contrast interface, Geophysical Journal International, 132, 159–166.

Zhang, Y., 1996. Nonhyperbolic converted wave velocity analysis and normal moveout, SEG Expanded Abstracts, 1555–1558.

Zhang, Y., and Robinson, E.A., 1992. Stacking of P-SV converted wave data with raypath velocity, SEG expanded abstracts. 1214–1217.

INDEX